Symbol	Definition	Section

Mass-dependent structures

Symbol	Definition	Section
r_m	Mass-dependent structure: Obtained from a fit of the structural parameters to the mass-dependent moments of inertia $I_m = 2I_s - I_0$, where I_s are the substitution moments of inertia that are calculated from the substitution coordinates r_s, and I_0 the ground-state moments of inertia. Generally no better than r_s.	5.5.1
r_c	Improvement of the r_m structure by using complementary sets of isotopologues.	5.5.2
r_m^ρ	r_m^ρ structure: Since the rovibrational contributions ε^0 vary less within a set of isotopologues than the inertial moments I^0, scaling of the inertial moments of all isotopologues by an appropriate common factor $(2\rho_g - 1)$ (for each principal axis g) calculated for the parent, and then submitting the scaled inertial moments to a least-squares fit to obtain the bond coordinates.	5.5.3
$r_m^{(1)}, r_m^{(2)}$	Mass-dependent structure: Based on the different dependence of inertial moments and their rovibrational contributions on the atomic masses (of one and half degrees). Models the rovibrational contributions by the (least possible number of) parameters: c_g (for each principal axis g), multiplied by the square root of the inertial moment of the individual isotopologue $\sqrt{I_g}$, and d_g (for $r_m^{(2)}$ only), multiplied by an isotopologue-dependent, but g-independent mass factor. Least-squares fitting to obtain bond coordinates and rovibrational parameters. The Laurie contraction of a X-H bond upon deuteration is modeled ($r_m^{(1L)}, r_m^{(2L)}$) by an additional parameter δ_H. Refined models $r_m^{(1r)}, r_m^{(2r)}$ assume that the rovibrational effects depend more on the overall shape or contour of the molecule (equal for all isotopologues) than on the principal axis systems whose orientations within the molecular shapes differ among the isotopologues.	5.5.4

Empirical structures

Symbol	Definition	Section
r_0	Effective structure: Least-squares fitting of experimental ground-state inertial moments I^0 of a set of isotopologues to obtain bond coordinates, neglecting rovibrational contributions ε^0 completely, can be realized in practice by means of different sets of observables: $r_0(I)$, $r_0(P)$, $r_0(B)$, and also $r_0(I,\Delta I)$, $r_0(P,\Delta P)$, where the moments of only the parent and the moment differences between parent and isotopologues are used.	5.3
r_s	Substitution structure: Aimed at obtaining Cartesian coordinates of individual atom, numerically dominated by inertial moment difference ΔI upon substitution, no least-squares used, though expandable on sets of several isotopologues (substituted atoms) by least-squares fitting: r_s-fit (determined via the Kraitchman equations).	5.4
r_s variants: $r_{\Delta I}, r_{\Delta P}$	ps-Kr ("pseudo-Kraitchman") structure: Attempts to compensate rovibrational contributions by least-squares, fitting exclusively differences of moments between parent and isotopologues ΔI^0 or ΔP^0 to obtain bond coordinates, realized by $r(\Delta I)$, $r(\Delta P)$.	5.3.2

Equilibrium Molecular Structures

From Spectroscopy to Quantum Chemistry

Equilibrium Molecular Structures

From Spectroscopy to Quantum Chemistry

Edited by Jean Demaison
James E. Boggs • Attila G. Császár

Foreword by Harry Kroto

CRC Press
Taylor & Francis Group
Boca Raton   London   New York

CRC Press is an imprint of the
Taylor & Francis Group, an **informa** business

CRC Press
Taylor & Francis Group
6000 Broken Sound Parkway NW, Suite 300
Boca Raton, FL 33487-2742

© 2011 by Taylor and Francis Group, LLC
CRC Press is an imprint of Taylor & Francis Group, an Informa business

No claim to original U.S. Government works

Printed in the United States of America on acid-free paper
10 9 8 7 6 5 4 3 2 1   *1006461632*

International Standard Book Number: 978-1-4398-1132-0 (Hardback)

Visit the Taylor & Francis Web site at
http://www.taylorandfrancis.com

and the CRC Press Web site at
http://www.crcpress.com

Contents

Foreword

At some point during the education process, which resulted in my becoming a professional researcher and teacher of chemistry, I must have made some sort of subliminal intellectual jump into thinking about molecules as realizable physical objects and indeed architectural/engineering structures. I became quite comfortable, essentially thinking "unthinkingly" about objects that I had never actually "seen." I started to take for granted that my new world was made up of networks of atoms. I do not know when my mind squeezed through this wormhole into what we now call "The Nanoworld," but it seems to have been quite painless, and only much later did I think about this as I became aware that scientists, chemists in particular, live in an abstract world in which we have a deep atomic/molecular perspective of the material world. Neither the sizes of molecules nor the numbers of atoms in a liter of water ever seemed to be amazing. Long ago, the number 6.023×10^{23} (now apparently $6.022 \times 10^{23} - 10^{20}$ seem to have disappeared!) was permanently inscribed on a piece of paper placed in a drawer labeled Avogadro's number in the chest of drawers of my mind. Over the years, some pieces of paper seem to have fallen down the back ending up in the wrong drawers without my knowledge or awareness of the fact—sometimes with dire consequences! In the early days, I do not remember wondering too much, about how this number had been determined, or how we "knew" the value of this number, or how the bond length of H_2 was determined to be 0.74 Å, or indeed, what we actually meant by the term "bond length."

While at school, I bought Fieser and Fieser's book at the suggestion of my chemistry teacher, Harry Heaney, who left the school a little later to become, ultimately, a professor of organic chemistry, and became fascinated by organic chemistry. My memory is that Harry and his wife had two Siamese cats, called Fieser and Fieser. Another friend had two Siamese cats called Schrödinger and Heisenberg. Gradually, I became quite fluent in the abstract visual/graphic language of chemistry, drawing hexagons for benzene rings and writing symbolic schemes to describe the intricate musical chair games that bunches of atoms perform during chemical reactions. At university (Sheffield), I suddenly became completely enamored with molecular spectroscopy during an undergraduate lecture by Richard Dixon. I was introduced to the electronic spectrum of the diatomic radical AlH in which elegant branch structure indicated that the molecule could count accurately—indeed certainly better than I could.

Spectroscopy is arguably the most fundamental of the experimental physical sciences. After all, we obtain most of our knowledge through our eyes and it is via the quest for an in-depth understanding of what light is, and what it can tell us, that almost all our deeper understanding of the universe has been obtained. Answers to these questions about light have led to many of our greatest discoveries, not least our present description of the way almost everything works both on a macroscopic and on a microscopic scale. In the deceptively simple question of why objects possess color at all—such an everyday experience that probably almost no one thinks it odd—lies the seed for the development of arguably our most profound and far-reaching theory—quantum mechanics.

I decided to do research on the spectra of small free radicals produced, detected, and studied by flash photolysis—the technique pioneered by George Porter who was then professor of physical chemistry at Sheffield University. In 1964, I went to the National Research Council (NRC) in Ottawa where Gerhard Herzberg, Alec Douglas, and their colleagues, such as Cec Costain, had created the legendary Mecca of Spectroscopy. While at NRC, I discovered microwave spectroscopy in Cec Costain's group and from that moment, the future direction of my career as a researcher was sealed. I gained a very high degree of satisfaction from making measurements at high resolution on the rotational spectra of small molecules and in particular from the ability to fit the frequency patterns with theory to the high degree of accuracy that this form of spectroscopy offered. Great intellectual satisfaction comes from knowing that the parameters deduced—such as bond lengths, dipole moments, quadrupole and centrifugal distortion parameters—are well-determined quantities both numerically and in a physically descriptive sense. Some sort of deep understanding seems to develop as one gains more-and-more familiarity with quantum mechanical (mathematical) approaches to spectroscopic analyses that add a quantitative perspective to the (subliminal?) classical descriptions needed to convince oneself that one really knows what is going on. I was to learn later that such levels of satisfying certitude of knowledge are a rarity in many other branches of science and in almost all aspects of life in general. It gives one a very clear view of how the scientific mindset develops and what makes science different from all other professions and within the sciences, a clear vision of what it means to really "know" something.

The equations of Kraitchman [1] and the further development of their application in the r_s substitution approach to isotopic substitution data in the 1960s by my former supervisor Cec Costain [2] resulted in a wealth of accurate structural information on small to moderate size molecules from rotational microwave measurements. Jim Watson took these ideas a step further in his development of the r_m method [3]. At Sussex, in 1974, my colleague David Walton and I put together a project for an undergraduate researcher, Andrew Alexander, to synthesize some long(ish) chain species starting with HC_5N and study their spectra—infrared and NMR as well as microwave [4]. This study was to lead to the discovery of long carbon chain molecules in interstellar space and stars [5] and ultimately the experiment that uncovered the existence of the C_{60} molecule.

I sometimes feel that as other scientists casually bandy about bond lengths, our exploits as spectroscopists are not appreciated—the hard work that is needed to obtain those simple but accurate numbers and the efforts needed to determine the molecular architectures as well as the deep understanding of the dynamic factors involved. Indeed, it took quite a significant amount of research before an understanding of what the experimentally obtained numbers really mean was gradually achieved. In particular, the realization that different techniques yield different values for the "bond lengths," for example, the average value of r is obtained by electron diffraction and this can differ significantly from the average values or $1/r^2$ for a particular vibrational state, which is obtained from rotational spectra [6]. Alas, it seems it is the particular lot of the molecular rotational microwave spectroscopy community to be so little appreciated! I sometimes feel that we should forbid the use our structural data by scientists who do not appreciate us in a way parallel to the way I feel about "creationists," who I suggest should be deprived of the benefits of the medications that have been developed on the basis of a clear understanding of Darwinian evolution.

Microwave measurements can reveal many important molecular properties. Internal rotation can give barriers heights, centrifugal distortion parameters can be analyzed to extract vibrational force-field data, and splittings due to the quadrupole moments can yield bond electron-density properties. Arguably, Jim Watson made the major final denouement in his classic paper on the vibration-rotation Hamiltonian—or "the Watsonian"—in which some issues involved in the Wilson–Howard Hamiltonian formulation were finally resolved [7]. Early on in my career I had wondered about the spectrum of acetylene studied by Ingold and King [8] and the way in which shape changes might affect the spectrum—in this case from linear to trans bent in the excited state. Later, I started to learn about quasi-linearity and quasi-planarity. Our present understanding of this phenomenon was due to the groundbreaking work of, among others, Richard Dixon [9] and Jon Hougen, and Phil Bunker and John Johns [10]. At Sussex, we obtained a truly delightful spectrum that afforded us great intellectual pleasure as well as a uniquely satisfying insight into the meaning of "quasi-linearity." This was to be found in the microwave spectrum of NCNCS which Mike King and Barry Landsberg studied [11]. As the angle bending vibration of this V-shaped molecule increases, the spectroscopic pattern observed at low v_{bend} changes to that of a linear one at ca. $v_{bend} = 4$, where the bending amplitude is so large that when averaged over the A axis it appears roughly linear. Brenda Winnewisser et al. have taken the study of this beautiful system to a further fascinating level of even deeper understanding in their elegant study of quantum monodromy [12].

As we now trek deeper into the twenty-first century, numerous ingenious researchers have resolved many fundamental theoretical spectroscopic problems. Molecular spectroscopy itself has become less of an intrinsic art form, but more of a powerful tool to uncover the ever more fascinating secrets of complex molecular behavior, and has become worthy of fundamental study in its own right. The compendium assembled in this monograph is one that helps a new generation of scientists, interested in understanding the deeper aspects of molecular behavior, to understand this fascinating subject. Even so, it is a fairly advanced textbook that even expert practitioners will find absorbing as it contains much of value as the articles deal with our state-of-the art understanding of, among other things: ab initio, Born–Oppenheimer, equilibrium, adiabatic and vibrationally averaged structures; Coriolis, Fermi, and other interactions; variational approaches as well as conformations of complexes and so on.

Of course, there is now a new twenty-first century buzzword—"nanotechnology" or as I prefer to call it, N&N (not to be confused with M&M!) or nanoscience and nanotechnology. There is much confusion in the mind of the public as to what N&N actually is. However, as it deals with molecules and atomic aggregates at nanoscale dimensions, it is really only a new name for chemistry with a twenty-first century "bottom-up" perspective. Our molecule C_{60} is, as it happens, almost exactly 1 nm in diameter, or to be more accurate, the center-to-center distance of C_{60} molecules in a crystal is 1 nm (to an accuracy of ca. 1%). C_{60} has become something of an iconic symbol representing N&N and therefore I cannot help feeling a bit like Monsier Jourdain in Moliére's *Bourgeois Gentilhomme* (*MJ*—Monsieur Jourdain, *PM*—Philosophy Master):

MJ   *I wish to write to my lady.*
PM   *Then without doubt it is verse you will need.*
MJ   *No. Not verse.*
PM   *Do you want only prose then?*
MJ   *No—neither.*
PM   *It must be one or the other.*
MJ   *Why?*
PM   *Everything that is not prose is verse and everything that is not verse is prose.*
MJ   *And when one speaks—what is that then?*
PM   *Prose.*
MJ   *Well by my faith! For more than forty years I have been speaking prose without knowing anything about it.*

My response is (preferably in London Cockney vernacular):

"Cor blimey, guv … I'm a spectroscopist so I must have been a nanotechnologist all my life!"

Harold Kroto
The Florida State University

REFERENCES

1. Kraitchman, J. 1953. *Am J Phys* 21:17–24.
2. Costain, C. C. 1951. *Phys Rev* 82:108.
3. Smith, J. G., and J. K. G. Watson. 1978. *J Mol Spectrosc* 69:47–52.
4. Alexander, J., H. W. Kroto, and D. R. M. Walton. 1976. *J Mol Spectrosc* 62:175–80.
5. Avery, L. W., N. W. Broten, J. M. MacLeod, T. Oka, and H. W. Kroto. 1976. *Astrophys J* 205:L173–5.
6. Kroto, H. W. 1974. *Molecular Rotation Spectra.* New York: Wiley. Then republished by Dover: New York in 1992 as a paperback, with an extra preface including many spectra. Now republished in Phoenix editions: New York, 2003.
7. Watson, J. K. G. 1968. *Mol Phys* 15:479–90.
8. Ingold, K., and G. W. King. 1953. *J Chem Soc* 2702–4.
9. Dixon, R. N. 1964. *Trans Faraday Soc* 60:1363–8.
10. Hougen, J. T., P. R. Bunker, and J. W. C. Johns. 1970. *J Mol Spectrosc* 34:136–72.
11. King, M. A., H. W. Kroto, and B. M. Landsberg. 1985. *J Mol Spectrosc* 113:1–20.
12. Winnewisser, B., M. Winnewisser, I. R. Medvedev, et al. 2005. *Phys Rev Lett* 95:243002/1–4.

Editors

Jean Demaison is a former Research Director at CNRS, University of Lille I. The research for his PhD, which he received in 1972, was performed in Freiburg and Nancy in the field of microwave spectroscopy. He was invited to be a Professor at the Universities of Ulm, Louvain-La-Neuve, and Brussels. In 2008, he received the International Barbara Mez-Starck prize for outstanding contribution in the field of structural chemistry. He has published over 300 papers in research journals and contributed to 17 books.

Professor Attila G. Császár is the head of the Laboratory of Molecular Spectroscopy at Eötvös University of Budapest, Hungary. He received his PhD in 1985 in theoretical chemistry at the same place. His research interests include computational molecular spectroscopy, structure determinations of small molecules, ab initio thermochemistry, and electronic structure theory. He has published more than 150 papers in these fields, mostly in leading international journals.

James E. Boggs is Professor Emeritus at the University of Texas at Austin. His PhD was received from the University of Michigan after working on the Manhattan District Project. He has spent sabbaticals at Harvard, Berkelcy, and the University of Oslo. Dr. Boggs has published over 325 papers, mostly on microwave spectroscopy and applications of quantum theory. He organized the first 23 biennial meetings of the Austin Symposium on Molecular Structure. In 2010, he received the International Barbara Mez-Starck prize for outstanding contributions in structural chemistry. He has been chosen as a Fellow of the American Chemical Society.

Contributors

W. D. Allen
Center for Computational Quantum
    Chemistry
Department of Chemistry
University of Georgia
Athens, Georgia

J. E. Boggs
Department of Chemistry and
    Biochemistry
Institute for Theoretical Chemistry
University of Texas
Austin, Texas

A. G. Császár
Laboratory of Molecular Spectroscopy
Institute of Chemistry
Eötvös University
Budapest, Hungary

J. Demaison
Department of Physics (PhLAM)
University of Lille I
Villeneuve d'Ascq, France

J. -M. Flaud
Laboratory of Atmospheric Systems
    (LISA)
Associated to the National Center for
    Scientific Research (CNRS) and to
    the Universities of Paris-Est and
    Paris-Diderot
Créteil, France

W. J. Lafferty
Optical Technology Division
National Institute of Standards and
    Technology
Gaithersburg, Maryland

A. C. Legon
School of Chemistry
University of Bristol
Bristol, United Kingdom

R. J. Le Roy
Department of Chemistry
University of Waterloo
Waterloo, Ontario, Canada

A. Perrin
Laboratory of Atmospheric Systems
    (LISA)
Associated to the National Center for
    Scientific Research (CNRS) and to
    the Universities of Paris-Est and
    Paris-Diderot
Créteil, France

H. D. Rudolph
Department of Chemistry
University of Ulm
Ulm, Germany

K. Sarka
Deparment of Physical Chemistry,
    Faculty of Pharmacy
Comenius University
Bratislava, Slovakia

J. F. Stanton
Department of Chemistry and Biochemistry
Institute for Theoretical Chemistry
University of Texas
Austin, Texas

J. Vázquez
Department of Chemistry and Biochemistry
Institute for Theoretical Chemistry
University of Texas
Austin, Texas

Introduction

James E. Boggs

In 1861 [1], the famous Russian chemist Aleksandr Mikhailovich Butlerov (1828–1886) used the term "chemical structure" for perhaps the first time in a modern sense. He argued as we argue today that molecular structure is perhaps the most basic information about a substance and it has very strong ties to most macroscopic physical and chemical properties.

The study of molecular structures has been hampered by the fact that every experimental method applies its own definition of "structure" and thus structural results corresponding to different sources are usually significantly different. For example, the distance between maxima in the electron density distribution as measured by X-ray diffraction is very different from the distance corresponding to the minima in the vibrational potential energy surface as measured by quantum chemical computations or the various vibrational averages of that distance as measured by different methods of molecular spectroscopy. The sophisticated protocols that have been developed to account for these differences, and render intercomparisons and the use of combined experimental and computational techniques possible, is the subject of this advanced textbook.

Most of our notions about structure arise from within the Born–Oppenheimer approximation. The potential energy surfaces that result from this venerable approximation are one of the most useful and ubiquitous paradigms in descriptive chemistry. They give rise to our notions of activation energies and transition states for chemical reactions, force constants to which the strength of various bonds can be related, and most important for the topic of this textbook, the equilibrium structure (r_e). The latter is defined by the geometry that the nuclei adopt when in a minimum on the potential energy surface. None of these common concepts "exists" in the context of more rigorous theory—they are in a sense artifacts of the Born–Oppenheimer approximation. However, forming the central paradigm of molecular structure and chemical dynamics, the Born–Oppenheimer approximation is a very good one, and knowing what the r_e structures really "are" is desirable.

This book is novel in several ways. To the best of our knowledge, the subject matter of equilibrium molecular structures has never before been treated in a book in a manner that provides balance between quantum theory and experiment. Another novel aspect of this textbook is that the editors have endeavored to bring together a number of distinguished educators and practitioners in this branch of science to write chapters on their own fields of expertise, starting with the basic elements and proceeding to the latest advances and current best practices. Reading the book may be compared to sitting in on a series of lectures by some of the best experts in the world on the subjects they address. This is a book on molecular structure, but it does not describe the instruments or details of the experimental methods that are used in determining the structure. Rather, it describes the theory involved in determining, and converting measured or computed data into the most accurate and best understood molecular structures possible from the available data set. This step is of vital importance in chemistry where most of the significant information in a structure is contained in differences of structural parameters amounting to less, often considerably less than one percent.

The book is not only intended to be a textbook suitable for advanced undergraduate or graduate courses but is also sufficiently complete for interested readers and active workers in the area who would like to learn about certain aspects of the field with which they are not familiar. As Linus Pauling pointed out in 1939 in the preface of his famous book, *The Nature of the Chemical Bond*, [2] "the ideas involved in modern structural chemistry are no more difficult and require for their understanding no more, or a little more, mathematical preparation than the familiar concepts of chemistry." Thus, while most chapters of our textbook do make extensive use of mathematics, it is never beyond the scope of a student who is in the last half of an undergraduate program in chemistry or physics. In keeping with its purpose to be used as a textbook, the chapters contain several examples and exercises, some given with solutions and some without. Each chapter is provided with a table of contents and an overall index is given at the end of the book. Important references are given in case the reader wants to look at the original presentation of the information discussed.

The book is organized in the following way:

Chapter 1 deals with quantum chemistry, introduces the concept of potential energy surfaces on which the idea of equilibrium molecular structures is built. It also discusses the quantum chemical computation of structures and anharmonic force fields, the two central quantities of this book.

Chapter 2 describes the method of least squares that is commonly used to calculate a structure from the moments of inertia. The dangers posed by the problem of ill-conditioning and the presence of outliers and leverage points are discussed in detail and some remedies are proposed.

Chapter 3 discusses certain uses of perturbation theory in the study of molecular structures as well as computational aspects related to the study of so-called semiexperimental equilibrium structures.

Chapter 4 deals with the determination of moments of inertia from experimental spectra. The resonances, which make difficult the determination of reliable equilibrium constants, are discussed in detail.

Chapter 5 derives the relationship between moments of inertia and structural parameters and discusses the different methods permitting derivation of the structure. Empirical structures which are obtained from ground-state moments of inertia and which are assumed to be a good approximation of the equilibrium structure are also presented.

Chapter 6 deals with the determination of the potential of a diatomic molecule. Semiclassical methods as well as quantum mechanical methods are discussed and the Born–Oppenheimer breakdown effects are also treated here.

Chapter 7 presents complementary sources of information, which can be used for at least partial structure analysis with particular emphasis on the structure of molecular complexes.

Chapter 8 defines temperature-dependent position and distance averages and how they can be computed in addition to equilibrium molecular structures, bridging the gap between usual quantum theory and experiment.

The table Principal Structure, which can be found on the inside cover and after the Introduction, gathers the structures that are discussed in the book and that are encountered in the literature. The book is accompanied by a CD that presents further examples and exercises and additional information on the methods that are discussed in the main text as well as more technical material.

The editors and the authors are grateful to Therese Huet for reading Chapter 7, to Francois Rohart for reading Chapter 2, and to Harald Møllendal for reading most of the chapters.

REFERENCES

1. A. M. Butlerov. 1861. *Z. Chem. Pharm.* 4:549
2. Pauling, L. 1960. *The Nature of the Chemical Bond and the Structure of Molecules and Crystals: An Introduction to Modern Structural Chemistry.* Third edition. Ithaca, NY: Cornell University Press.

Principal Structures

Symbol	Definition	Section
Equilibrium structures		
r_e^{BO}	Born–Oppenheimer equilibrium structure: Corresponds to a minimum of the potential energy hypersurface defined within the Born–Oppenheimer separation of electronic and nuclear motion and determined by techniques of electronic structure theory.	1.1
r_e^{ad}	Adiabatic equilibrium structure: Mass-dependent equilibrium structure corresponds to the adiabatic potential energy hypersurface obtained after adding a small, first-order, so-called diagonal Born–Oppenheimer correction (DBOC) to r_e^{BO}.	1.6
r_e^{SE}	Semiexperimental equilibrium structure: Determined from a fit of the structural parameters to the equilibrium moments of inertia, obtained from the experimental effective, ground-state rotational constants corrected by the rovibrational contribution calculated using a cubic force field usually determined first principles (ab initio).	3.3
r_e^{exp}	Experimental equilibrium structure: Obtained from a fit of the structural parameters to the experimental equilibrium moments of inertia.	4.3
Average structures		
Position averages		
$r_z = r_{\alpha,0}$	Zero-point average structure: A temperature-independent average structure belonging to the average nuclear positions in the ground vibrational state.	8.1
$r_{\alpha,T}$	r_α-structure: Distance between the nuclear positions averaged at a given temperature T assuming thermal equilibrium.	8.1
Distance (and angle) averages		
$r_{g,T}$	Mean (average) internuclear distance (angle): Average internuclear distance (angle), related to the expectation value $<r>$, at temperature T assuming thermal equilibrium ("g" stands for center of gravity of the distance distribution function).	8.1
$r_{a,T}$	Inverse internuclear distance (angle): Average related to electron scattering intensities.	8.1
$\langle r^2 \rangle_T^{1/2}$	Root-mean-square (rms) internuclear distance (angle): Related to the expectation value $<r^2>$, at temperature T assuming thermal equilibrium.	8.1
$\langle r^{-2} \rangle_T^{-1/2}$	Effective internuclear distance (angle): Related to the expectation value $<r^{-2}>$, at temperature T assuming thermal equilibrium.	8.1
$\langle r^3 \rangle_T^{1/3}$	Cubic internuclear distance (angle): Related to the expectation value $<r^3>$, at temperature T assuming thermal equilibrium.	8.1
$\langle r^{-3} \rangle^{-1/3}$	Inverse cubic: Average, appears in dipolar coupling constants.	8.1, 7.5

(Continued)

Symbol	Definition	Section
Mass-dependent structures		
r_m	Mass-dependent structure: Obtained from a fit of the structural parameters to the mass-dependent moments of inertia $I_m = 2I_s - I_0$, where I_s are the substitution moments of inertia that are calculated from the substitution coordinates r_s, and I_0 the ground-state moments of inertia. Generally no better than r_s.	5.5.1
r_c	Improvement of the r_m structure by using complementary sets of isotopologues.	5.5.2
r_m^ρ	r_m^ρ structure: Since the rovibrational contributions ε^0 vary less within a set of isotopologues than the inertial moments I^0, scaling of the inertial moments of all isotopologues by an appropriate common factor $(2\rho_g - 1)$ (for each principal axis g) calculated for the parent, and then submitting the scaled inertial moments to a least-squares fit to obtain the bond coordinates.	5.5.3
$r_m^{(1)}$, $r_m^{(2)}$	Mass-dependent structure: Based on the different dependence of inertial moments and their rovibrational contributions on the atomic masses (of one and half degrees). Models the rovibrational contributions by the (least possible number of) parameters: c_g (for each principal axis g), multiplied by the square root of the inertial moment of the individual isotopologue $\sqrt{I_g}$, and d_g (for $r_m^{(2)}$ only), multiplied by an isotopologue-dependent, but g-independent mass factor. Least-squares fitting to obtain bond coordinates and rovibrational parameters. The Laurie contraction of a X-H bond upon deuteration is modeled ($r_m^{(1L)}$, $r_m^{(2L)}$) by an additional parameter δ_H. Refined models $r_m^{(1r)}$, $r_m^{(2r)}$ assume that the rovibrational effects depend more on the overall shape or contour of the molecule (equal for all isotopologues) than on the principal axis systems whose orientations within the molecular shapes differ among the isotopologues.	5.5.4
Empirical structures		
r_0	Effective structure: Least-squares fitting of experimental ground-state inertial moments I^0 of a set of isotopologues to obtain bond coordinates, neglecting rovibrational contributions ε^0 completely, can be realized in practice by means of different sets of observables: $r_0(I)$, $r_0(P)$, $r_0(B)$, and also $r_0(I,\Delta I)$, $r_0(P,\Delta P)$, where the moments of only the parent and the moment differences between parent and isotopologues are used.	5.3
r_s	Substitution structure: Aimed at obtaining Cartesian coordinates of individual atom, numerically dominated by inertial moment difference ΔI upon substitution, no least-squares used, though expandable on sets of several isotopologues (substituted atoms) by least-squares fitting: r_s-fit (determined via the Kraitchman equations).	5.4
r_s variants: $r_{\Delta I}$, $r_{\Delta P}$	ps-Kr ("pseudo-Kraitchman") structure: Attempts to compensate rovibrational contributions by least-squares, fitting exclusively differences of moments between parent and isotopologues ΔI^0 or ΔP^0 to obtain bond coordinates, realized by $r(\Delta I)$, $r(\Delta P)$.	5.3.2

1 Quantum Theory of Equilibrium Molecular Structures

Wesley D. Allen and Attila G. Császár

CONTENTS

1.1 CONCEPT OF THE POTENTIAL ENERGY SURFACE

Molecular quantum mechanics, as embodied in the time-independent Schrödinger equation $\hat{H}\Psi = E\Psi$, is the physical foundation of chemistry. For systems containing atoms no heavier than Ar, highly accurate results are obtained from the standard nonrelativistic Hamiltonian involving only Coulombic interactions:

$$
\hat{H} = -\frac{\hbar^2}{2}\sum_{\alpha}\frac{\nabla^2_{\alpha}}{M_{\alpha}} - \frac{\hbar^2}{2m_e}\sum_i\nabla^2_i + \sum_{\alpha}\sum_{\beta>\alpha}\frac{Z_{\alpha}Z_{\beta}e^2}{4\pi\varepsilon_0 r_{\alpha\beta}}
$$
$$
-\sum_{\alpha}\sum_i\frac{Z_{\alpha}e^2}{4\pi\varepsilon_0 r_{i\alpha}} + \sum_i\sum_{j>i}\frac{e^2}{4\pi\varepsilon_0 r_{ij}}
\tag{1.1}
$$

in which Greek (α and β) indices refer to nuclei with masses M_{α} and charges Z_{α}, and Latin (i and j) indices refer to electrons with mass m_e and charge e, while the corresponding interparticle distances are denoted by $r_{\alpha\beta}$, $r_{i\alpha}$, and r_{ij}. The Laplacian operator for each particle takes the simple form $\Delta \equiv \nabla^2 = \partial^2/\partial x^2 + \partial^2/\partial y^2 + \partial^2/\partial z^2$ in rectilinear Cartesian coordinates but generally is considerably more complicated if curvilinear internal coordinates are used.

The five operators in order of appearance in Equation 1.1 represent nuclear kinetic energy $(\hat{T}_N)$, electronic kinetic energy $(\hat{T}_e)$, nuclear–nuclear repulsion $(\hat{V}_{NN})$, electron–nuclear attraction $(\hat{V}_{eN})$, and electron–electron repulsion $(\hat{V}_{ee})$. Because exact, analytic solutions to the Schrödinger equation built on $\hat{H}$ are not possible for many-particle systems, effective approximation methods must be employed. The development of algorithms for such methods has been one of the main goals of modern computational quantum chemistry. Rigorous approaches that do not resort to empirical parameterization and only invoke the fundamental constants are termed ab initio (from the beginning) or first-principles methods.

Nuclear and electronic motions in molecular systems have greatly different time-scales and a wide separation in classical velocities (at least three orders of magnitude) that has profound consequences for chemistry. Because electrons are much lighter than nuclei $(m_e/m_H \approx 1/1836)$, they move much more vigorously. In effect, the light, fast electrons adjust instantaneously to the motions of the slow, heavy nuclei. Therefore, the nuclear and electronic degrees of freedom can be separated adiabatically, as in the highly accurate Born–Oppenheimer (BO) approximation,* whereby the electronic part of the Schrödinger equation is solved repeatedly with nuclei clamped at various positions. The purely electronic equation is

$$\hat{H}_e \psi_e(\mathbf{r}_i ; \mathbf{r}_\alpha) = E_e(\mathbf{r}_\alpha) \psi_e(\mathbf{r}_i ; \mathbf{r}_\alpha) \tag{1.2}$$

in which the electronic Hamiltonian is $\hat{H}_e = \hat{T}_e + \hat{V}_{ee} + \hat{V}_{eN}$, and the nuclear coordinates $\mathbf{r}_\alpha$ are fixed parameters. Adding the nuclear–nuclear repulsion energy to the electronic energy eigenvalues $E_e(\mathbf{r}_\alpha)$ that depend parametrically on the nuclear positions yields a potential energy surface (PES) for nuclear motion,

$$V(\mathbf{r}_\alpha) = E_e(\mathbf{r}_\alpha) + V_{NN}(\mathbf{r}_\alpha) \tag{1.3}$$

The nuclear Schrödinger equation resulting from the BO approximation is

$$[\hat{T}_N + V(\mathbf{r}_\alpha)]\psi_N(\mathbf{r}_\alpha) = E_N \psi_N(\mathbf{r}_\alpha) \tag{1.4}$$

This equation can be solved for the vibrational-rotational states that occur within a given electronic state. Derivatives of the electronic wave function with respect to the nuclear coordinates, namely $\nabla_\alpha \psi_e$ and $\nabla_\alpha^2 \psi_e$, are neglected in the BO approximation and are usually very small.

The PESs $V(\mathbf{r}_\alpha)$, illustrated by a model function in Figure 1.1, are fundamental to most modern branches of chemistry, especially spectroscopy and kinetics. The topography of the surface $V(\mathbf{r}_\alpha)$ constitutes the basis for ascribing geometric structures to

---

* The BO separation of electronic and nuclear degrees of freedom was introduced in Born, M., and J. R. Oppenheimer. 1927. *Ann Phys* 84:457. However, a better, more contemporary and accessible reference is Born, M., and K. Huang. 1954. *Dynamical Theory of Crystal Lattices*, appendix VIII. London: Oxford University Press.

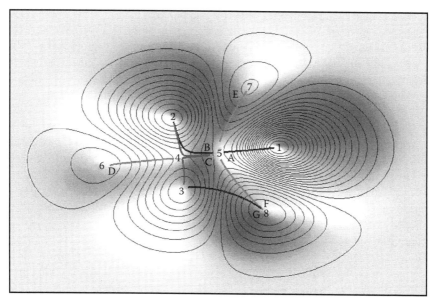

FIGURE 1.1   A two-dimensional model of a molecular potential energy surface and characteristic features and paths on it: points (**1**, **2**, **3**) are local minima; (**4**, **5**) are transition states (first-order saddle points); and (**6**, **7**, **8**) are second-order saddle points that appear as local maxima in this cross section of the PES. Paths A and B comprise the intrinsic reaction path of steepest descent that connects reactant **1** to product **2** via transition state **5**. Path C starts at a valley-ridge inflection point; small perturbations about such a point can cause a bifurcation of steepest descent paths and instability in the final products, in this case either **2** or **3**. Paths D, E, and F are gradient extremum paths descending from points **6**, **7**, and **8**, respectively, along ridges to minimize the steepness of the route. Path G is a corresponding steepest descent path that starts out coincident with F but falls off the ridge into the basin of minimum **3**.

molecules.* The local minima occurring on this multidimensional PES correspond to the equilibrium (r_e) structures of molecules, on which virtually all chemical intuition is built. Accordingly, it is the BO approximation that allows equilibrium structures to be defined as special points among the instantaneous configurations (geometries) that nuclei may exhibit. Depictions of static molecular frameworks are pervasively used to describe and understand chemical phenomena, and the implicit assumption therein is that the nuclei are localized in potential energy wells centered about the corresponding r_e structures and execute only small-amplitude vibrations away from their equilibrium positions. Without the BO separation of nuclear and electronic motions, the traditional concept of molecular structure would be lost, and only a murky quantum soup of delocalized particles would exist.

---

* As indicated by the notation $V(\mathbf{r}_\alpha)$, PESs are inherently hypersurfaces for all molecules larger than triatomics, involving $3N - 6$ internal degrees of freedom for a nonlinear N-atomic molecule. Even in the case of a nonlinear triatomic molecule, a four-dimensional plot (V vs. three degrees of freedom) would be required to fully represent the potential energy function.

The variation of the total energy of the chemical system as a function of the internal coordinates of the constituent nuclei is described by PESs. Internal coordinates describe the vibrations of N-atomic molecules, and thus their number is 6(5) less than the total number ($3N$) of Cartesian variables for nonlinear (linear) molecules. Because an equilibrium structure is a local minimum of the corresponding PES, the associated quadratic force constant matrix must be positive definite.*

In the conventional BO separation of nuclear and electronic motions, the resulting PES is isotope independent, because the masses of the nuclei are assumed to be infinitely heavy. For example, the BO PESs of molecules containing deuterium (D) instead of hydrogen (H) are identical. By means of first-order perturbation theory (PT), the diagonal Born—Oppenheimer correction (DBOC) may be used to relax this strict assumption somewhat while keeping the concept of a PES intact. Appending the DBOC to $V(\mathbf{r}_\alpha)$ gives rise to adiabatic PESs (APESs) that are dependent on the masses of the nuclei and are slightly different for a series of isotopologues or isotopomers.[†]

It is important to realize that many PESs exist for any given molecule, each corresponding to a different electronic state solution of Equation 1.2. Of course, the most fundamental PESs and r_e structures are those of ground electronic states. Nevertheless, well-defined r_e structures are also generally exhibited for the PESs of excited electronic states. Frequently, the r_e structures of excited states are markedly different from those of ground states, as in the case of CO_2, for which bent equilibrium structures are found for the lowest excited states. Equilibrium structures are most useful for interpretive purposes if the PESs of excited electronic states are well separated and not highly coupled, but their mathematical basis is retained even if such circumstances are not met. In special cases where nonadiabatic nuclear–electronic interactions occur, as in the Jahn–Teller or Renner–Teller effects,[‡] multiple PESs that are strongly coupled must be considered simultaneously to understand the motion of the nuclei. However, to maintain focus, we are concerned neither with such cases where multiple electronic states are coupled nor with the evaluation of nonadiabatic coupling matrix elements.

Much of contemporary experimental physical chemistry, through spectroscopic, scattering, and kinetic studies, is directed toward the elucidation of salient features of potential energy hypersurfaces (Figure 1.1). One can obtain details of the PES most easily, including structural and spectroscopic signatures of its minima, from an analysis of well-resolved vibrational-rotational (often abbreviated as rovibrational) spectra or from scattering experiments. When characterization of local minima of

---

* A square matrix is called "positive definite" if all of its eigenvalues are larger than zero. A square matrix is called "positive semidefinite" if all of its eigenvalues are nonnegative.

[†] According to the International Union of Pure and Applied Chemistry (IUPAC), *isotopologues* are molecular entities that differ only in isotopic composition (number of isotopic substitutions), for example, CH_4, CH_3D, and CH_2D_2. On the other hand, an *isotopomer*, where the term comes from the contraction of "isotopic isomer," refers to an isomer having the same number of each isotopic atom in a molecule but differing in positions.

[‡] The interested reader can find details about the Renner–Teller and Jahn–Teller effects, related to degeneracies forced by symmetry at linear and nonlinear molecular geometries, respectively, in part 4 of the book Jensen, P., and P. R. Bunker, eds. 2000. *Computational Molecular Spectroscopy*. Chichester: Wiley.

the PES is the goal, the best spectroscopic techniques possess several advantages over scattering measurements: (1) they can provide results of higher intrinsic accuracy and (2) there is less need to average over the usually somewhat loosely defined experimental conditions. Generally, experiments, through well-defined modeling approaches, yield parameters, including molecular structures, in more or less local representations of potential surfaces.

Much of modern quantum chemistry is also aimed at mapping out given portions or the whole of potential energy hypersurfaces of molecular species or reactive (scattering) systems by computational, rather than experimental, means. The availability of analytic gradients and higher derivative methods in standard electronic structure programs,* for reasons discussed in Sections 1.3 and 1.4, has substantially increased the utility of quantum chemistry for the exploration of PESs. For structural studies, the PES is needed mostly in the vicinity of a minimum. Therefore, techniques based on power series expansions around a single stationary point can be highly useful. Indeed, locating r_e structures and evaluating attendant (anharmonic) force fields based on series expansions of rather large molecules is now commonplace in quantum chemistry.

1.2 INTERPLAY OF ELECTRONIC AND NUCLEAR CONTRIBUTIONS TO THE POTENTIAL ENERGY SURFACE

Equilibrium structures, transition states, and other stationary points of chemical systems occur when the gradient of the PES with respect to nuclear coordinates is zero. Force fields for molecular vibrations are constituted by the higher-order derivatives of the PES at these stationary points. According to Equation 1.3, all derivatives of the PES can be decomposed into electronic energy $[E_e(\mathbf{r}_\alpha)]$ and nuclear–nuclear repulsion $[V_{NN}(\mathbf{r}_\alpha)]$ terms. Both of these contributions are large and almost always of opposite signs. Thus, it is the interplay of these competing terms that determine the positions of equilibrium structures and the strength and sign of the force constants for molecular vibrations. The V_{NN} contribution and its derivatives can be calculated exactly by simple algebraic expressions involving Coulombic terms, whereas the E_e contribution and its derivatives can be determined only approximately by means of computationally intensive electronic structure theory. This situation creates an imbalance of errors that must be appreciated to understand the effects that govern the accuracy of ab initio theoretical predictions of structures and force fields.

The N_2 and F_2 diatomic molecules provide paradigms for the interplay of the E_e and V_{NN} contributions to molecular PESs. Experimental potential energy curves $[V_{RKR}(r)]$ for N_2 and F_2 (Figure 1.2) can be extracted from rovibrational spectroscopic data by means of the Rydberg–Klein–Rees (RKR) inversion technique. In particular, the classical turning points for each quantized vibrational level are known from RKR inversion up to vibrational quantum numbers $v = 22$ and $v = 23$ for N_2 and F_2, respectively. Details of the RKR method are available in the related literature and Chapter 6 of this book. For comparison to experiment, we also consider the potential

---

* Yamaguchi, Y., Y. Osamura, J. D. Goddard, and H. F. Schaefer III. 1994. *A New Dimension to Quantum Chemistry: Derivative Methods in* Ab Initio *Molecular Electronic Structure Theory.* New York: Oxford University Press.

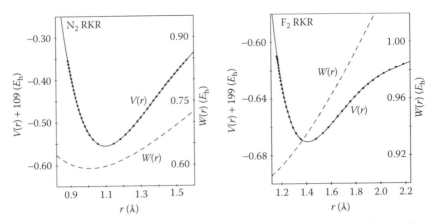

FIGURE 1.2 Rydberg–Klein–Rees (RKR) potential energy curves of N_2 and F_2 and the corresponding functions $W(r) = V_{RHF}(r) - V_{RKR}(r)$.

energy curves obtained from a beginning level of ab initio electronic structure theory, namely, the restricted Hartree–Fock (RHF) method with a Gaussian double-ζ plus polarization (DZP) basis set.* In Figure 1.2, the difference function $W(r) = V_{RHF}(r) - V_{RKR}(r)$ is plotted alongside $V_{RKR}(r)$ for N_2 and F_2, showing the variation of the magnitude of the electron correlation energy with bond distance.

Robust analytical representations[†] of both the RKR and RHF potential curves allow derivatives of $V(r)$ and hence $E_e(r) = V(r) - V_{NN}(r)$ to be determined analytically through fourth order as a function of the bond distance for our diatomic paradigms. Thus, at any specified point within a given range, the RHF/DZP[‡] theoretical predictions for the potential energy derivatives of various orders can be compared to "exact experimental" values. Specific numerical comparisons are made in Tables 1.1 and 1.2 at the distinct equilibrium bond distances of the RHF and RKR curves. In addition, the RKR derivative functions are plotted in Figure 1.3, and the corresponding RHF curves are virtually indistinguishable on the scale of the plots. The N_2 and F_2 examples are chosen not only because accurate experimental data are available but also because they exhibit very different levels of agreement between theoretical and experimental equilibrium structures. In particular, for N_2 the RHF/DZP equilibrium distance is 0.015 Å too short, within typical ranges of error, whereas for F_2 this difference is 0.077 Å, which is very large even for this introductory level of electronic structure theory.

The ab initio and experimental data for N_2 and F_2 in Tables 1.1 and 1.2 clearly demonstrate that the $E_e(r)$ and $V_{NN}(r)$ derivatives are sizable at all orders and opposite in sign. The E_e and V_{NN} contributions to the gradient obviously cancel each

---

* See Section 1.5 for a description of Gaussian basis sets and electronic structure methods.
† For a detailed account, see Allen, W. D., and A. G. Császár. 1993. *J Chem Phys* 98:2983.
‡ In ab initio electronic structure theory, it is customary to employ the notation "level/basis," where "level" denotes a particular wave function method and "basis" a particular one-particle basis set (see Figure 1.5).

TABLE 1.1

A Comparison of RHF/DZP Theoretical and RKR Experimental Data for the Electronic (E_e), Nuclear–Nuclear Repulsion (V_{NN}), and Total (V) Energies of N_2 and Their Geometric Derivatives through Fourth Order[a]

	At r_e(RHF/DZP) = 1.082707 Å			At r_e(Expt) = 1.097685 Å		
	RHF/DZP	RKR	Percentage Error	RHF/DZP	RKR	Percentage Error
$E_e(N_2)$	−132.907896	−133.503919	−0.45	−132.580357	−133.177747	−0.45
E_e'	96.437	96.074	0.38	94.255	93.823	0.46
E_e''	−147.88	−152.54	−3.1	−143.55	−148.01	−3.0
E_e'''	294.1	308.4	−4.6	283.9	297.6	−4.6
E_e''''	−692.1	−733.8	−5.7	−665.5	−704.9	−5.6
$V_{NN}(N_2)$	23.948932	23.948932	0	23.622147	23.622147	0
V_{NN}'	−96.437	−96.437	0	−93.823	−93.823	0
V_{NN}''	178.14	178.14	0	170.95	170.95	0
V_{NN}'''	−493.6	−493.6	0	−467.2	−467.2	0
V_{NN}''''	1823.6	1823.6	0	1702.5	1702.5	0
$V(N_2)$	−108.958964	−109.554988	−0.54	−108.958210	−109.555600	−0.55
V'	0.00	−0.3632	–	0.4315	0.00	–
V''	30.26	25.60	18.2	27.40	22.94	19.4
V'''	−199.5	−185.2	7.7	−183.3	−169.6	8.1
V''''	1131.4	1089.8	3.8	1037.0	997.6	3.9

[a] All energies are given in hartrees, whereas all derivatives correspond to energies measured in attojoules and distances in Å. The percentage errors are given as 100(RHF/RKR − 1). The RKR data are based on Lofthus, A., and P. H. Krupenie. 1977. *J Phys Chem Ref Data* 6:113.

other completely at equilibrium. What is less appreciated is that the cancellation is almost as great for the quadratic force constants. For the higher-order force constants, the derivatives of V_{NN} become increasingly dominant. To be precise, for N_2 at the experimental geometry the ratios are $E_e'/V_{NN}' = -1.00$, $E_e''/V_{NN}'' = -0.87$, $E_e'''/V_{NN}''' = -0.63$, and $E_e''''/V_{NN}'''' = -0.38$, whereas in the F_2 case these ratios are −1.00, −0.96, −0.87, and −0.74, respectively. Figure 1.3 shows that this behavior is not restricted to the experimental bond distance alone, because the $V(r)$ derivative curves shift away from the r axis as the order of the derivative is increased as a consequence of the growing importance of the V_{NN} contributions. In brief, the higher-order bond stretching derivatives depend strongly on core–core nuclear repulsions, and the cancellation of the E_e and V_N derivative terms decreases substantially in higher order.

The accuracy of the RHF/DZP electronic energy derivatives of both N_2 and F_2 is remarkably good for such a modest level of theory. The errors in the $E_e(r)$ derivatives through fourth order are under 6% for both molecules over bond-length intervals of at least 0.5 Å surrounding r_e. However, the theoretical values for the second derivatives

TABLE 1.2

A Comparison of RHF/DZP Theoretical and RKR Experimental Data for the Electronic (E_e), Nuclear–Nuclear Repulsion (V_{NN}), and Total (V) Energies of F_2 and Their Geometric Derivatives through Fourth Order[a]

	At r_e(RHF/DZP) = 1.334980 Å			At r_e(Expt) = 1.411930 Å		
	RHF/DZP	RKR	Percentage Error	RHF/DZP	RKR	Percentage Error
$E_e(F_2)$	−230.847255	−231.773494	−0.40	−229.092257	−230.027524	−0.41
E_e'	104.859	104.371	0.47	94.277	93.741	0.57
E_e''	−148.28	−148.88	−0.4	−127.42	−128.08	−0.5
E_e'''	298.1	296.8	0.4	245.9	245.7	0.08
E_e''''	−755.4	−745.8	1.3	−604.6	−588.0	2.8
$V_{NN}(F_2)$	32.107853	32.107853	0	30.357979	30.357979	0
V_{NN}'	−104.859	−104.859	0	−93.741	−93.741	0
V_{NN}''	157.094	157.094	0	132.784	132.784	0
V_{NN}'''	−353.03	−353.03	0	−282.13	−282.13	0
V_{NN}''''	1057.8	1057.8	0	799.3	799.3	0
$V(F_2)$	−198.739402	−199.665641	−0.46	−198.734278	−199.669545	−0.47
V'	0.00	−0.4873	–	0.5365	0.00	–
V''	8.818	8.217	7.3	5.365	4.703	14.1
V'''	−54.95	−56.24	−2.3	−36.18	−36.39	−0.6
V''''	302.3	311.9	−3.1	194.7	211.3	−7.9

[a] All energies are given in hartree, whereas all derivatives correspond to energies measured in attojoules and distances in Å. The percentage errors are given as 100(RHF/RKR − 1). The RKR data are based on Colboum, E. A., M. Dagenais, A. E. Douglas, and J. W. Raymonda. 1976. *Can J Phys* 54:1343.

of $V(r)$ are much less accurate than those of $E_e(r)$—a disparity that becomes smaller for higher-order derivatives. Because the errors in the $E_e(r)$ derivatives are comparable at all orders, the fact that the V'' predictions are much poorer than the V''' and V'''' results is a direct consequence of the cancellation of nuclear repulsion and electronic energy effects. As a specific example, the theoretical E_e'' value for N_2 is in error by only 3.0% at the experimental r_e distance, but the corresponding discrepancy for V'' is 19.4% (see Table 1.1). In contrast, E_e'''' and V'''' for N_2 are predicted by the RHF/DZP method with comparable accuracies of 5.6% and 3.9%, respectively.

It must be recognized that slight inaccuracies in the evaluation of $E_e'(r)$ by theoretical methods may lead to substantial errors in the value of $V'(r)$. This is the reason correlation effects (see Section 1.5) are prominent in computing gradients and, consequently, equilibrium structures. In this sense, it is fundamentally more difficult to determine accurate r_e parameters by electronic structure techniques than force constants (especially higher-order ones). The case of F_2 demonstrates the situation

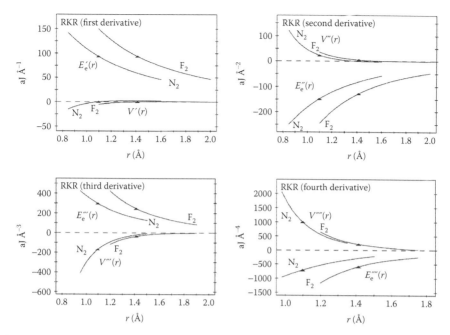

FIGURE 1.3 Derivative functions of the Rydberg–Klein–Rees (RKR) potential energy curves of N_2 and F_2. Solid triangles indicate the equilibrium distances.

dramatically. In Table 1.2, it is seen that RHF/DZP theory predicts $E'_e(r)$ with an error of only about 0.5%, regardless of the geometry sampled. However, the r_e (RHF/DZP) value of 1.3350 Å is a gross underestimation of the 1.4119 Å experimental distance. This example is often cited to highlight the possible extent of electron correlation effects on equilibrium bond distances. As to the $V'(r)$ curves of N_2 and F_2 (Figure 1.3), at the shortest distances the force is large and acts to separate the atoms. At $r = r_e$ the force vanishes. Linearity of the curve around r_e is connected with the harmonic character of the oscillator. For $r > r_e$ the force is of opposite sign than for $r < r_e$ and helps to establish the chemical bond.

Our analysis of N_2 and F_2 demonstrates the necessity of using an accurate reference geometry for evaluating molecular force fields. If theoretical derivatives of the total energy are compared to experimental values at the same geometry, despite the problem of cancellation of V_{NN} and E_e terms, even RHF/DZP theory is quite successful in predicting force constants. Note in Table 1.2 that for F_2 the "pure" theoretical quadratic force constant of 8.82 aJ·Å^{-2} is 87% larger than the experimental value of 4.70 aJ·Å^{-2}, but the error comes almost exclusively from the drastically different reference geometries upon which the force constants are based. A direct comparison at the experimental r_e structure reveals a much smaller error of 14.1%, and at the theoretical r_e distance the discrepancy is only 7.3%.

The agreement between the RHF/DZP and RKR values for V''' and V'''' is even better, provided once again that a direct comparison of quantities at the same geometry is made. For example, at the experimental r_e structure, the RHF/DZP cubic force constant for F_2 (-36.18 aJ·Å^{-3}) differs from the RKR value (-36.39 aJ·Å^{-3}) by only 0.6%. Because F_2 is recognized as a pathological case for computational electronic structure theory, it is remarkable that the errors in the second, third, and fourth derivatives of $E_e(r)$ are considerably smaller for F_2 than for N_2 (cf. Tables 1.1 and 1.2). This comparison emphasizes that the quality of the reference geometry is critical in the ab initio prediction of force constants.

EXERCISE 1.1

The following model diatomic potential energy function correctly describes the interplay of electronic and nuclear–nuclear repulsion energy:

$$V(r) = \varepsilon\left(\frac{e^{-ar}}{r} - be^{-cr^2}\right)$$

where a, b, and c are adjustable, dimensionless parameters; ε is the nuclear–nuclear repulsion energy at the equilibrium distance R_e; and $r = R/R_e$ is a scaled bond-length variable. The parameters a, b, and c can be determined by requiring $V(r)$ to reproduce known spectroscopic values for R_e, the dissociation energy (D_e), and the harmonic vibrational frequency (ω_e).

(a) Show that

$$b = e^c(d + e^{-a})$$

$$c = \frac{a+1}{2(de^a+1)}$$

and

$$\frac{d(a^2+3a+3)+e^{-a}(a+2)}{(de^a+1)} = f$$

where

$$d = \frac{D_e}{\varepsilon}$$

and

$$f = c_f\left(\frac{\mu R_e^3 \omega_e^2}{Z_A Z_B}\right)$$

are dimensionless quantities, and

$$c_f = \frac{4\pi^2}{\bar{m}_e a_0 \alpha^2 10^{16}}$$

is a unit conversion factor. The equation for f employs the reduced mass μ for diatomic vibrations in atomic mass units (u), R_e in Å, and ω_e in cm^{-1}, whereas c_f

involves the fine-structure constant $\alpha = 1/137.0359997$, the relative mass of the electron $\bar{m}_e(u) = 1/1822.8849$, and the Bohr radius $a_0 = 0.529\ 177\ 209$ Å.

(b) Given the following spectroscopic constants, determine the parameters a, b, and c for the diatomic potential energy curves of H_2, HF, N_2, O_2, F_2, and I_2.

	R_e (Å)	ω_e (cm^{-1})	D_e (cm^{-1})
H_2	0.7412	4403.2	38297
HF	0.95706	4138.7	49314
N_2	1.0977	2358.0	79868
O_2	1.2074	1580.2	42046
F_2	1.41193	916.6	13395
I_2	2.667	214.52	12560

(c) Use the diatomic potential curves to compute and interpret the derivative ratios $\rho_n = E_e^{(n)}(R_e)/V_{NN}^{(n)}(R_e)$ for $n = 2\text{--}6$, where the superscript (n) denotes the order of the derivative.

1.3 OPTIMIZATION ALGORITHMS

The geometric stationary points on a molecular PES include local and global minima, maxima, and saddle points of various orders (Figure 1.1). Minima display positive curvature for distortions along any direction and are characterized by a positive-definite second-derivative matrix (quadratic force constant or Hessian matrix). A local minimum is simply a minimum near an input or reference geometry. The lowest-energy minimum that exists on a given PES is called the *global minimum*. In general, minima represent the BO equilibrium structures (r_e^{BO}) of different conformers and isomers corresponding to a given molecular formula and are thus paramount in molecular applications. In contrast, genuine local maxima are unimportant and rarely encountered. Among saddle points, those characterized by a single negative eigenvalue of the Hessian matrix are of special significance. These stationary points are the classic (first-order) *transition states* for chemical reactions. In any one-step process, the reactant and product minima will be connected by a transition state via an intrinsic reaction path (IRP) of steepest descent, for example, paths A and B in Figure 1.1. The kinetic stability of a molecule and the activation energies for its reactions are determined by the relative energies of the surrounding transition states. Transition state theory in its various forms allows rates of chemical reactions to be computed simply from local properties of transition states.

Because finding and characterizing stationary points on a molecular PES is fundamental to much of chemistry, there is a great need for mathematical optimization algorithms. It would be desirable to readily perform both local and global geometry optimizations on a PES. However, global optimizations are rarely feasible, and the completeness of such searches usually cannot be guaranteed. Therefore, we focus on practical techniques for local optimizations in this discussion.

The optimization algorithms most frequently employed by quantum chemistry programs can be categorized as (1) those using energy points alone, (2) those using numerical or analytic gradients and approximate second-derivative information, and (3) those using (analytic) techniques to obtain both first and second derivatives of the PES. If no derivative information is available, the two common optimization techniques applied are the univariate method and the simplex method. Here, we give only a brief description of the univariate method. In this exceedingly simple approach, one-dimensional optimizations are performed in a sequential and repetitive manner over the geometric coordinates $\{q_i\}$ of the system. Starting from a current point $\mathbf{q}_m$, the ith coordinate (q_i) is displaced by preselected amounts $\{0, a_i, -b_i\}$ and the energy points $\{E(0), E(a_i), E(-b_i)\}$ are computed. These three points are fit to a parabola to minimize the energy with respect to q_i and obtain a new point $\mathbf{q}_{m+1}$. If $\mathbf{q}_{m+1}$ is not interpolated by the existing q_i data, then energies for additional q_i displacements may be computed to ensure a reliable geometry update. The same minimization procedure is then performed along the next coordinate q_{i+1}. The process is continued by cyclically passing through all the coordinates until the energy changes are acceptably small along each direction and the desired local minimum is reached. Convergence of the method is usually very slow, especially if the coordinates are strongly coupled or the PES is flat along some direction (which is often the case for at least some internal degrees of freedom for all but the smallest molecules).

Energy gradients, computed by either efficient analytic techniques or more expensive numerical procedures, greatly facilitate the optimization of stationary points on a PES. Analytic gradients are available for many electronic structure methods at costs comparable to the corresponding energy computations. Quasi-Newton methods are generally the preferred choice among methods that employ energy gradients. To understand such techniques, consider a set of equations $\{f_k(\mathbf{q}) = 0\}$ that must be solved to optimize a set of variables $\mathbf{q}$. The multidimensional Newton–Raphson (NR) approach employs a series expansion of the functions $f_k(\mathbf{q})$ about some reference point $\mathbf{q}_m$ to obtain

$$f_k(\mathbf{q}) = f_k(\mathbf{q}_m) + \nabla f_k(\mathbf{q}_m) \cdot (\mathbf{q} - \mathbf{q}_m) + \cdots = 0 \qquad (1.5)$$

Truncating the expansion after first order and solving for $\mathbf{q}$ gives the iterative update equation

$$\mathbf{q}_{m+1} = \mathbf{q}_m - \mathbf{H}_m^{-1} \mathbf{g}_m \qquad (1.6)$$

in which the components of the vector $\mathbf{g}_m$ are

$$(\mathbf{g}_m)_k \equiv f_k(\mathbf{q}_m) \qquad (1.7)$$

and the elements of the matrix $\mathbf{H}$ are

$$(\mathbf{H}_m)_{kj} \equiv \left(\frac{\partial f_k}{\partial q_j} \right)_{\mathbf{q}_m} \qquad (1.8)$$

In a standard optimization problem, the $f_k(\mathbf{q})$ functions comprise the gradient of the potential energy function $V(\mathbf{q})$, and $\mathbf{H}$ is the matrix of second derivatives $(\partial^2 V/\partial q_k \partial q_j)_{\mathbf{q}_m}$. The PES is thus represented locally as a quadratic function of the coordinates $\mathbf{q}$, which is usually a good approximation near a stationary point. The NR scheme may also be used in problems other than energy minimizations, such as the determination of valley-ridge inflection points (Exercise 1.3).

Quasi-Newton optimization methods employ Equation 1.6 by explicitly evaluating gradients but only estimating the Hessian matrix $\mathbf{H}$ to cut down on computational costs. Quasi-Newton methods are in principle applicable to the optimization of all types of stationary points, not just minima. However, it is critical for the approximate Hessian matrix to accurately represent the shape of the PES in the region of concern by exhibiting the correct number of negative eigenvalues and properly describing soft versus stiff degrees of freedom. Numerous procedures have been developed for approximating $\mathbf{H}$ and updating the Hessian matrix as the geometry optimization proceeds. The efficiency of quasi-Newton optimizations depends on how close the starting point is to the target stationary point, how well the coordinate system describes the natural features of the chemical system and uncouples the degrees of freedom, how valid the quadratic approximation of the PES is, how accurately the Hessian matrix elements are approximated and improved, and what controls are placed on the size of the geometry update by means such as line searches and trust radius schemes.

If both the gradient and the second-derivative matrix are explicitly computed and employed in the coordinate update formula (Equation 1.6), the optimization algorithm is a proper NR method. If the PES were truly a quadratic function in the vicinity of some stationary point, the NR method would require only one step for a complete geometry optimization, regardless of the starting position. Of course, for real chemical systems, more NR steps will be required, but the convergence will be very rapid once the quadratic region surrounding the stationary point is reached. To be precise, if the error in optimizing the energy is ε at one point, it will be reduced to roughly ε^2 at the next point, meaning that the NR algorithm is quadratically convergent. Thus, only one NR optimization step would be necessary to take a $10^{-5}\,E_h$ error down to $10^{-10}\,E_h$. In quantum chemistry, the explicit computation of the Hessian matrix is usually too expensive for levels of theory that produce highly accurate equilibrium structures. Fortunately, efficient geometry optimizations with high levels of theory are often achieved by substituting a Hessian matrix computed at a cost-effective lower level of theory.

EXERCISE 1.2

The surface depicted in Figure 1.1 was generated from the model potential energy function $V(x,y) = (-8x^3 + 17xy^2 - 9x^2y - 10x^2 - 2xy - 1)\exp(-x^2 - y^2)$. Derive the polynomial equations that determine the stationary points **1–8**. Find analytic expressions for the elements of the Hessian matrix (**H**) of $V(x, y)$. Implement an NR algorithm based on analytic gradient and Hessian matrix formulas to precisely locate stationary points **1–8**. Compute the eigenvalues and eigenvectors of **H** at each of these points and interpret the results. Determine the relative energies of **1–8**. What is the reaction energy and barrier height for transformation **1 → 2** and **2 → 3**? Is there a transition state connecting **1** and **3**?

EXERCISE 1.3

A *valley-ridge inflection point* may be defined as a point on a PES at which there is a zero eigenvalue of the Hessian matrix whose corresponding eigenvector is orthogonal to the gradient vector. For the model potential energy function given in Exercise 1.2, derive the polynomial equations that determine the valley-ridge inflection point depicted in Figure 1.1. Use an NR scheme to find the proper root of these equations and to compute the position and energy where the valley-ridge inflection occurs.

EXERCISE 1.4

A gradient extremum path connecting two stationary points is one for which the gradient vector is always an eigenvector of the Hessian matrix. For the model potential energy function given in Exercise 1.2, derive the polynomial equation $f(x, y) = 0$ that implicitly determines the gradient extremum paths that exist on the surface shown in Figure 1.1. Note that gradient extremum paths do not require the solution of a differential equation and that points on these paths can be located independently, unlike IRPs. Find numerical solutions for the gradient extremum paths D, E, and F traced in Figure 1.1.

1.4   ANHARMONIC MOLECULAR FORCE FIELDS

Force field representations of PESs, as mentioned in Section 1.1, provide a general and effective means of determining the low-lying vibrational states of semirigid molecules and quantifying vibrational effects on geometric structures and spectroscopic parameters. The expansion of the PES around a reference geometry, usually chosen as an equilibrium structure of the molecular system, can be written as follows:

$$V = V_0 + \sum_i f^i R_i + \frac{1}{2}\sum_{ij} f^{ij} R_i R_j + \frac{1}{6}\sum_{ijk} f^{ijk} R_i R_j R_k + \frac{1}{24}\sum_{ijkl} f^{ijkl} R_i R_j R_k R_l$$

$$+ \frac{1}{120}\sum_{ijklm} f^{ijklm} R_i R_j R_k R_l R_m + \frac{1}{720}\sum_{ijklmn} f^{ijklmn} R_i R_j R_k R_l R_m R_n + \cdots$$

(1.9)

where $\mathbf{R}$ or $\{R_i\}$ denotes a set of nuclear *displacement* coordinates, defined to be zero at the reference structure. Common choices for $\mathbf{R}$ include internal, Cartesian, and normal coordinates (see Table 1.3 for their usual symbols and units).

The unrestricted summations preceded by a factor $(1/n!)$ at each order n in Equation 1.9 ensure that the expansion coefficients (force constants) are equal to the true derivatives of V with respect to $\mathbf{R}$ taken at the reference configuration, $f^{ijk\cdots} = \left(\dfrac{\partial^n V}{\partial R_i \partial R_j \partial R_k \cdots}\right)_0$. In some publications, especially in the older literature, the expansion coefficients correspond to restricted summations and the $(1/n!)$ factor is absorbed directly into the numerical values. Caution is thus warranted when comparing "force constants" from different studies. It is important to understand that the force constants $f^{ijk\cdots}$ may not be well-defined mathematically unless a *complete* and *nonredundant* set of coordinates is specified. The reason is that there can be

TABLE 1.3

Names, Symbols, and Units for Vibrational Coordinates and Force Constants

Name	Symbol	SI Unit	Customary Unit
Vibrational coordinates			
Cartesian	X	m	Å
Internal			
Bond stretching	R_i, l_i	m	Å
Angle bending	α_i, Θ_i	radian	radian
Linear angle bending	λ_i	radian	radian
Out-of-plane bending	γ_i	radian	radian
Torsion	τ_i	radian	radian
Symmetry	S_i	(Varies)	(Varies)
Normal			
Mass adjusted	Q_r	$kg^{1/2} \cdot m$	$u^{1/2} \cdot Å$
Dimensionless	q_r	1	1
Vibrational force constants			
In internal coordinates	f^{ijk}	(Varies)	(Varies)
In symmetry coordinates	F^{ijk}	(Varies)	(Varies)
In dimensionless normal coordinates	Φ^{rst}	m^{-1}	cm^{-1}

considerable sensitivity to the choice of coordinates being held fixed whenever partial derivatives of a multivariate function are taken.

The quadratic, cubic, quartic, quintic, and sextic force constants will have 2, 3, 4, 5, and 6 superscripts, respectively, corresponding to the indices for the coordinates with respect to which the derivatives of the PES are taken. Retaining only quadratic force constants in the PES expansion provides a harmonic vibrational analysis, which is the most widely employed approximation. The next most common approach treats vibrational anharmonicity by means of a quartic force field representation. The need to simultaneously include both cubic and quartic force constants to describe anharmonicity is revealed by second-order vibrational perturbation theory (VPT2; for details see Chapter 3), wherein the quartic terms contribute in first order while the cubic terms appear only in second order, placing these contributions on an equal footing. The use of sextic and higher-order force fields to provide local representations of PESs is much less common than the use of quartic force fields.

A depiction of quadratic through sextic force field representations in comparison to an exact PES is shown in Figure 1.4, in which contour plots are shown of the same model potential energy function appearing in Figure 1.1, but with focus on the region surrounding minimum **1**. Successive improvement is seen as the order of the force field is increased, especially in describing the valley leading to the transition state on the left of the plot. However, it is also apparent that the higher-order force

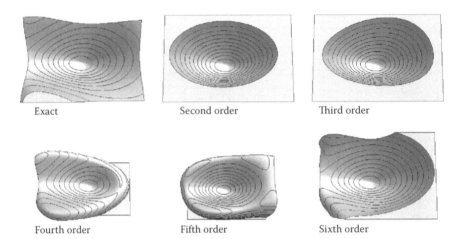

Exact	Second order	Third order
Fourth order	Fifth order	Sixth order

FIGURE 1.4 Comparison of an exact PES with force field representations of it from second through sixth order: The model potential energy function is that of Figure 1.1 and Exercise 1.2; contour plots are shown in the region surrounding minimum **1**.

fields can exhibit physically incorrect chasms in the surface if the distance from the reference point is too large. Such problems are common for molecular PESs, and the limited range of validity of a force field expansion must always be appreciated. Although any complete and nonredundant set of coordinates may be used in principle for a force field expansion, some representations may have more desirable properties than others. A dramatically better radius of convergence of the expansion may be achieved by choosing one set of coordinates over another, leading to a more accurate representation of the PES at lower orders. A fine example is the use of Simons–Parr–Finlan variables $(1 - r_e/r)$ for bond stretching motions rather than simple bond distances (r).

Several procedural issues must be considered when determining (anharmonic) force fields by methods of electronic structure theory. After an appropriate coordinate system has been selected, one should identify all unique force constants to be determined. A reference geometry must then be adopted, following the principles outlined in Section 1.2. Once an appropriate level of electronic structure theory has been chosen, the necessary quantum chemical computations can be performed. Careful consideration must be given to the method of computing the high-order force constants from low-order analytic information without much loss in numerical precision. Checks on the computed force constants may be provided from an understanding of the underlying chemical principles. For example, higher-order diagonal stretching force constants almost always follow the patterns of relative signs and magnitudes expected for simple diatomic Morse oscillators (see Chapter 8). Finally, there is often a need to transform the computed force fields from one coordinate system to another. Analytic force field transformations are preferred over numerical fitting approaches, but the necessary mathematical formulas are complex, requiring so-called **B** tensors that contain higher derivatives of the first set of geometric variables with respect to the second. Extensive research already exists on analytic force field transformations, which can most easily be derived, programmed, and visualized by means of a *brace notation* technique.*

The recommended and customary usage of symbols and units for coordinates and force constants in vibrational analyses is summarized in Table 1.3. In practice, most researchers avoid the use of SI units, instead defining energies in attojoules (aJ), bond stretching internal or symmetry coordinates in Å (1 Å = 10^{-10} m = 100 pm), and angle bending internal or symmetry coordinates in radians. Therefore, the force constant units become, at nth order, aJ·Å^{-n} (mdyn·Å^{-n+1}) for stretching coordinates and aJ·rad^{-n} for bending coordinates. The units of interaction force constants follow from these definitions.

EXERCISE 1.5

Compute the quartic vibrational force field for minima **1–3** of the model potential energy function depicted in Figure 1.1, $V(x, y) = (-8x^3 + 17xy^2 - 9x^2y - 10x^2 - 2xy - 1) \exp(-x^2 - y^2)$. Determine the *normal coordinates* for each minimum that yield a diagonal quadratic force constant matrix. Transform the quartic force field representations from the (x, y) space to the system of normal coordinates for minima **1–3**.

1.5   A HIERARCHY OF ELECTRONIC STRUCTURE METHODS

Electronic structure theory and nuclear motion theory are the two main areas of quantum chemistry, as explained in Section 1.1. Several excellent introductory and advanced textbooks are available on the subject of electronic structure theory (often referred to as "quantum chemistry" by itself) and some on nuclear motion theory. Some of these books are listed at the end of this chapter.

Nuclear motion theory is not considered in this section. A few remarks about the time-independent picture of nuclear motion theory are given in Chapter 8. Due to the availability of the reviews mentioned and space limitations, a detailed treatment of the advanced field of molecular electronic structure theory is not attempted here. Furthermore, it is assumed that the reader is familiar with the elements of quantum mechanics (e.g., Hilbert space, operators, spherical harmonics, the Pauli exclusion principle) and with the most basic concepts of quantum chemistry (e.g., atomic and molecular orbitals [MOs], Slater determinants, Rayleigh–Schrödinger PT, the variational principle). In this section, a necessarily rudimentary and nontechnical description is given of theoretical methods for studies on molecular structures and vibrational force fields in order to highlight the most important issues facing readers interested in the ab initio determination of these quantities.

For all systems of chemical interest, the exact solution to the (nonrelativistic, time-independent) electronic Schrödinger equation cannot be obtained; thus, a hierarchy of increasingly accurate wave function approximation methods is needed beyond the BO separation of nuclear and electronic motions. Basic to the understanding of

---

* This technique is described in detail in Allen, W. D., and A. G. Császár. 1993. *J Chem Phys* 98:2983 and Allen, W. D., A. G. Császár, V. Szalay, and I. M. Mills. 1996. *Mol Phys* 89:1213.

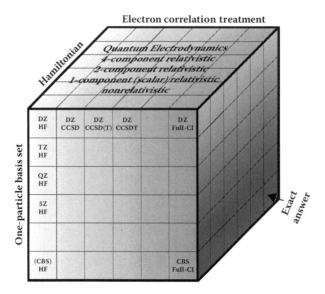

FIGURE 1.5 Computational cube of ab initio electronic structure theory indicating quality of the one-particle space (basis set), quality of the n-particle space (wave function methodology), and quality of the electronic Hamiltonian (for the abbreviations employed, see Section 1.5).

this hierarchy is the computational cube depicted in Figure 1.5. It demonstrates that there are three fundamental approximations in electronic structure theory: choice of the electronic Hamiltonian, truncation of the one-particle basis, and the extent of the electron correlation treatment. The "exact answer" is approached as closely as possible by choosing an appropriate Hamiltonian and extending both the atomic-orbital (one-particle) basis set and the many-electron correlation method (n-particle basis) to technical limits. For lighter elements, perhaps up to Ar, the effects of special relativity will not be consequential, except in electronic structure studies seeking ultimate precision. Thus, computations are usually performed using the nonrelativistic Hamiltonian, introduced in Section 1.1.

It is customary to use atom-centered Gaussian functions for the one-particle basis set. Many-electron wave functions are usually obtained by PT, configuration interaction (CI), or coupled-cluster (CC) approaches. The usefulness, quality, and reliability of any given approximation, whether in the Hamiltonian or in the one- or n-particle spaces, must be assessed from comprehensive studies on a large number of systems. It is highly advantageous if the error introduced by the different approximations (1) is controllable, (2) can be cancelled in some systematic way, (3) is comparable for similar systems and physical situations, and (4) is balanced over a large region of the geometrical space.

A key concept in quantum chemistry is the electron correlation energy of a chemical system,

$$\varepsilon_{corr} = E_{exact} - E_{HF} \tag{1.10}$$

where E_{HF} is the electronic energy obtained from a Hartree–Fock (HF) computation in which the electrons move in the mean fields of one another, and E_{exact} is the exact energy resulting when the instantaneous electronic interactions are completely reckoned.

A distinction must be made between the exact correlation energy, defined with respect to the HF and full configuration interaction (FCI) energies in the complete one-particle basis set limit, and the computed correlation energy, defined in a given finite basis set as the FCI–HF energy difference. At geometric minima, the HF method typically recovers some 99% of the total electronic energy, but its performance can deteriorate considerably when one moves away from equilibrium. Because both the HF and FCI procedures obey the variational theorem, the correlation energy is always negative.

Fermi correlation arises from the Pauli antisymmetry principle; it is not part of the electron correlation defined in Equation 1.10 and is already taken into account at the HF level. Dynamic correlation (DC) serves to keep electrons apart instantaneously. It is a cumulative effect built up from myriad small contributions and usually forms the largest part of ε_{corr}. It originates primarily from the failure of most reference wave functions to describe the short-range interactions in the electron–electron cusp regions. Nondynamic correlation (NDC) arises when an electronic state is not adequately described by a single Slater determinant of MOs, generally due to near-degeneracy effects. NDC is a long-range effect. In diverse chemical systems, the correlation energy per electron pair remains roughly constant at its value for the ground electronic state of He-like ions and H_2, about $-0.04\ E_h$.

There are two computational strategies possible if NDC is significant. In the single-reference (SR) route, NDC is accounted for along with DC merely by sufficiently increasing the highest order of electronic excitations included in the correlation treatment (PT, CI, or CC). In the multiconfiguration/multireference (MC/MR) route, NDC is accounted for at the start in the zeroth-order wave function, and DC is added subsequently via MR CI, CC, or PT schemes. It is of considerable importance to determine whether single-configuration and SR methods, which are applicable with relative ease even for large molecular systems and have a "black box" nature, are sufficient, or whether the conceptually more involved MC and MR methods need to be applied. For this purpose, different tests have been developed that allow estimation of the MR character of an electronic state at a given geometry.

Electronic structure techniques used to study PESs should provide comparable accuracy for all subsystems (fragmentation products) investigated. Such considerations lead to the concepts of size consistency and size extensivity. In accord with traditional thermodynamic concepts, a "size-extensive" method scales correctly with the number of particles in the system, as in the pedagogical example of an electron gas or N noninteracting H_2 molecules. A "size-consistent" method is one that more specifically leads during molecular fragmentation to a wave function that is multiplicatively separable and an energy that is additively separable. There is some disagreement over the precise definition of size consistency, and many use the terms size-extensive and size-consistent synonymously, despite the fact that in some cases the former property is exhibited but not the latter. Neither property alone ensures that a molecule and its dissociation products are described with precisely the same accuracy. While variational computations (like CI) are size consistent only if an exponentially growing direct-product space of the fragments is employed for their construction, CC theory provides natural *ansätze* for multiplicatively separable approximate wave functions.

1.5.1   PHYSICALLY CORRECT WAVE FUNCTIONS

In order to account for electron correlation successfully, a good (model) reference wave function must be chosen. The simplest standard model offered by wave-function-based ab initio electronic structure theory is the HF mean-field theory, often called *self-consistent-field* (SCF) *theory*. An appealing feature of HF theory is that it defines MOs as delocalized one-electron functions describing the movement of an electron in an average (effective) field of all the other electrons. In HF methods, the MOs are variationally optimized in order to obtain an energetically "best" many-electron function of a single-configuration form. Of course, energy optimization does not necessarily imply a similar favorableness for properties, like structures or force fields. The HF model is size extensive, but the HF wave function is not an eigenfunction of the exact Hamiltonian. There are several HF techniques, like RHF, spin-unrestricted Hartree–Fock (UHF), spin-restricted open-shell HF (ROHF), and generalized HF (GHF), which differ in the form of the orbitals used for electrons of different spin. The results of an HF computation are the total energy, the wave function consisting of MOs (canonical, localized, or other), and the electron density, from which various properties can be calculated.

When qualitative electronic structure analysis, chemical intuition, or simple tests indicate a serious breakdown of the HF (single-configuration) approximation, it is necessary to turn to multiconfiguration SCF (MCSCF) methods (e.g., complete-active-space [CAS] SCF). Multiconfiguration SCF methods are to be used when an HF description of the electronic state under consideration is qualitatively incorrect, when proper space and spin eigenfunctions must be constructed for linear/atomic systems/fragments, or when problematic radicals or transition metals are studied. A basic premise of many MCSCF approaches is that the important chemical aspects of most molecular systems can be represented by just a limited number of carefully chosen configurations. A significant problem with the MCSCF technique, aimed to recover important NDC, is that the selection of the configurations to be included in the wave function is not always straightforward.

Selection of individual configurations for an MCSCF treatment can be based on chemical intuition and/or perturbative estimates. The most rigorous strategy is to employ the CASSCF technique, in which the orbitals are divided into three subspaces: core, active, and virtual. Core and virtual orbitals have fixed occupation numbers of 2 and 0, respectively. Active orbitals have varying occupation: all configurations are included in the wave function that are obtained by distribution of the active electrons among the active orbitals in all possible ways, satisfying the relevant symmetry and spin requirements of the total electronic wave function. Therefore, within the active space, a CASSCF wave function is an FCI wave function. The framework of the CASSCF method is general, its application is straightforward, and it can be employed for the most difficult problems of electronic structure theory, but within rather severe size limitations.

Although the HF description of the electronic structure of most molecular species is surprisingly accurate overall (see Section 1.2), the HF wave function fails to exhibit the requisite Coulomb hole, that is, substantially reduced electron density around the instantaneous position of each electron. In the wave function models that go beyond the HF description, techniques of varying sophistication are employed to represent the long- and short-range electron–electron interactions, especially the Coulomb hole. While long-range DC effects are described adequately by most techniques, it has become clear over the years that it is difficult to arrive at an accurate representation of the short-range electron–electron interactions. This difficulty has led to the development of so-called *explicitly correlated* approaches for the treatment of electron correlation, in which interelectronic distances are directly incorporated in the wave function.

As in many branches of physics and chemistry, it is often expedient to use some form of PT in electronic structure computations. Møller–Plesset (MP) perturbation theory is a form of Rayleigh–Schrödinger PT in which the unperturbed Hamiltonian is taken as a sum of one-particle Fock operators. When this PT is carried out to second order (MP2), it defines the simplest method (besides density functional theory) that incorporates electron correlation, and it provides size-extensive energy corrections at low cost. The present-day standard of electronic structure theory is the CC method. Efficient single-reference coupled-cluster (SR-CC) procedures have been developed, which are based on several types of reference wave functions, including closed-shell RHF, open-shell UHF and ROHF, and quasi-restricted HF wave functions. In the vicinity of equilibrium structures, these methods have now reached a high degree of sophistication and accuracy.

The fundamental equation of CC theory is

$$\left| \psi \right\rangle_{CC} = e^{\hat{T}} \left| \Phi_0 \right\rangle \tag{1.11}$$

where $\left| \psi \right\rangle_{CC}$ is the correlated molecular electronic wave function based on the exponential CC ansatz, and in most applications $\left| \Phi_0 \right\rangle$ is a normalized HF wave function. Nothing in CC theory is fundamentally limited, however, to the HF choice for the reference function. The $\exp(\hat{T})$ operator obeys the usual Taylor series expansion

$$e^{\hat{T}} = 1 + \hat{T} + \frac{\hat{T}^2}{2!} + \cdots = \sum_{k=0}^{\infty} \frac{\hat{T}^k}{k!} \tag{1.12}$$

The cluster operator $\hat{T}$ is defined as the sum of n-tuple excitation operators $\hat{T}_n$ that promote n electrons from the occupied to the virtual orbitals of the reference wave function. The maximum value of n equals the number of electrons (in this case, CC becomes equivalent to FCI), but practical restrictions almost always dictate $n \leq 4$. Accordingly, only certain excitation operators are included in the usual applications of CC theory. Restricting $\hat{T}$ just to $\hat{T}_2$ results in the coupled-cluster doubles (CCD) method. Inclusion of $\hat{T}_1$ and $\hat{T}_2$ gives the widely employed CC singles and doubles (CCSD) method. This method is capable of recovering typically 95% or more of the correlation energy for molecules in the vicinity of their equilibrium structures.

Simplification in CC theory can be accomplished by restricting the evaluation of (connected) contributions corresponding to higher excitations to certain lead terms. The most important technique is CCSD(T), the gold standard of electronic structure theory. The well-balanced CCSD(T) approach, which includes a noniterative, perturbative accounting of the effect of connected triple excitations, is able to provide total and relative electronic energies of chemical accuracy or better, as well as excellent results for molecular structures and a wide range of molecular properties. In SR cases and for most properties of general interest, the HF, MP2, CCSD, and CCSD(T) methods provide the most useful and practical hierarchy of approximations of increasing accuracy (see Section 1.6).

The conceptually simple SR CI methods, which are variational in nature in contrast to CC methods, have been in use from the early days of computational electronic structure theory. If we take the HF wave function as the zeroth-order wave function in PT, then all triple and higher excitations make no contribution to the correlated wave function in first order. Accordingly, CI including all single and double excitations, that is, CISD, became the first standard method for the treatment of electron correlation. Nevertheless, CISD is seldomly used today for ground electronic state computations. It has been shown, for example, that for the prediction of equilibrium geometries, complete basis set (CBS) CISD performs rather poorly. The lack of size extensivity of truncated CI treatments is another weakness of this methodology. While multireference coupled-cluster (MR-CC) theories have been difficult to formulate and implement until very recently, multireference configuration interaction (MR-CI) methods have been employed for a long time. The MR-CI and MR-CC techniques are considered to be the most accurate ab initio electron correlation procedures that can be employed for reasonably large molecular systems over an extended range of nuclear configurations.

Achieving the FCI limit with a complete orbital basis has been a persistent goal of molecular quantum mechanics. However, for most chemical systems, explicit FCI computations are intractable due to their factorial growth with respect to the one-particle basis and the number of electrons. Nonetheless, numerous schemes have been developed to estimate the FCI limit from series of truncated CI computations conjoined with, in many cases, PT. Numerous applications have demonstrated the viability of generating PES information by such composite methods for small molecules.

1.5.2 ONE-PARTICLE BASIS SETS

MOs can be constructed either numerically or by expansion techniques. For polyatomic systems, the usual choice is to expand MOs in a set of simple analytical one-electron functions. Although arbitrarily accurate numerical HF procedures have been developed, electron correlation computations still require the use of one-electron basis sets. Indeed, all traditional quantum chemical procedures, whether HF, CI, MP, or CC, start with the selection of a one-particle basis set. The accuracy and dependability of any quantum chemical computation depend supremely on this basis set.

The usual choice for the form of the one-particle basis functions is the Cartesian Gaussian-type orbital (GTF):

$$g^{a,b,c}(x, y, z; \alpha, \mathbf{r}_A) = N_a N_b N_c (x - x_A)^a (y - y_A)^b (z - z_A)^c \exp(-\alpha \,|\, \mathbf{r} - \mathbf{r}_A \,|^2),$$

$$\text{with } N_i = \left[\frac{(2i-1)!!}{\alpha^i} \left(\frac{\alpha}{\pi} \right)^{1/2} \right]^{1/2} \tag{1.13}$$

where the N_i ($i = a, b, c$) are normalization constants; a, b, and c are nonnegative integers; the orbital exponents α are taken to be positive; and the basis function is centered on atom A at $\mathbf{r}_A = (x_A, y_A, z_A)$.

Frequently, linear transformations of Cartesian Gaussians are invoked to yield manifolds of pure spherical harmonics (Y_{lm}, $l = a + b + c$ plus lower contaminants) or real combinations thereof.* The GTFs neither satisfy the nuclear cusp condition† nor show the proper exponential decay at long range. Nevertheless, GTFs give one- and two-electron integrals that can be computed very efficiently, allowing the actual number of primitive functions in the basis to be increased to mitigate any short- and long-range deficiencies. The ease with which integrals over GTFs may be computed is related to two important analytic properties of Gaussian distributions: their separability in the Cartesian directions and the Gaussian product rule. The Gaussian product rule states that a product of two Gaussians of arbitrary exponents centered at arbitrary spatial positions can be represented as a third Gaussian with an exponent determined by the sum of the original exponents and located at a point that lies on the line connecting the original centers.

Molecular electron correlation procedures, as opposed to HF, require not only a set of atomic orbitals that resemble the occupied orbitals of the constituent atoms but also a set of spatially compact, orthogonal, virtual orbitals, into which electrons can be excited and hence correlated. Therefore, it is generally expected that optimal basis sets for uncorrelated and correlated calculations will be quite different. Electron correlation computations demand the use of much better one-electron basis sets; for example, basis sets of double-ζ (DZ) quality (two sets of basis functions for each shell) in the valence region augmented with polarization functions, designated as DZP, are considered to be of the lowest acceptable quality.

Numerous hierarchical Gaussian basis sets have been developed for the efficient computation of energies and molecular properties. We use the term hierarchical to indicate that these basis sets allow approach to the CBS limit in a systematic fashion. Perhaps the most successful approach in this regard is the family of

---

* Most basis sets used in electronic structure theory contain either single Gaussians or a linear combination (contraction) of several Gaussians with fixed coefficients. It is typical to fix the Gaussian exponents in the basis sets.

† In the limit of an electron approaching a nucleus of charge Z very closely ($r \to 0$), all other particles and interactions in the quantum system can be neglected, and solution of the corresponding Schrödinger equation for the radial function $R = R(r)$ yields the "cusp condition" $\lim_{r \to 0} dR/dr = -ZR(0)$.

correlation-consistent basis sets.* Correlation-consistent basis sets result from convergence studies of the correlation energy with respect to saturation of both the radial and angular spaces. These basis sets are correlation consistent in the sense that each basis set contains all functions that contribute to the correlation energy by at least a set amount. The resulting correlation-consistent (cc), polarized valence (pV) basis sets are denoted as (aug)-cc-p(C)VXZ (X = D, T, Q, 5, 6, ...), in which (aug) denotes the addition of diffuse (low-exponent) functions, (C) signifies the addition of (high-exponent) functions to describe core correlation, and X is the cardinal number of the basis. The cardinal number also represents the highest spherical harmonic contained in the basis set. Atomic natural orbital basis sets form another family of hierarchical basis sets of considerable practical utility.

For conventional wave function methods of electronic structure theory, the convergence of electron correlation energies with the expansion of the one-particle basis set is intrinsically and painfully slow. The difficulty is that Slater determinants built from MOs are fundamentally the products of one-particle functions, which cannot properly describe the region around the two-particle Coulomb singularities that occur at electron–electron coalescence points. Therefore, very high angular momentum functions must be included in the one-particle basis set to adequately recover the electron correlation energy. The performance of correlation-consistent basis sets has been thoroughly tested in this regard. Although the correlation energy convergence is slow, it is quite systematic, allowing extrapolations to the CBS limit using simple asymptotic formulas. We demonstrate the success of this approach in Section 1.6.

Another solution to the electron correlation convergence problem is to invoke explicitly correlated electronic structure techniques. We may impose the correct electronic Coulomb-cusp condition on any determinant-based wave function by multiplying the orbital product expansion by some *correlation factor* Γ. The term "explicit correlation" refers to methods that employ such correlation factors or otherwise make explicit use of interelectronic distances in the wave function. If a single term Φ_0 dominates the conventional Slater-determinant expansion Φ of the wave function, one can write an improved many-electron wave function (Ψ) as

$$\Psi = \Gamma\Phi = \Gamma(\Phi_0 + \omega) = \Gamma\Phi_0 + \chi \qquad (1.14)$$

Then, a practical approach is to represent χ in the usual manner as a linear combination of determinants for excited configurations. Several choices for the correlation factor have been investigated. Correlation factors containing only linear r_{12} terms proved successful in the 1990s, but more recent work has demonstrated the superiority of writing Γ in terms of Slater functions $\exp(-\gamma\, r_{12})$, usually represented by an expansion in Gaussian geminals. Many molecular applications now show the dramatic acceleration in correlation energy convergence attainable by explicitly correlated methods.

---

* The principal reference for the correlation-consistent family of atom-centered Gaussian basis sets is Dunning Jr., T. H. 1989. *J Chem Phys* 90:1007.

1.6 PURSUIT OF THE AB INITIO LIMIT

Ab initio *limits* can be closely approached by *composite schemes* that employ multiple electronic structure computations at different levels of theory to arrive at a single energy at a given molecular geometry. A general composite scheme that is highly successful is the focal-point analysis (FPA) approach.* A fundamental characteristic of the FPA approach is the dual extrapolation to the one- and *n*-particle limits of electronic structure theory. The process leading to these limits can be characterized as follows: (1) use of a family of basis sets, such as (aug)-cc-pVXZ, which systematically approaches completeness through an increase in a cardinal number (X); (2) application of lower levels of theory (typically, HF and MP2 computations) with very extensive basis sets; (3) execution of a sequence of higher-order correlation treatments with the largest possible basis sets; and (4) layout of a two-dimensional extrapolation grid based on the assumed additivity of correlation *increments*, that is, the differences between correlation energies given by successive levels of theory in the adopted hierarchy.

A key aspect of FPA is the assumption that the higher-order correlation increments show diminishing basis set dependence. From the large number of FPA studies, it is clear that the first correlation increment that results from double excitations is by far the most important and also the hardest to determine in a converged manner. Thus, the computation of MP2 – HF or CCSD – HF energy differences requires the largest basis sets, and sometimes convergence is not reached even with $X = 6$. Contributions from triple excitations are generally important, and their inclusion is necessary to obtain chemical accuracy or better. Fortunately, smaller basis sets are usually sufficient here, and CCSD(T) – CCSD increments may be satisfactorily converged with $X = 3$ or 4. Occasionally, the contribution from quadruple excitations is significant, requiring the evaluation of increments such as CCSDT(Q) – CCSD(T) with $X = 2$ and 3. For example, a composite (c~) approach to estimating the energy at the CBS limit of CCSDT(Q) theory would be

$$E^{\text{CBS}}_{\text{c~CCSDT(Q)}} = E^{\text{CBS}}_{\text{CCSD(T)}} + E^{\text{cc-pVDZ}}_{\text{CCSDT(Q)}} - E^{\text{cc-pVDZ}}_{\text{CCSD(T)}} \tag{1.15}$$

The CBS values required in the FPA approach are usually obtained by extrapolating HF energies with an exponentially decaying function of X and extrapolating the correlation energies by means of an inverse power dependence such as X^{-3}. These choices are based on mathematical analyses of the respective wave functions and reflect the fact that correlation energies converge much more slowly than HF energies, as discussed in Section 1.5. The energy of a composite method can be written as a simple function (often linear) of the energies from the input levels of theory; thus, it is straightforward to obtain energy gradients for the composite method, provided gradients are available for the constituent levels. Otherwise, high-precision composite energies could be used to obtain numerical gradients and force constants. In

---

* For original publications on the FPA approach, see Allen, W. D., A. L. L. East, and A. G. Császár. 1993. *Structures and Conformations of Non-Rigid Molecules*, ed. J. Laane, M. Dakkouri, B. van der Veken, and H. Oberhammer, 343. Dordrecht: Kluwer; Császár; A. G., W. D. Allen, and H. F. Schaefer III. 1998. *J Chem Phys* 108:9751.

this manner, composite techniques can be used to determine equilibrium molecular structures and vibrational force fields at the ab initio limit.

The virtues of the FPA approach can be appreciated best by way of examples. Consider the diatomic CO molecule, for which the experimental equilibrium bond length is $r_e^{BO} = 1.1283$ Å. The optimized all-electron (AE) CCSD(T)/aug-cc-pCVQZ result is 1.1293 Å, a full 0.001 Å greater than the well-established experimental value, showing the difficulty in precisely describing multiple bonds even with the gold standard CCSD(T) method. By using the cc-pCVQZ and cc-pCV5Z basis sets to extrapolate to the CBS limit of AE-CCSD(T) theory, one arrives at $r_e^{BO} = 1.1276$ Å, now 0.0007 Å smaller than the experimental value. Finally, by employing the composite scheme of Equation 1.15 to estimate the AE-CCSDT(Q)/CBS bond distance, one arrives at $r_e^{BO} = 1.1285$ Å, in remarkable agreement with experiment. A similar composite approach for estimating the CCSDTQ/CBS value gives an even better result, $r_e^{BO} = 1.1284$ Å.

A more complete example of the FPA approach is the analysis shown in Table 1.4 of the barrier to linearity of the ketenyl radical (HCCO).* Note that the basis set convergence of the entries in Table 1.4 going down the columns is excellent, suggesting that the CBS limit has been reached to the level of about 0.01 kcal $\cdot$ mol^{-1} at each step in the electron correlation series. Nonetheless, in accord with the FPA principles stated earlier, the most difficult increment to converge is the lowest-order correlation contribution δ[ZAPT2], which is given by a particular open-shell variant of second-order perturbation theory. Starting from the HF/CBS barrier of 1.06 kcal$\cdot$mol^{-1}, the successive ZAPT2, CCSD, CCSD(T), CCSDT, and CCSDT(Q) corrections are +0.81, −0.56, +0.60, +0.07, and +0.10 kcal$\cdot$mol^{-1}, respectively. Therefore, the valence-correlated ΔE_e[CCSDT(Q)] barrier is placed at 2.08 kcal$\cdot$mol^{-1}, and the error with respect to the n-particle (FCI) limit is indicated to be less than 0.1 kcal$\cdot$mol^{-1}. The core electron correlation (Δ_{core}), relativistic (Δ_{rel}), and diagonal BO (Δ_{DBOC}) contributions to the barrier are −0.23, +0.04, and −0.10 kcal$\cdot$mol^{-1}, respectively, yielding a final classical (vibrationless) barrier to linearity of 1.79 ± 0.10 kcal$\cdot$mol^{-1}. Overall, Table 1.4 shows the detailed FPA procedure, including extrapolation formulas, by which an energetic quantity or spectroscopic property can be pinpointed, together with sound uncertainty estimates.

As illustrated by the HCCO example, the FPA approach includes the consequences of several small physical effects: core electron correlation, special relativity, and adiabatic corrections to the BO approximation. Quantum electrodynamics (QED) also provides electronic radiative corrections (or Lamb shifts)† arising from the interaction of the electron with the fluctuation of the electromagnetic field in vacuum. Studies of atoms and simple molecules have indicated that QED effects are generally orders of magnitude smaller than relativistic corrections, and thus further discussion of them is not warranted here.

---

* Simmonett, A. C., N. J. Stibrich, B. N. Papas, H. F. Schaefer, and W. D. Allen. 2009. *J Phys Chem A* 113:11643.

† Lamb shifts have virtually no effect on chemistry. Nevertheless, for highly accurate spectroscopic applications, their influence may be detected once the underlying quantum chemical approaches are pushed to their limits (Pyykkö, P., K. G. Dyall, A. G. Császár, G. Tarczay, O. L. Polyansky, and J. Tennyson. 2001. *Phys Rev A* 63:024502).

TABLE 1.4

Focal-Point Analysis[a] of the Barrier to Linearity (in kcal·mol^{-1}) for the HCCO Radical

Basis set	ΔE_e[ROHF]	δ[ZAPT2]	δ[CCSD]	δ[CCSD(T)]	δ[CCSDT]	δ[CCSDT(Q)]	ΔE_e[CCSDT(Q)]
aug-cc-pVDZ	+1.85	+1.71	−0.50	+0.68	+0.07	+0.10	[+3.91]
aug-cc-pVTZ	+1.08	+1.21	−0.60	+0.61	+0.07	[+0.10]	[+2.47]
aug-cc-pVQZ	+1.08	+0.96	−0.61	+0.60	[+0.07]	[+0.10]	[+2.20]
aug-cc-pV5Z	+1.07	+0.89	−0.59	+0.60	[+0.07]	[+0.10]	[+2.14]
aug-cc-pV6Z	+1.06	+0.85	−0.58	+0.60	[+0.07]	[+0.10]	[+2.10]
CBS limit	[+1.06]	[+0.81]	[−0.56]	[+0.60]	[+0.07]	[+0.10]	[+2.08]
Fit function	$a + be^{-cX}$	$a + bX^{-3}$	$a + bX^{-3}$	$a + bX^{-3}$	Additive	Additive	
Points (X)	4, 5, 6	5, 6	5, 6	5, 6			

$$\Delta E_e(\text{final}) = \Delta E_e[\text{CBS CCSDT(Q)}] + \Delta_{core}[\text{CCSD(T)/aug-cc-pCVQZ}] + \Delta_{rel}[\text{CCSD(T)/cc-pCVTZ}]$$
$$+ \Delta_{DBOC}[\text{HF/aug-cc-pVQZ}] = 2.08 - 0.23 + 0.04 - 0.10 = \mathbf{+1.79\ kcal·mol^{-1}}$$

[a] The symbol δ denotes the increment in the relative energy (ΔE_e) with respect to the preceding level of theory in the hierarchy ROHF→ZAPT2→CCSD→CCSD(T)→CCSDT→CCSDT(Q). Square brackets signify results obtained from basis set extrapolations or additivity assumptions. Final predictions are boldfaced.

The core correlation energy is defined as the difference between the AE correlation energy and the valence-valence (VV) correlation energy (usually, simply called the valence correlation energy). Size-extensive methods should be employed to estimate core correlation effects because a consistent accuracy needs to be maintained when the number of active electrons is changed. While computing the core correlation energy, special attention must also be paid to the one-particle basis set, because most basis sets are designed to describe only the VV interaction of electrons. The compactness of core MOs requires inclusion in the basis set of higher angular momentum functions with very large exponents.

In diatomic molecules containing first-row atoms, the following core correlation effects have been established: (1) equilibrium bond distances experience a substantial contraction, 0.002 Å or more for multiple bonds and 0.001 Å for single bonds; (2) the *direct* effect of core correlation is a correction function to the valence diatomic potential energy curve that has negative curvature at all bond lengths in the vicinity of the equilibrium position; and (3) core correlation decreases all higher-order force constants as well. The direct effect of core correlation means that under the constraint of fixed internuclear distance, core correlation decreases the quadratic force constant and serves to lower the harmonic vibrational frequency. Nevertheless, if the total effect is considered in a conventional, phenomenological sense as the difference between the harmonic frequencies from AE and valence-electron treatments, the harmonic frequency actually increases due to electron correlation, in accord with the shift anticipated from the attendant bond-length contractions. The core correlation effect on the r_e^{BO} bond length and bond angle of water are ~0.001 Å and +0.13°, respectively, which are well within the realm of modern equilibrium structure determinations.

For extremely accurate computations, the DBOC cannot be neglected, especially if hydrogen or other light atoms are present in the system. As discussed in Section 1.1, by appending the DBOC to $V(\mathbf{r}_\alpha)$ one gets APESs that are dependent on the masses of the nuclei. Mass-dependent equilibrium structures, which can be designated as r_e^{ad}, result from this process. In the case of water, the differences between r_e^{BO} and r_e^{ad} OH bond lengths are minuscule: 3×10^{-5} and 1×10^{-5} Å for $H_2^{16}O$ and $D_2^{16}O$, respectively. The corresponding adiabatic change in the bond angle is only $0.01°$.

The relativistic effect is defined as the difference in an observable property that arises from the true velocity of light as opposed to the assumed infinite velocity in traditional treatments of quantum chemistry. The importance of accounting for relativistic effects in quantum-chemical computations on systems containing heavy atoms is well recognized. Relativistic phenomena indeed have striking chemical and physical consequences in the lower part of the periodic table.* Nevertheless, for systems containing atoms no larger than Ar, relativistic effects are quite small. In the water example, the relativistic changes in bond length and bond angle are only 1.6×10^{-4} Å and $-0.07°$, respectively.[†]

REFERENCES AND SUGGESTED READING

The fundamental modern source on quantum chemistry is:
Schleyer, P. v. R., ed. 1998. *The Encyclopedia of Computational Chemistry*. vols 1–6. Chichester, UK: Wiley.
The six volumes of this encyclopedia contain a lot of useful information on all fields discussed in this chapter.

An excellent source on the elementary aspects of electronic structure theory is:
Szabo, A., and N. S. Ostlund. 1996. *Modern Quantum Chemistry*. New York: Dover.

Further recommended sources on computational quantum chemistry include:
Helgaker, T., P. Jørgensen, and J. Olsen. 2000. *Molecular Electronic-Structure Theory*. New York: Wiley.
Mayer, I. 2003. *Simple Theorems, Proofs, and Derivations in Quantum Chemistry*. New York: Kluwer.
McWeeny, R. 1992. *Methods of Molecular Quantum Mechanics*. 2nd ed. London: Academic Press.
Pilar, F. L. 2001. *Elementary Quantum Chemistry*. 2nd ed. New York: Dover.
Tannor, D. J. 2007. *Introduction to Quantum Mechanics*. Sausalito, CA: University Science Books.

Some aspects of nuclear motion theory and the bridge between quantum chemistry and spectroscopy are covered in the following monograph:
Jensen, P., and P. R. Bunker, eds. 2000. *Computational Molecular Spectroscopy*. Chichester, UK: Wiley.

---

* Entire books, like Balasubramanian, K. 1997. *Relativistic Effects in Chemistry*. New York: Wiley; and Dyall, K. G., and K. Faegri Jr. 2007. *Introduction to Relativistic Quantum Chemistry*. Oxford: Oxford University Press, have been devoted to chemical applications of relativistic electronic structure theory and to the elucidation of relativistic effects.
† The structural data presented here for the water isotopologues were taken from Császár, A. G., G. Czakó, T. Furtenbacher, J. Tennyson, V. Szalay, S. V. Shirin, N. F. Zobov, and O. L. Polyansky. 2005. *J Chem Phys* 122:214305.

2 The Method of Least Squares

Jean Demaison

CONTENTS

2.1 INTRODUCTION

This chapter is meant to be a practical introduction to the use of the *method of least squares*, which is important for the following chapters. This method is indeed extremely important to analyze a spectrum or to determine a structure. This chapter is not intended to be comprehensive. In particular, the proofs for the theorems have been omitted. The method of least squares can be automated, but it is nevertheless important to understand the details of what is done and how it is done in order to do it correctly. This method has already been reviewed with an emphasis on structure determination [1] and on analysis of spectra [2]. In this chapter, we follow the notations of these references. Many textbooks are devoted to the least squares method; some of these are referenced in the text.

Section 2.2 describes the method of ordinary least squares (OLS). OLS is used to solve an overdetermined system of linear equations for which the measurements have the same precision and the associated errors have a mean of zero. When some parameters are not well-defined by the observations, the rounding errors and tiny changes in the experimental data may strongly affect these parameters. A method to identify this problem is discussed. Errors in measurements or data that do not fit the model (both are called outliers) occur frequently and they often remain unidentified. Some data, called leverage points, are particularly dangerous because they have a strong influence although they can be outliers. Diagnostics to detect the outliers are proposed.

Section 2.3 presents the method of nonlinear least squares. In most cases, a linearization by a first-order Taylor series expansion is possible and we are brought back to OLS. Section 2.4 is dedicated to weighted least squares and Section 2.5 to correlated least squares. In both cases, a simple transformation of parameters brings us back to OLS. Section 2.6 describes the method of mixed regression, which permits to improve the conditioning as well as the accuracy of the determined parameters. Finally, systematic errors are discussed in Section 2.7. Additional examples are given in Appendix II, where some applications and special topics are also discussed.

2.2   ORDINARY LEAST-SQUARES METHOD

2.2.1   PRINCIPLE OF THE METHOD

If we want to determine p parameters β_j ($j = 1, \ldots, p$) from n experimental data y_i ($i = 1, \ldots, n, n > p$), the starting equation of the linear least-squares methods is

$$\mathbf{y} = \mathbf{X}\boldsymbol{\beta} + \boldsymbol{\varepsilon} \tag{2.1}$$

where $\mathbf{X}$ is an $n \times p$ matrix (called design matrix) whose coefficients are supposed to be known exactly and $\boldsymbol{\varepsilon}$ is the vector of (unknown) errors, assumed to be random. If $\hat{\boldsymbol{\beta}}$ is the least-squares estimate (or estimator) of the unknown vector $\boldsymbol{\beta}$, the vector of residuals of length n is

$$\mathbf{r} = \mathbf{y} - \mathbf{X}\hat{\boldsymbol{\beta}} \tag{2.2}$$

The method of least squares minimizes the sum of squared residuals

$$S = \mathbf{r}^{\mathrm{T}}\mathbf{r} = \sum_{i=1}^{n} r_i^2 = \min \tag{2.3}$$

where the superscript T means the transpose of the vector.

Minimizing Equation 2.3 gives the best estimator $\hat{\beta}$ as

$$(\mathbf{X}^T\mathbf{X})\hat{\beta} = \mathbf{X}^T\mathbf{y} \tag{2.4a}$$

$$\hat{\beta} = (\mathbf{X}^T\mathbf{X})^{-1}\mathbf{X}^T\mathbf{y} \tag{2.4b}$$

The practical solution of the system of Equation 2.4a will be discussed in Section 2.2.2. It is recommended to avoid the inversion of $(\mathbf{X}^T\mathbf{X})$. It is better to calculate the eigenvalues of this matrix [3] or it is still better to decompose $\mathbf{X}$ into singular values [4].

The residual standard deviation s, which is an estimate of σ (i.e., of the unknown errors ε), is given by

$$s = \hat{\sigma} = \sqrt{\frac{\sum_{i=1}^{n} r_i^2}{n-p}} \tag{2.5}$$

and the estimated standard errors of the regression coefficients are equal to the square roots of the diagonal elements of the variance–covariance matrix Θ

$$\Theta(\hat{\beta}) = s^2(\mathbf{X}^T\mathbf{X})^{-1} \tag{2.6}$$

It is then possible to define the correlation matrix ρ with elements

$$\rho_{ij} = \frac{\Theta_{ij}}{\sqrt{\Theta_{ii}\Theta_{jj}}} \tag{2.7}$$

The solution given by Equation 2.4 is optimal when the following so-called Gauss–Markov conditions are obeyed:

- The errors have a mean of zero (no bias): $E(\varepsilon) = 0$, where E is the expectation value of ε*.
- The experimental data are of equal precision (equal variances): $E(\varepsilon_i^2) = \sigma^2$ $E(\varepsilon\varepsilon^T) = \sigma^2\mathbf{I}_n$.
- The errors are independent (uncorrelated): the variance–covariance matrix Θ is diagonal.

$$\text{Var}(\mathbf{y}) = \Theta(\mathbf{y}) = \text{Var}(\varepsilon) = \Theta(\varepsilon) = \sigma^2\mathbf{I}_n \tag{2.8}$$

where $\mathbf{I}_n$ is the n-dimensional unit matrix.

---

* The expectation value (also called population mean) of any function $f(x)$ is the weighted average value of the function averaged over all possible values of the variable x, with each value of $f(x)$ weighted by the probability distribution: $E[f(x)] = \langle f(x) \rangle = \sum f(x_j)P(x_j)$.

If these conditions are met, it is possible to determine a confidence interval for the parameters β_i, that is, an interval which is likely to include the true value of the parameter

$$\hat{\beta}_i - ts(\hat{\beta}_i) \leq \beta_i \leq \hat{\beta}_i + ts(\hat{\beta}_i) \tag{2.9}$$

where t is given by the Student distribution [5] and $s(\hat{\beta}_i)$ is the standard deviation of the parameter $\hat{\beta}_i$ that is, the square root of $\left[\boldsymbol{\Theta}(\hat{\boldsymbol{\beta}}) \right]_{ii}$ from Equation 2.6. How likely the interval is to contain the parameter is determined by the confidence level. For instance, if $n \gg p$, $t = 1.96$ gives a confidence level of 95%. Generally, the uncertainties $s(\hat{\beta}_i)$ are reported with two-figure accuracy. However, in case of ill-conditioning (Section 2.2.2), this is not always sufficient. The rounding of parameters is discussed in Appendix II.1.4.

When the variance–covariance matrix $\boldsymbol{\Theta}(\hat{\boldsymbol{\beta}})$ is known, it is possible to calculate the variance–covariance matrix of any function of $\boldsymbol{\beta}$. Suppose the relationship

$$\boldsymbol{\gamma} = \mathbf{D}\boldsymbol{\beta} \tag{2.10}$$

where $\mathbf{D}$ is a matrix whose coefficients are exactly known. The variance–covariance matrix of $\boldsymbol{\gamma}$ may be written

$$\begin{aligned} \boldsymbol{\Theta}(\hat{\boldsymbol{\gamma}}) = E\left[(\boldsymbol{\gamma} - \hat{\boldsymbol{\gamma}})(\boldsymbol{\gamma} - \hat{\boldsymbol{\gamma}})^{\mathrm{T}} \right] = \mathbf{D}E\left[(\boldsymbol{\beta} - \hat{\boldsymbol{\beta}})(\boldsymbol{\beta} - \hat{\boldsymbol{\beta}})^{\mathrm{T}} \right]\mathbf{D}^{\mathrm{T}} \\ = \mathbf{D}\boldsymbol{\Theta}(\hat{\boldsymbol{\beta}})\mathbf{D}^{\mathrm{T}} \end{aligned} \tag{2.11}$$

Equation 2.11 is called the law of propagation of errors. It is worth noting that even if the errors on $\hat{\boldsymbol{\beta}}$ are not correlated, the errors on $\hat{\boldsymbol{\gamma}}$ will be correlated. This equation will be used in Appendix II.1.5 to discuss what happens when the parameters or the data are linearly transformed. The importance of correlations in the calculation of standard deviations is also discussed.

2.2.2 PRACTICAL SOLUTION AND ILL-CONDITIONING

2.2.2.1 Solution

To solve the system of normal equations (2.4a), one of the best techniques is the *singular value decomposition* (SVD), which, at the same time, diagnoses possible numerical problems such as high correlations between the parameters and ill-conditioning.

First, the columns of the Jacobian matrix $\mathbf{X}$ are scaled to have unit length, that is, each term of the vector column $\mathbf{X}_i$ is divided by the norm $\|\mathbf{X}_i\|$), where

$$\left\| \mathbf{X}_i \right\| = \sqrt{\sum_{j=1}^{n} X_{ji}^2} \tag{2.12}$$

Then, the singular values (square roots of the eigenvalues of $\mathbf{X}^T\mathbf{X}$) of the scaled $\mathbf{X}$ matrix, $\mu_1, \mu_2, \ldots, \mu_p$, are calculated. The SVD of $\mathbf{X}$ may be written [4] as

$$\mathbf{X} = \mathbf{U}\mathbf{D}\mathbf{V}^T \tag{2.13}$$

where $\mathbf{U}$ is a $n \times p$ matrix with orthonormal columns containing the p eigenvectors associated to the largest eigenvalues of $\mathbf{X}\mathbf{X}^T$, $\mathbf{D}$ is a $p \times p$ diagonal matrix whose elements are the singular values μ_i of $\mathbf{X}$, and $\mathbf{V}$ is a $p \times p$ orthonormal matrix, which contains the eigenvectors of $\mathbf{X}^T\mathbf{X}$. For this decomposition, which can always be done, several computer codes are available, for instance SVDCMP in *Numerical Recipes* [6].

The solution is

$$\hat{\boldsymbol{\beta}} = \mathbf{V}\mathbf{D}^{-1}\mathbf{U}^T\mathbf{y} \tag{2.14}$$

Likewise, the variance–covariance matrix of the parameters may be written as

$$\hat{\Theta}(\hat{\boldsymbol{\beta}}) = s^2(\mathbf{X}^T\mathbf{X})^{-1} = s^2\mathbf{V}\mathbf{D}^{-2}\mathbf{V}^T \tag{2.15a}$$

$$\mathrm{var}(\hat{\beta}_k) = s^2\sum_j\left(\frac{V_{kj}}{\mu_j}\right)^2 \tag{2.15b}$$

2.2.2.2 Ill-Conditioning

It often happens that some parameters cannot be estimated with accuracy and that they are very sensitive to small perturbations in the data. This is due to the correlation of parameters. This increases the variances of the estimated parameters and is responsible for important round-off errors. There are diagnostics that determine whether a correlation exists and that can identify the parameters affected. The correlation matrix, Equation 2.7, is often employed for this purpose. Note that the absence of high correlations does not imply the absence of collinearity. Therefore, many different procedures have been proposed to detect the presence of ill-conditioning.

Lees [7] suggested we calculate the eigenvalues and eigenvectors of the square matrix $\mathbf{X}^T\mathbf{X}$: for each vanishing eigenvalue, there is a linear dependency among the columns of $\mathbf{X}$. Therefore, a small eigenvalue is an indication of collinearity. Watson et al. [8] advocated the use of the diagonal elements of the inverse of the correlation matrix. This notion has later been generalized by Femenias [9] and then by Grabow et al. [10].

Belsley [11] has critically reviewed these procedures and has concluded that "None is fully successful in diagnosing the presence of collinearity and variable involvement or in assessing collinearity's potential harm" (p. 37). To palliate the weaknesses of the existing diagnostics, he has introduced the condition indices and has shown that they can be easily used to determine the strength and the number of near-dependencies.

The *scaled condition indices* of the scaled matrix $\mathbf{X}$ are defined as

$$\eta_k = \frac{\mu_{max}}{\mu_k} \, k = 1, \ldots, p \tag{2.16}$$

where μ_{max} is the largest value of μ_k. The highest condition index, $\max(\eta_k)$, is the condition number $\kappa(\mathbf{X})$. It is an error magnification factor, and it is used to determine whether a matrix is ill-conditioned: if the data are known to d significant figures and if the condition number of $\mathbf{X}$ is 10^r, then a small change in the data in its least significant digit can affect the solution in the $(d - r)$th place. If we consider a perturbation $\delta\mathbf{y}$ in $\mathbf{y}$, then

$$\frac{\left\|\delta\hat{\boldsymbol{\beta}}\right\|}{\left\|\hat{\boldsymbol{\beta}}\right\|} \leq \kappa(\mathbf{X}) \frac{\left\|\delta\mathbf{y}\right\|}{\left\|\mathbf{y}\right\|} \tag{2.17}$$

where $\hat{\boldsymbol{\beta}}$ is the least-squares solution and $\delta\hat{\boldsymbol{\beta}}$ the perturbation in $\hat{\boldsymbol{\beta}}$ due to $\delta\mathbf{y}$. Thus, a large κ may be responsible for a large error.

Likewise, a perturbation $\delta\mathbf{X}$ in $\mathbf{X}$ leads to

$$\frac{\left\|\delta\hat{\boldsymbol{\beta}}\right\|}{\left\|\hat{\boldsymbol{\beta}}\right\|} \leq \kappa(\mathbf{X}) \frac{\left\|\delta\mathbf{X}\right\|}{\left\|\mathbf{X}\right\|} \tag{2.18}$$

It shows that, when the condition number is large, the result may be sensitive to the accuracy of $\mathbf{X}$. In other words, in the case of the nonlinear least-squares method (Section 2.3), it is important to calculate the derivatives accurately (for more details, see Chapter 5 and Appendices V.2.1 and Va.3).

More generally, in the case of an inexact linear system (i.e., a least-squares regression), it is possible to show that [11]

$$\frac{\left\|\delta\hat{\boldsymbol{\beta}}\right\|}{\left\|\hat{\boldsymbol{\beta}}\right\|} \leq \kappa(\mathbf{X})\hat{R}^{-1}\left[2 + \sqrt{1 - \hat{R}^2}\kappa(\mathbf{X})\right]\max\left(\frac{\left\|\delta\mathbf{X}\right\|}{\left\|\mathbf{X}\right\|}, \frac{\left\|\delta\mathbf{y}\right\|}{\left\|\mathbf{y}\right\|}\right) \tag{2.19}$$

where $\hat{R}^2$ is the squared multiple correlation coefficient

$$\hat{R}^2 = 1 - \frac{\mathbf{r}^T\mathbf{r}}{\mathbf{y}^T\mathbf{y}} \tag{2.20}$$

The number of near-dependencies is equal to the number of large scaled condition indices. To determine which parameters are involved in the collinearities, Belsley defines the variance-decomposition proportions. Starting from the variance–covariance matrix $\Theta(\hat{\boldsymbol{\beta}})$ of the least-squares estimator $\hat{\boldsymbol{\beta}}$ of $\boldsymbol{\beta}$, Equation 2.15, the variance-decomposition proportions are defined as

$$\pi_{jk} = \frac{\left(\dfrac{V_{kj}}{\mu_j}\right)^2}{\displaystyle\sum_{i=1}^{p}\left(\dfrac{V_{ki}}{\mu_i}\right)^2} \quad k,j = 1,\ldots,p \tag{2.21}$$

Belsley proposes the following rule of thumb: estimates are degraded when two or more variances have at least half of their magnitude (i.e., >0.5) associated with a scaled condition index of 30 or more. Evidently, the probability of encountering the problem of ill-conditioning rapidly increases with the number of parameters to be determined.

EXAMPLE 2.1: EQUILIBRIUM STRUCTURE OF FLUOROPHOSPHAETHYNE

The first illuminating example is given by the equilibrium structure of the fluorophosphaethyne (FCP) molecule, which was determined independently several times. The equilibrium structure of FC≡P was first determined using the ground-state moments of inertia of the $F^{12}CP$ and $F^{13}CP$ isotopic species, the vibrational correction being calculated from an experimental anharmonic force field [12]. Later the equilibrium structure was redetermined using the experimental equilibrium moment of inertia of $F^{12}CP$ and an estimated equilibrium moment of inertia for the $F^{13}CP$ species [13]. The two determinations are obviously incompatible (see Table 2.1). The reason is that the carbon atom is near the center of mass; therefore, the system of equations is ill-conditioned (condition number $\kappa = 450$), and it is not possible to determine accurately both interatomic distances which are fully correlated (variance-decomposition proportions equal to 1). However, note that their sum, $r(FP)$, is accurate. It is possible to improve the conditioning by using the "corrected" ab initio distances as additional data, the discussion of this example is continued in Section 2.6.

EXAMPLE 2.2: SEMIEXPERIMENTAL EQUILIBRIUM STRUCTURE OF CYANOBUTADIYNE, HC₅N

The equilibrium structure of the linear molecule cyanobutadiyne (HC_5N) was calculated [14] using 14 experimental ground-state rotational constants and the

TABLE 2.1
Comparison of Different Equilibrium Structures (in Å) of FC≡P

Method	r(C–F)	r(C≡P)	r(FP)	References
Force field + exp.	1.272	1.548	2.820	12
Experimental r_e^{exp}	1.284	1.538	2.820	13
Mixed	1.276	1.544	2.820	27
Ab initio[a], r_e^{BO}	1.2757	1.5431	2.819	32
Semiexperimental, r_e^{SE}	1.2759(4)	1.5445(2)	2.819	32

[a] CCSD(T)/CBS, see Section 1.5 for the explanation of this level of electronic structure theory.

α-constants (rotation-vibration interaction constants defined in Sections 3.2.2 and 4.3.1) calculated at the CCSD(T)/cc-pVQZ level. The results are given in Table 2.2.

The first observation is that the standard deviations of the parameters are not reliable indicators of their precision because the number of parameters (6) is not small compared to the number of data (14). The condition number, $\kappa = 1186$, is quite large, and the three central bond lengths [$r(C_b–C_c)$, $r(C_c\equiv C_d)$, and $r(C_d–C_e)$] are highly correlated ($\pi > 0.5$) and may thus be sensitive to small errors, in other words they might not be accurate. This is easy to check by increasing successively each rotational constant by 0.1 MHz (the typical order of magnitude of the error) and by repeating the least-squares fit. The variation of the molecular parameters is given in Table 2.3. The three bond lengths [$r(C_b–C_c)$, $r(C_c\equiv C_d)$, and $r(C_d–C_e)$] are indeed the most sensitive to errors.

TABLE 2.2
Semiexperimental Equilibrium Structure (in Å) of
H–C$_a$≡C$_b$–C$_c$≡C$_d$–C$_e$≡N

Parameter	Value	π^a
r(H–C$_a$)	1.06220(3)	0.042
r(C$_a$≡C$_b$)	1.20949(5)	0.023
r(C$_b$–C$_c$)	1.36518(12)	**0.696**
r(C$_c$≡C$_d$)	1.21284(17)	**0.992**
r(C$_d$–C$_e$)	1.37079(12)	**0.705**
r(C$_e$≡N)	1.16209	0.014

a Variance-decomposition proportions, Equation 2.21 corresponding to the condition number $\kappa = 1186$. Potentially harmful values are bold.

TABLE 2.3
Variation of the Bond Lengths (in Å) of H–C$_a$≡C$_b$–C$_c$≡C$_d$–C$_e$≡N When the Equilibrium Rotational Constant of Each Isotopologue Is Increased Successively by 0.1 MHz

	N	^{13}C$_a$	^{13}C$_b$	^{13}C$_c$	^{13}C$_d$	^{13}C$_e$	^{15}N	D
r(H–C$_a$)	−0.0016	−0.0083	−0.0002	0.0003	−0.0012	−0.0007	−0.0004	0.0018
r(C$_a$≡C$_b$)	0.0000	0.0090	−0.0119	0.0020	−0.0003	0.0004	0.0006	0.0000
r(C$_b$–C$_c$)	−0.0003	0.0078	**0.0230**	−0.0213	−0.0117	−0.0007	0.0014	0.0003
r(C$_c$≡C$_d$)	0.0002	**−0.0132**	−0.0135	**0.0292**	**0.0279**	−0.0124	**−0.0148**	−0.0007
r(C$_d$–C$_e$)	0.0001	0.0014	−0.0005	−0.0122	−0.0199	**0.0218**	0.0088	0.0003
r(C$_e$≡N)	0.0002	0.0005	0.0004	−0.0003	0.0015	−0.0116	0.0090	0.0002

Note: Largest variation is outlined in bold.

Further examples of ill-conditioning are given in Appendix II.1. In particular, it is shown in II.1.5 that a linear transformation of the variables will not improve the conditioning. On the other hand, a nonlinear transformation may improve the conditioning, a typical example being the different reductions and representations of the molecular rovibrational Hamiltonian (see Section 4.2.3).

2.2.3 OUTLIER ANALYSIS

It is important to check that each experimental datum is not affected by a systematic error. An outlier in a data set is an observation that is numerically distant from its true value. Although outliers are common, they often remain unnoticed when the least-squares method is used as a "black box" without a careful analysis of the results. Furthermore, outliers sometimes remain hidden even to a conscientious user because they do not show in the residuals r_i, which should be of the order of magnitude of the experimental accuracy. Indeed, by definition, the least-squares method tries to avoid large residuals and, thus, tries to accommodate the outlying data at the expense of the remaining observations. Therefore, an outlier may have a small residual, especially when it is a leverage point, that is, a point that contributes an enormous amount to the solution.

2.2.3.1 Leverage

The most common measure of leverage is given by the diagonal elements of the so-called hat matrix **H** defined as

$$\mathbf{H} = \mathbf{X}(\mathbf{X}^T\mathbf{X})^{-1}\mathbf{X}^T \tag{2.22}$$

H is a square matrix of size $n \times n$. The hat matrix is a projection matrix, which transforms the vector **y** of experimental data into the vector $\hat{\mathbf{y}}$ of predicted values.

$$\hat{\mathbf{y}} = \mathbf{H}\mathbf{y} \ (\text{from } \hat{\mathbf{y}} = \mathbf{X}\hat{\boldsymbol{\beta}} = \mathbf{X}(\mathbf{X}^T\mathbf{X})^{-1}\mathbf{X}^T\mathbf{y}) \tag{2.23}$$

It can be calculated from the **U** matrix (see Equation 2.13).

$$\mathbf{H} = \mathbf{U}\,\mathbf{U}^T \tag{2.24}$$

and hence $h_{ii} = \sum_{j=1}^{n} U_{ij}^2$

It is easy to check that **H** is idempotent (**H H** = **H**) and symmetric ($\mathbf{H}^T = \mathbf{H}$). Hence, the following properties

$$0 \le h_{ii} \le 1 \tag{2.25}$$

and

$$\mathrm{Tr}(\mathbf{H}) = \sum_{i=1}^{n} h_{ii} = p \tag{2.26}$$

hold, and so the average value of h_{ii} is p/n. From Equation 2.23, $h_{ii} = \dfrac{\partial \hat{y}_i}{\partial y_i}$.

Thus, a leverage of zero indicates an observation without influence on the fit. On the other hand, if h_{ii} is close to one, $\hat{y}_i \approx y_i$ and the corresponding residual will be close to zero whatever the value of y_i. In this case, one parameter has been devoted to fitting this observation and it indeed sometimes happens that one parameter is determined from only one observation (which may be further incorrect). It is quite important to avoid data with "large" h_{ii}. The definition of large is not easy. Many authors believe that $h_{ii} > 2p/n$ may be considered large. However, this definition is often too strict; one may consider that the data for which $h_{ii} < 0.6$ are rather safe. When h_{ii} is larger, more observations have to be introduced in the fit in order to reduce the value of h_{ii}; otherwise, the offending observation has to be eliminated unless there are very strong reasons to keep it in the fit.

A simple example will illuminate the usefulness of the h_{ii} diagnostics. Assume that we try to determine two unknown parameters, α and β, by measuring several y_i values, which obey the following relation:

$$y_i = \alpha f_i + \beta k_i + \varepsilon_i \tag{2.27}$$

The f_i and k_i, which are the coefficients of the $\mathbf{X}$ matrix, are known, but it may happen that the following conditions hold:

$$\begin{cases} f_i \neq 0 & \text{for all} \quad i \neq n \; f_n = 0 \\ k_i = 0 & \text{for all} \quad i \neq n \; k_n \neq 0 \end{cases} \tag{2.28}$$

This corresponds, for instance, to the case in which we measure many rotational transitions $\Delta J = 0$ and only one $\Delta J \neq 0$ (see Appendix IV.2). The parameter β is obviously determined only by y_n and its estimate is

$$\hat{\beta} = \frac{y_n}{k_n} \tag{2.29}$$

but, more importantly, its standard deviation is given by

$$s(\hat{\beta}) = \frac{s}{k_n} \tag{2.30}$$

where s is the standard deviation of the fit, Equation 2.5, which may be quite small even if y_n is an outlier because $r_n = 0$. Thus, we have the wrong impression that the fit is good and β is well determined. In this case, obviously, $h_{nn} = 1$. It may also happen that one $f_m \gg f_i$ for all $i \neq m$. In this case

$$h_{mm} = \frac{f_m^2}{\sum f_i^2} = \frac{f_m^2}{f_m^2 + \delta} \approx 1 \tag{2.31}$$

Here, δ means a small quantity. This situation often occurs when one tries to determine higher-order terms. Finally, note that this measure of leverage is not "robust." In other words, it is not safe to try to eliminate all the leverage points in one step.

h_{ii} indicates that there is a problem with the ith measurement, but it does not indicate which parameter is affected. Another diagnostic that gives more specific information is the Difference in Betas (DFBETAS)

$$\text{DFBETAS}_j(i) = \frac{\hat{\beta}_j - \hat{\beta}_j(i)}{s(\hat{\beta}_j)} \frac{s}{s(i)} \tag{2.32}$$

where $\hat{\beta}_j(i)$ is the estimate of the jth parameter when the ith data is omitted, $s(\hat{\beta}_j) = \left[\sqrt{\Theta(\hat{\beta})} \right]_{jj}$ the estimated standard error of $\hat{\beta}_j$ and $s(i)$ is the estimated standard deviation of the fit when the ith measurement is dropped. $s(i)$ is easy to calculate using the following relation:

$$s^2(i) = \frac{\left[(n-p)s^2 - \frac{r_i^2}{(1-h_{ii})} \right]}{(n-p-1)} \tag{2.33}$$

If

$$\mathbf{C} = (\mathbf{X}^T\mathbf{X})^{-1}\mathbf{X}^T \tag{2.34}$$

then

$$\text{DFBETAS}_j(i) = \frac{c_{ji}}{\sqrt{\sum_{k=1}^{n} c_{jk}^2}} \frac{r_i}{s(i)(1-h_{ii})} \tag{2.35}$$

The DFBETAS test shows us how much the estimation of the parameter β_j would change if the ith measurement were to be deleted. The critical or cutoff value for DFBETAS is $2/\sqrt{n}$. For more details about these tests, see references [15,16].

2.2.3.2 Analysis of Residuals

Several textbooks describe the different outlier diagnostics; see for instance [5,15–17]. To check whether a particular datum is compatible with the other data, the "jackknifed" residual (also called "studentized" residual) has proven to be particularly efficient.

$$t(i) = \frac{r_i}{s(i)\sqrt{1-h_{ii}}} \tag{2.36}$$

It is based on the fact that the variance of a particular residual r_i is $s^2(1 - h_{ii})$. In this diagnostic, the standard deviation s is replaced by $s(i)$, which is the estimated standard deviation of the fit when the ith measurement is dropped (Equation 2.33). $t(i)$ follows a Student distribution (t) if the errors are Gaussian and has a near t-distribution under a wide range of circumstances. In other words, if $t(i)$ is large ($t(i) > 3$–3.5), the ith data is likely to be an outlier.

EXAMPLE 2.3: ROTATIONAL SPECTRUM OF MONODEUTERATED FORMALDEHYDE (HDCO)

A detailed analysis of the rotational spectrum of HDCO was published in 1978 [18]. A more recent and more extensive analysis [19] obtained incompatible parameters. In particular, the value of the sextic centrifugal distortion constant Φ_J is $-31(1)$ Hz in the first analysis, whereas it is only 0.1671(16) Hz in the second one. Inspection of the fit of the transitions of Dangoisse et al. [18] reveals nothing wrong. However, using the diagnostics defined in Equations 2.22, 2.32, and 2.36, the $10_{2,8} \leftarrow 11_{1,11}$ transition* at 33,989.68 MHz appears extremely dubious. Although its residual, $r = 4$ kHz, is extremely small (compared to the experimental accuracy of about 100 kHz), its $t(i) = 4.7$ is too large and, especially, its $h_{ii} \approx 1$ indicates that it is a very strong influential point that mainly affects Φ_J because DFBETAS(Φ_J) = 6004 (compared to a cutoff value of 27). The removal of this transition solves the problem and its right frequency is found to be about 134 MHz lower. This example shows the usefulness of the diagnostics introduced in this section because neither the quality of the fit nor the smallness of the residuals permits to conclude that the derived parameters are correct. Further examples are discussed in Appendix II.2. One way to improve the conditioning and to decrease the leverage is discussed in Section 2.6.

2.3 NONLINEAR LEAST SQUARES

Often, the relationship between $\mathbf{y}$ and $\boldsymbol{\beta}$, $\mathbf{y} = f(\boldsymbol{\beta})$, is not linear. When it is not possible to linearize the equation by a coordinate transformation, the parameters must be refined iteratively as

$$\beta_j^{k+1} = \beta_j^k + \Delta\beta_j \tag{2.37}$$

where k is the iteration number. At each iteration, the function f is linearized by a first-order Taylor series expansion about $\boldsymbol{\beta}^k$.

$$f_i(\boldsymbol{\beta}) = f_i(\boldsymbol{\beta}^k) + \sum_{j=1}^{p} \frac{\partial f_i(\boldsymbol{\beta})}{\partial \beta_j}(\beta_j - \beta_j^k) \tag{2.38}$$

With $r_i = y_i - f_i(\boldsymbol{\beta}^k)$, $X_{ij} = \dfrac{\partial f_i(\boldsymbol{\beta})}{\partial \beta_j}$, and $\Delta\beta_j = \beta_j - \beta_j^k$ the solution may be written as

$$\mathbf{r} = \mathbf{X}\,\Delta\boldsymbol{\beta} + \boldsymbol{\varepsilon} \tag{2.39}$$

---

* For the definition of the rotational quantum numbers, see Chapter 4 and Appendix IV.1.

This equation is formally identical to Equation 2.1 and may be solved in the same way. Here, the design matrix $\mathbf{X}$ is called the Jacobian matrix. The particular case of a structure determination is discussed in Appendix V.1.

When the problem is strongly nonlinear or when the starting values of the parameters are quite far from the solution, this method may fail to converge. In this case, it may be helpful to use the Levenberg–Marquardt method [20], for which the solution of the OLS method is replaced by

$$\Delta\hat{\boldsymbol{\beta}} = \left[\mathbf{X}^{\mathrm{T}}\mathbf{X} + c\,\mathrm{diag}(\mathbf{X}^{\mathrm{T}}\mathbf{X})\right]^{-1}\mathbf{X}^{\mathrm{T}}\mathbf{r} \tag{2.40}$$

and c is first set to a small number (e.g., 10^{-3}). If the sum of squares of residuals, Equation 2.3, increases, c is scaled up (e.g., by a factor of 10), otherwise c is scaled down (e.g., by 0.1) for the next iteration. Routines based on this method are widely available, for instance, in the *Numerical Recipes* series [6].

A simpler method is to use a damping factor, d, where $0 < d < 1$, which is used to reduce the corrections to the parameters β_i to a fraction of the value calculated, as follows:

$$\beta_j^{k+1} = \beta_j^k + d\Delta\beta_j \tag{2.41}$$

2.4 WEIGHTED LEAST SQUARES

2.4.1 METHOD

Quite often, the observations have different uncertainties and their variance is written as follows:

$$\mathrm{Var}(\mathbf{y}) = \mathrm{Var}(\boldsymbol{\varepsilon}) = \sigma^2\mathbf{W}^{-1} \tag{2.42}$$

where $\mathbf{W}$ is the diagonal weight matrix.

Left-multiplying Equation 2.1 by $\mathbf{W}^{1/2}$, we get

$$\mathbf{W}^{1/2}\mathbf{y} = \mathbf{W}^{1/2}\mathbf{X}\boldsymbol{\beta} + \mathbf{W}^{1/2}\boldsymbol{\varepsilon} \tag{2.43}$$

The variance–covariance matrix of $\mathbf{W}^{1/2}\boldsymbol{\varepsilon}$ is $\sigma^2\mathbf{I}_n$ because

$$\Theta(\mathbf{W}^{1/2}\boldsymbol{\varepsilon}) = E(\mathbf{W}^{1/2}\boldsymbol{\varepsilon}\,\boldsymbol{\varepsilon}^{\mathrm{T}}\mathbf{W}^{1/2}) = \mathbf{W}^{1/2}E(\boldsymbol{\varepsilon}\boldsymbol{\varepsilon}^{\mathrm{T}})\mathbf{W}^{1/2} = \sigma^2\mathbf{I}_n \tag{2.44}$$

In other words, the substitutions $\mathbf{y} \rightarrow \mathbf{y}' = \mathbf{W}^{1/2}\mathbf{y}$ and $\mathbf{X} \rightarrow \mathbf{X}' = \mathbf{W}^{1/2}\mathbf{X}$ allow us to solve the system of equations by the standard OLS. The estimate is given by the solution of the set of normal equations, Equation 2.4a,

$$(\mathbf{X}^T\mathbf{W}\mathbf{X})\hat{\boldsymbol{\beta}} = \mathbf{X}^T\mathbf{W}\mathbf{y} \qquad (2.45)$$

The residual standard deviation is given by

$$s = \sqrt{\frac{\sum_{i=1}^{n} W_i r_i^2}{n - p}} \qquad (2.46)$$

and the variance–covariance matrix of the parameters Θ is

$$\Theta(\hat{\boldsymbol{\beta}}) = s^2 (\mathbf{X}^T\mathbf{W}\mathbf{X})^{-1} \qquad (2.47)$$

Usually, the weight of an observation is taken as the inverse square of the experimental uncertainty. However, in a structure determination, the inherent model limitations due to the lack of sufficiently well-known vibrational corrections generally result in errors, which are much larger than the experimental errors of the inertial moments, although the latter may themselves differ greatly in magnitude, usually being larger for the less abundant isotopologues (see Appendix II.3.1).

Furthermore, two different weighting schemes may be used.

1. The absolute weighting method, in which each weight is assumed to be the inverse square of the uncertainty
2. The relative weighting method, in which the weights are normalized using, for instance, the average value of the weights W_A, and dimensionless weights are defined as

$$W'_i = \frac{W_i}{W_A} \qquad (2.48)$$

where

$$W_A = \frac{1}{n} \sum_{i=1}^{n} W_i$$

In the first case, the residual standard deviation, Equation 2.46, should be close to 1, but it is not enough to prove the quality of the fit because it is possible to obtain $s = 1$ by just scaling the weights. On the other hand, in the second case (relative weighting), the residual standard deviation has the dimension of the observed quantity and its value is between the smallest and the largest errors.

An example showing the advantage of the relative weighting is given in Appendix II.3.2.

Another point worth mentioning is that the values of the estimated parameters do not depend on the weighting provided that the number of data is much larger than the number of parameters. Indeed, from Equation 2.45

$$E(\hat{\beta}) = (\mathbf{X}^{\mathsf{T}}\mathbf{W}\mathbf{X})^{-1}\mathbf{X}^{\mathsf{T}}\mathbf{W}E(\mathbf{y}) = (\mathbf{X}^{\mathsf{T}}\mathbf{W}\mathbf{X})^{-1}\mathbf{X}^{\mathsf{T}}\mathbf{W}\left[X\beta + E(\varepsilon)\right] = \beta \quad (2.49)$$

However, in the particular case of a structure determination, where the number of data is not much larger than the number of parameters, this is not true, for an example see Appendix II.3.1 and Appendix V.2.2, Table V.3. On the other hand, the standard deviations are sensitive to the weighting.

2.4.2 ITERATIVELY REWEIGHTED LEAST SQUARES

Finding appropriate weights is sometimes difficult, especially when the errors are mainly due to the approximations of the model (as is the case in a structure determination) or when there are outliers (due to misassignments or perturbations). The iteratively reweighted least squares (IRLS) method [21] can sometimes be used to estimate the weights. It has the advantage of being a *robust* method, that is, it mitigates the influence of outliers. The first step is performed with the standard least-squares method (with no weighting or with an approximate weighting), then the weight matrix is updated to some nonnegative function $g(r)$ of the residuals r. With these new weights, the weighted least-squares equation is solved. The process is iterated in the following way:

Step 0: Initial residuals r_i and leverage h_{ii} are calculated using the OLS.

Step 1: The median* absolute deviations (MAD) of the residuals is calculated.

$$\text{MAD} = \text{median}|r_i - \text{median}(r_i)| \quad (2.50)$$

and the standard deviation is estimated from MAD as

$$s = 1.4826\left(1 + \frac{5}{n-p}\right)\text{MAD} \quad (2.51)$$

(This estimate of the standard deviation is more robust than the one obtained with Equation 2.5.)

The residuals are scaled by dividing them by s from Equation 2.51

$$u_i = \frac{|r_i|}{s} \quad (2.52)$$

and the weights are calculated first using the Huber function, Equation 2.53, and then, if possible in a second round of iterations, the biweight function, Equation 2.54. The Huber weights are calculated in the following way:

$$\begin{vmatrix} W_i = 1 & \text{if} & u_i \leq 1.345 \\ W_i = 1.345/u_i & \text{if} & u_i > 1.345 \end{vmatrix} \quad (2.53)$$

---

* If the n data is listed in the order of increasing magnitude, the median is the value at position $(n+1)/2$. If n is odd, the median is the middle value and, if n is even, it is the mean of the two middle values.

The biweight function gives

$$
\begin{vmatrix}
W_i = W_i^H \left[1 - \left(\dfrac{u_i}{4.685} \right)^2 \right]^2 & \text{if} \quad u_i \le 4.685 \\
W_i = 0 \quad \text{if} \quad u_i > 4.685
\end{vmatrix}
\tag{2.54}
$$

where

$$
\begin{vmatrix}
W_i^H = 1 \quad \text{if} \quad h_{ii} \le 0.3 \\
W_i^H = \left(\dfrac{0.3}{h_{ii}} \right)^2 \quad \text{if} \quad h_{ii} > 0.3
\end{vmatrix}
\tag{2.55}
$$

The leverage-based weights, W_i^H, are introduced to reduce the influence of leverage points.

Step 2: The weighted least-squares method is used with the weights, Equation 2.53 or 2.54, to obtain a new set of regression parameters and new residuals.

Step 3: Steps 1 and 2 are repeated until there is negligible change from one iteration to the next.

The iteration is stopped when the maximum change in weights from one iteration to the next becomes negligible (typically, less than 0.1).

Note that the weights in IRLS are random variables, contrary to the standard weighted least squares, where weights are fixed numbers. For this reason, the standard errors of the parameters cannot be calculated in the usual way. A procedure to calculate robust standard errors is described in reference [21], but it often leads to a small increase of the standard deviation, which may be considered negligible. An example of the use of this method is the determination of the ground-state constants of ethane [22].

This method often gives good results when the uncertainties of all data are similar. On the other hand, when one deals with data of very different uncertainties, another reweighting scheme should be used. Watson [23] proposed the following formula to calculate the weights:

$$
W_i = \frac{1}{\sigma_i^2 + \alpha r_i^2}
\tag{2.56}
$$

where σ_i is the estimated uncertainty of the ith data and α a positive constant, which is 1/3 in many cases. This weighting scheme was used with success to calculate the "experimental" energy levels of water [24].

2.5   CORRELATED LEAST SQUARES

Frequently, particularly in structure determination, the errors of the observables are not independent. In this case, the variance–covariance matrix of the errors is nondiagonal.

$$\text{Var}(\mathbf{y}) = \text{Var}(\boldsymbol{\varepsilon}) = \sigma^2 \mathbf{M} \qquad (2.57)$$

where $\mathbf{M}$ is a symmetric and positive definite matrix. Its inverse, $\mathbf{M}^{-1}$, is called the general weight matrix. When $\mathbf{M}$ is known, the solution is easy. If $\mathbf{L}$ is the matrix of eigenvectors of $\mathbf{M}$ and $\boldsymbol{\Lambda}$ the diagonal matrix of associated eigenvalues, we can define the matrix $\mathbf{P}$ such that

$$\mathbf{P} = \mathbf{L}\boldsymbol{\Lambda}^{-1/2} \qquad (2.58)$$

If we left-multiply $\mathbf{y} = \mathbf{X}\boldsymbol{\beta} + \boldsymbol{\varepsilon}$ by $\mathbf{P}^{\mathsf{T}}$, we are again brought back to the case of the OLS with $\mathbf{y}' = \mathbf{P}^{\mathsf{T}}\mathbf{y}$ and $\mathbf{X}' = \mathbf{P}^{\mathsf{T}}\mathbf{X}$ because it is easy to show that $\text{Var}(\mathbf{P}^{\mathsf{T}}\boldsymbol{\varepsilon}) = \sigma^2 \mathbf{I}_n$ (see Equation 2.44).

The solution may be written as

$$(\mathbf{X}^{\mathsf{T}}\mathbf{M}^{-1}\mathbf{X})\hat{\boldsymbol{\beta}} = \mathbf{X}^{\mathsf{T}}\mathbf{M}^{-1}\mathbf{y} \qquad (2.59)$$

$$s^2 = \frac{1}{n-p}\mathbf{r}^{\mathsf{T}}\mathbf{M}^{-1}\mathbf{r} \qquad (2.60)$$

and

$$\text{Var}(\hat{\boldsymbol{\beta}}) = s^2 (\mathbf{X}^{\mathsf{T}}\mathbf{M}^{-1}\mathbf{X})^{-1} \qquad (2.61)$$

This method can easily be used in the term value approach or to analyze the hyperfine structure of a spectrum because, in these particular cases, it is easy to calculate the elements of $\mathbf{M}$ (see Appendix II.4), but in most cases $\mathbf{M}$ is not known and, because there are $n(n + 1)/2$ distinct elements in $\mathbf{M}$, we cannot estimate them on the basis of n observations.

2.6 MIXED REGRESSION (METHOD OF PREDICATE OBSERVATIONS)

2.6.1 METHOD OF PREDICATE OBSERVATIONS

When the system of normal equations is ill-conditioned or when there are leverage points, the simplest method is the introduction of new data. Unfortunately, this is rarely easy. A better method is to introduce prior or auxiliary information [11,25].

It is assumed that it is possible to construct q linear prior restrictions on the elements of $\boldsymbol{\beta}$ in the form

$$\mathbf{c} = \mathbf{R}\boldsymbol{\beta} + \boldsymbol{\xi} \qquad (2.62)$$

Here $\boldsymbol{\xi}$ is a random vector with $\text{Var}(\boldsymbol{\xi}) = \sigma^2\mathbf{N}$, $\mathbf{R}$ is a matrix of rank q of known constants, and $\mathbf{c}$ is a vector of specifiable values.

y and X are then augmented to give

$$\begin{bmatrix} y \\ c \end{bmatrix} = \begin{bmatrix} X \\ R \end{bmatrix} \beta + \begin{bmatrix} \varepsilon \\ \xi \end{bmatrix} \tag{2.63}$$

with

$$\mathrm{Var}\begin{bmatrix} \varepsilon \\ \xi \end{bmatrix} = \sigma^2 \begin{bmatrix} M & 0 \\ 0 & N \end{bmatrix} \tag{2.64a}$$

From Equation 2.59, the solution is

$$\hat{\beta} = (X^T M^{-1} X + R^T N^{-1} R)^{-1} (X^T M^{-1} y + R^T N^{-1} c) \tag{2.64b}$$

This method proved to be particularly useful when determining harmonic force constants from experimental data [26]. Example 2.4 discusses its application to the improvement of structural parameters.

EXAMPLE 2.4: EQUILIBRIUM STRUCTURE OF FLUOROPHOSPHAETHYNE, FCP

It is possible to improve the conditioning (see Example 2.1) by using the "corrected" ab initio distances as additional data. The offsets used to correct the ab initio distances are estimated from the structurally similar molecules HC≡P and FC≡N using the same method and the same basis set: CCSD(T)/6-311+G(2d, p) [27]. The input data are given in Table 2.4.

The results of this fit are collected in Table 2.1 (line "mixed"). The condition number drops from 450 to 58 and the structure obtained with the mixed fit is almost identical with the best ab initio r_e^{BO} structure.

EXAMPLE 2.5: ANALYSIS OF ROTATIONAL SPECTRA

At the beginning of the analysis of a rotational spectrum, it is customary to fix some (or all) centrifugal distortion constants at their ab initio computed values, or at the value of another isotopologue. This method does not always permit to obtain a good fit and makes the assignment of the spectrum more difficult.

For the parameters that cannot be determined but that may have an effect on the fit, it is much better to use the ab initio value (or the value of another isotopologue) as predicate observation with an uncertainty of about ten to twenty percent. This gives more flexibility to the fit and improves the accuracy of the fitted parameters, thus permitting a more accurate prediction.

A typical example is the analysis of the ground-state rotational spectrum of proline-*II* [28]. The quartic centrifugal distortion constants Δ_K and δ_K are highly correlated. For this reason, δ_K was set to zero in the original experimental analysis. This turned out to be somewhat unfortunate because δ_K is large. Using $\Delta_K = 8.95(300)$ kHz and $\delta_K = 0.65(30)$ kHz as predicate observations (from the ab initio calculations) permits to obtain a good fit and reliable molecular constants, which

TABLE 2.4
Input Data for the Mixed Structure of FCP

Parameter	Input Value	Exp – Calc[a]
B_e(FCP)(MHz)	5273.178(250)	0.0090
B_e(F^{13}CP) (MHz)	5265.226(1000)	−0.0812
r_e(FC) (Å)[b]	1.2770(20)	0.0009
r_e(CP) (Å)[b]	1.5455(20)	0.0010

Source: The data B_e(FCP) and B_e(F^{13}CP) are from McNaughton, D., and D. N. Bruget. 1993. *J Mol Spectrosc* 161:336–50.

FCP = fluorophosphaethyne.

[a] Residuals of the mixed fit.

[b] Scaled CCSD(T)/6-311+G(2d, p) value; see text.

TABLE 2.5
Rotational Parameters of Proline II

	Original Fit	Ab Initio[a]	With Predicate
A (MHz)	3923.5648(53)	3837.02	3923.5847(27)
B (MHz)	1605.87630(62)	1693.76	1605.87807(46)
C (MHz)	1279.79761(46)	1339.05	1279.79622(34)
Δ_J (kHz)	1.0198(105)	0.88	1.0252(56)
Δ_{JK} (kHz)	−4.645(85)	−3.32	−4.585(43)
Δ_K (kHz)	−0.0119(49)	8.95	8.2(25)
δ_J (kHz)	0.2512(63)	0.14	0.2471(31)
δ_K (kHz)	[0.0]	0.65	0.91(14)

Source: Allen, W. D., E. Czinki, and A. G. Császár. 2004. *Chem Eur J* 10:4512–7.

[a] MP2(AE)/cc-pVTZ.

are in better agreement with the ab initio values (see Table 2.5). Furthermore, the predictive power of these new constants is much better. Another example is discussed in Appendix II.5. Further examples may be found in reference [29].

The effect of fixing a parameter to an arbitrary value is discussed in Appendix II.6. This is particularly important in structure determination in which it is sometimes difficult to determine all the parameters.

2.6.2 CONSTRAINED LEAST SQUARES

An important case is when Equation 2.62 is a constraint, that is, $\xi = 0$ (or N = 0). In this case, the minimization of the sum of squared residuals, Equation 2.3 or Equation 2.60 in its general form, is minimized under the constraint, Equation 2.62. The solution is different from Equation 2.64a,b and is obtained using the method of Lagrange multipliers [5]

$$\hat{\beta} = \mathbf{b} + (\mathbf{X}^T \mathbf{M}^{-1} \mathbf{X})^{-1} \mathbf{R}^T [\mathbf{R}(\mathbf{X}^T \mathbf{M}^{-1} \mathbf{X})^{-1} \mathbf{R}^T]^{-1} (\mathbf{c} - \mathbf{Rb}) \qquad (2.65)$$

where $\mathbf{b}$ is the solution of the system without constraint, Equation 2.59.

2.7 SYSTEMATIC ERRORS

2.7.1 DEFINITION

The errors ε_i in Equation 2.1 are assumed to be random errors, their expectation value is zero. However, it may happen that a large (even dominant) part of the error is systematic. In this case, the error (called "bias") is about the same for each observation with a given apparatus and method. This may be due to a defect in the apparatus. For instance, Robert Boyle, in his famous experiments on the origin of Boyle's law, used a tube that was not perfectly cylindrical. For this reason, the measurement of the volumes was affected by a systematic error [30].

A more frequent source of systematic error is due to deviations of the real system from the laws assumed for it. For instance, Robert Millikan had to correct his measurements of the electronic charge for deviation from Stokes's law of fall as applied to the oil drops. A third common source of bias is due to the use of auxiliary quantities, which are incorrect. A typical example is the calibration of spectra. The problem can be easily pointed out and corrected when several independent methods are used to determine the same parameters. For instance, the vibrational ground-state spectra of OH were investigated by UV, infrared, and microwave spectroscopies [31]. It is permitted to conclude that the infrared measurements are affected by a systematic error of about 0.08 cm^{-1} (to be compared to an estimated uncertainty of 0.05 cm^{-1}).

Such errors must be identified and reduced until they are much less than the required precision. However, this goal is not always achievable and such errors are often hard to detect. There is an important exception discussed in Section 2.7.2.

2.7.2 AUTOCORRELATION OF THE ERRORS

The errors often become correlated due to the approximations of the model. For instance, when the transition frequencies are measured for successive quantum numbers, the r_i become correlated if a higher-order term is omitted (see for instance Figure 2.1). In this particular example, the solution is obvious: one has to free the higher-order term.

However, the situation is often more complicated. For instance, if we try to determine a structure from the ground-state moments of inertia, the errors will be correlated because the rovibrational contribution is neglected (see Figure 2.2). If instead of the coarse r_0-model (direct fit of the ground-state moments of inertia; see Section 5.3), the more sophisticated $r_{\mathrm{m}}^{(2)}$-model (see Section 5.5.4) is used, the residuals are decreased by more than one order of magnitude, but the residuals remain correlated (see Figure 2.3). Actually, even in a true r_e-fit (experimental or semiexperimental), the rovibrational corrections are affected by systematic errors, which leads to the problem of autocorrelation, see Figure 2.4 where the residuals r_i of a fit of the semiexperimental moments of inertia of CO_2 are plotted as a function of the

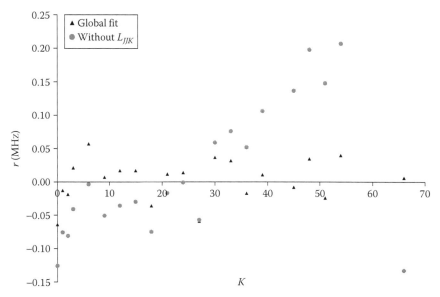

FIGURE 2.1 Residuals (MHz) as a function of the quantum number K of the fit of the rotational spectrum ($J = 74$) of the symmetric top tertiobutyl isocyanide, $(CH_3)_3CNC$, when the octic centrifugal distortion constant L_{JJK} is omitted. In the global fit, all octic constants are fitted. (Data from Cazzoli, G., G. Cotti, and Z. Kisiel. 1993. *J Mol Spectrosc* 162:467–73.)

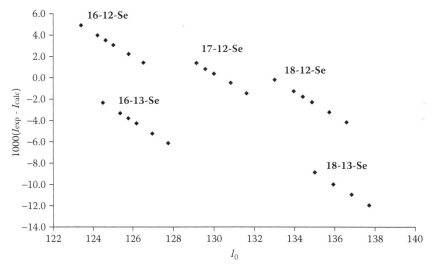

FIGURE 2.2 Residuals of the r_0-fit of the ground-state moments of inertia of carbonyl selenide (OCSe), each segment corresponds to the different isotopes of Se: ^{74}Se, ^{76}Se, ^{77}Se, ^{78}Se, ^{80}Se, and ^{82}Se (unit = u$Å^2$). (Reprinted from Le Guennec, M., G. Wlodarczak, J. Demaison, H. Bürger, M. Litz, and H. Willner. 1993. *J Mol Spectrosc* 157:419–46. copyright (1993). With permission from Elsevier.)

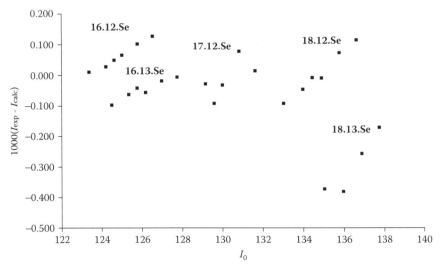

FIGURE 2.3 Residuals of the $r_m^{(2)}$-fit of the ground-state moments of inertia of OCSe (unit = uÅ^2). Source: Demaison, J. 2007. *Mol Phys* 105:3109–38. With permission.

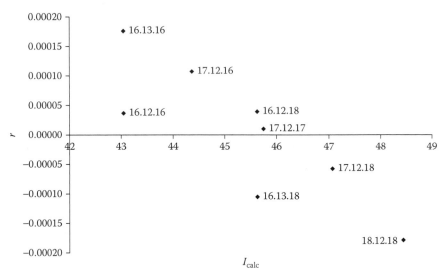

FIGURE 2.4 Residuals of the r_e-fit of the semiexperimental equilibrium moments of inertia of CO_2 (unit = uÅ^2). (Data from Puzzarini, C., M. Heckert, and J. Gauss. 2008. *J Chem Phys* 128:194108/1–9.)

calculated moments of inertia. It is obvious that almost all the points are aligned. If the number of degrees of freedom is large (number of data n much larger than the number of parameters p), a very exceptional occurrence in structure determination, the estimates are not much biased, but neglecting the correlation causes the variances of the parameters to increase substantially. There are diagnostics that permit to detect the autocorrelation of errors (briefly discussed in Appendix II.6), but the simplest and safest method is the plot of the residuals r_i as a function of the predicted $\hat{y}_i$.

REFERENCES

1. Rudolph, H. D. 1995. Accurate molecular structure from microwave rotational spectroscopy. In *Advances in Molecular Structure Research,* ed. M. Hargittai and I. Hargittai, Vol. 1, 63–114. Greenwich, CT: JAI Press.
2. Albritton, D. L., A. L. Schmeltekopf, and R. N. Zare. 1976. An introduction to the least-squares fitting of spectroscopic data. In *Molecular Spectroscopy: Modern Research*, ed. K. Narahari Rao, Vol. II, 1–67. New York: Academic Press.
3. Curl, R. F. 1970. *J Mol Spectrosc* 6:367–77.
4. Golub, G. H., and C. Reinsch. 1970. *Numer Math* 14:403–20.
5. Sen, A., and M. Srivastava. 1990. *Regression Analysis*. New York: Springer.
6. Press, W. H., B. P. Flannery, S. A. Teukolsky, and W. T. Vetterling. 1986. *Numerical Recipes.* Cambridge: Cambridge University Press.
7. Lees, R. M. 1970. *J Mol Spectrosc* 33:124–36.
8. Watson, J. K. G., S. C. Foster, A. R. W. McKellar, P. F. Bernath, T. Amano, F. S. Pan, et al. 1984. *Can J Phys* 62:1875–85.
9. Femenias, J. L. 1990. *J Mol Spectrosc* 144:212–23.
10. Grabow, J. -U., N. Heineking, and W. Stahl. 1992. *J Mol Spectrosc* 152:168–73.
11. Belsley, D. A. 1991. *Conditioning Diagnostics*. New York: Wiley.
12. Whiffen, D. H. 1985. *J Mol Spectrosc* 111:62–5.
13. McNaughton, D., and D. N. Bruget. 1993. *J Mol Spectrosc* 161:336–50.
14. Bizzocchi, L., C. Degli Esposti, and P. Botschwina. 2004. *J Mol Spectrosc* 225:145–51.
15. Belsley, D. A., E. Kuh, and R. E. Welsch. 1980. *Regression Diagnostics*. New York: Wiley.
16. Cook, R. D., and S. Weisberg. 1982. *Residuals and Influence in Regression*. London: Chapman & Hall.
17. Rousseeuw, P. J., and A. M. Leroy. 1987. *Robust Regression and Outlier Detection*. New York: Wiley.
18. Dangoisse, D., E. Willemot, and J. Bellet. 1978. *J Mol Spectrosc* 71:414–29.
19. Bocquet, R., J. Demaison, J. Cosléou, A. Friedrich, L. Margulès, S. Macholl, et al. 1999. *J Mol Spectrosc* 195:345–55.
20. Marquardt, D. W. 1963. *SIAM J Appl Math* 2:431–41.
21. Hamilton, L. C. 1992. *Regression with Graphics*. Belmont, CA: Duxbury Press.
22. Lin, K. F., W. E. Blass, and N. M. Gailar. 1980. *J Mol Spectrosc* 79:151–7.
23. Watson, J. K. G. 2003. *J Mol Spectrosc* 219:326–8.
24. Furtenbacher, T., A. G. Császár, and J. Tennyson. 2007. *J Mol Spectrosc* 245:115–25.
25. Bartell, L. S., D. J. Romanesko, and T. C. Wong. 1975. Augmented analyses: method of predicate observations. In *Chemical Society Specialist Periodical Report N° 20: Molecular Structure by Diffraction Methods*, ed. G. A. Sim, and L. E. Sutton, Vol. 3, 72–9. London: The Chemical Society.
26. Hedberg, L., and I. M. Mills. 1993. *J Mol Spectrosc* 160:117–42.
27. Dréan, P., J. Demaison, L. Poteau, and J. -M. Denis. 1996. *J Mol Spectrosc* 176:139–45.
28. Allen, W. D., E. Czinki, and A. G. Császár. 2004. *Chem Eur J* 10:4512–7.
29. Demaison, J., G. Wlodarczak, and H. D. Rudolph. 1997. Determination of reliable structures from rotational constants. In *Advances in Molecular Structure Research*, ed. M. Hargittai and I. Hargittai, Vol. 3, 1–51. Greenwich, CT: JAI Press.

30. Mandel, J. 1964. *The Statistical Analysis of Experimental Data*, 295–303. New York: Dover.
31. Destombes, J. L., C. Marlière, and F. Rohart. 1977. *J Mol Spectrosc* 67:93–116.
32. Bizzocchi, L., C. Degli Esposti, and C. Puzzarini. 2006. *Mol Phys* 104:2627–40.

CONTENTS OF APPENDIX II (ON CD-ROM)

3 Semiexperimental Equilibrium Structures
Computational Aspects

Juana Vázquez and John F. Stanton

CONTENTS

3.1 INTRODUCTION

Although the equilibrium structure, as defined by the Born–Oppenheimer approximation [1], can, in principle, be extracted from purely experimental information, the procedure is quite arduous. In particular, one needs to know the rotational constants in each and every vibrationally excited state that corresponds to a fundamental vibration—a task that becomes progressively more unwieldy as the molecular size increases—and for not just the molecule of interest but also several isotopologues.

Second, it is essential that the molecule under study does not have any appreciable Coriolis resonances. In recent years, workers in several laboratories have capitalized on a strategy first put forward by Pulay, Meyer, and Boggs [2] more than three decades ago. In that pioneering work, the equilibrium structure of methane (which had been estimated from electron diffraction studies [3] but was controversial) was estimated by applying computed "vibrational corrections" to the measured rotational constants of various isotopologues; the equilibrium structure was then deduced by the techniques discussed in Chapter 5 of this book.

These computations, which require one to determine the force field of the molecule (see Chapter 1) through cubic terms, were difficult to perform at the time of the original work on methane, but are now performed routinely. Advances in analytic derivative techniques in the field of quantum chemistry (electronic structure theory) has made the evaluation of molecular force fields straightforward, even at very high levels of theory with extended basis sets. Consequently, there has been a revolution of sorts in the determination of equilibrium structures by this technique; they are known as *semiexperimental equilibrium structures* because they contain input from both experiment (the ground-state rotational constants) and theory (the force fields that are necessary to obtain the corrections). The number of molecular equilibrium structures that are known with some precision (ca. 0.001–0.002 Å for bond lengths, and ca. 0.1–0.2° for bond angles) has grown exponentially in the last twenty years.

This chapter discusses the basic aspects involved in these computations and summarizes the basic perturbation theory [4] that relates the measured ground-state rotational constants [designated here as $A(B,C)_0$] to those that are related (in a simple way) to the principal moments of inertia of the equilibrium structure $A(B,C)_e$. In addition, several examples are given that illustrate the rich amount of information that has been generated by this procedure. Experimental aspects of determining semiexperimental equilibrium structures are discussed in Chapter 4.

3.2   GENERAL PROCEDURE FOR OBTAINING STRUCTURES

3.2.1   RIGID ROTATORS

In the simplest (rigid-rotator) treatment of rotational energy levels, the rotational constants are inversely and precisely proportional to the principal moments of inertia of the fixed structures [5], namely,

$$A = \frac{\hbar^2}{2I_A}$$

$$B = \frac{\hbar^2}{2I_B} \tag{3.1}$$

$$C = \frac{\hbar^2}{2I_C}$$

with the constants expressed in energy units. More typical, however, is a case in which the rotational constants are written in units of inverse time,

$$A = \frac{\hbar}{4\pi I_A}$$

$$B = \frac{\hbar}{4\pi I_B} \tag{3.2}$$

$$C = \frac{\hbar}{4\pi I_C}$$

or cm^{-1}, in which the constants above (A, B, C) are divided by the speed of light in cm · s^{-1}. The full form of the rotational Hamiltonian and the way in which these constants are extracted from experiment can be found in several chapters of this book and also in the classic literature of the field [6,7].

Suppose, for example, that one has a C_{2v} [8] symmetric triatomic molecule of the form AX$_2$ (water, F$_2$O, and CH$_2$ are examples). Let us further assume, for the moment, that the molecule behaves as an idealized rigid rotator [5–7] and that the structure of the rigid molecule under consideration is equal to its equilibrium structure. Like all asymmetric tops, such a molecule has three different rotational constants, all of which can be determined experimentally. However, in the case of a *planar* rigid molecule, only two of the three moments are independent, as is easily seen from the relations

$$I_A = \sum_i m_i (b_i^2 + c_i^2)$$

$$I_B = \sum_i m_i (a_i^2 + c_i^2) \tag{3.3}$$

$$I_C = \sum_i m_i (a_i^2 + b_i^2)$$

where the coordinates of atom i in the principal axis system (see Chapter 5) are given by (a_i, b_i, c_i). When the molecule is planar, the value of either a, b, or c is zero for all atoms. By convention, $I_C \geq I_B \geq I_A$, and by consequence this coordinate must be c. One can then straightforwardly derive the constraint as follows:

$$I_C - I_B - I_A = 0 \tag{3.4}$$

This establishes the fact that only two moments of inertia for a planar rigid rotator are independent. For our AX$_2$ model system, these constants are simple (nonlinear) functions of the principal axis coordinates, which are in turn functions of the two independent geometrical degrees of freedom of the molecule. Thus, if one knows the rotational constants from experiment, this information completely suffices to determine the structure of a rigid rotator.

If, instead of a simple AX$_2$ molecule, we have a more elaborate planar molecule, such as HFCO, it is somewhat less straightforward to determine the structure. For HFCO (a substituted form of formaldehyde), there are five geometrical degrees of freedom—conveniently chosen as the FCO and HCO bond angles, as well as the CO, CF, and CH bond distances. Thus, the two independent rotational constants for HFCO are ultimately functions of five geometrical coordinates and the straightforward inversion of the rotational constants to yield a structure is clearly impossible, as there are more equations than unknowns.

The way in which this problem is circumvented in practice is to take advantage of isotopically substituted forms of the molecule. Although ^{19}F is the only available isotope of fluorine, deuterium-, ^{18}O-, and ^{13}C-substituted forms of the compound can be studied conveniently. While the ^{13}C-substituted form is measurable in natural abundance, the others often require isotopically enriched samples. If the two independent rotational constants of all four isotopic species are obtained, then there are eight rotational constants to work with each of which is a function (assumed equal for each isotopic species, as is consistent with the Born–Oppenheimer picture) of the five geometrical coordinates. Thus, there are eight equations in five unknowns, and the structural parameters can be obtained by a nonlinear least-squares procedure [9], as discussed in Chapter 2 of this book. If the constants are exact (no experimental uncertainty and no systematic error), one can use any five of the eight rotational constants to determine the structure results in precisely the same distances and angles. However, in reality, where some uncertainty is always present, slightly different structures would result from the 56 possible choices of 5 constants, and the nonlinear least-squares approach is preferred. The magnitude of the residual then offers some estimate as to the uncertainty in geometrical parameters.

EXERCISE 3.1

Suppose you endeavor to calculate a semiexperimental equilibrium structure for the HFCO molecule, which has C_s point-group symmetry. How many independent molecular structure parameters exist for HFCO? For how many isotopic species would you need to find experimental data? Given that only one isotope, ^{19}F, is naturally occurring for fluorine, what do you think would be the likely isotopologues used in your study?

Procedures precisely analogous to the one mentioned here can be applied to other types of rigid-rotator molecules. For symmetric tops (which have two identical moments of inertia by definition) [10], there is one independent constant for planar systems (like BF_3, benzene) and two for nonplanar molecules like ammonia. Asymmetric tops [10] that are nonplanar have three independent rotational constants. The structures of all such molecules can be extracted from rotational constant data, given that a sufficiently large number of isotopic species are investigated so as to provide a number of independent constants that equals or exceeds the number of geometrical degrees of freedom in the system. This number, as dictated by symmetry, is always equal to the number of totally symmetric molecular vibrations for the molecule in question.

There are some subtleties that will not be discussed in detail in this chapter. For example, if one is interested in the structure of a particular molecule, some parameters can be obtained with greater confidence than others, depending on the choice of the isotopically substituted forms studied. If the rotational constants are determined exactly (which, again, is never possible), then any normal or overdetermined set of equations (in which the number of constants equals or exceeds the number of parameters) is actually equivalent, but when an experimental error bar can be assigned to each constant, the choice of isotopically substituted species can become critical. For example, if one was studying the species AXXY that has no nontrivial elements of symmetry (belonging to point group C_1), then one would have to obtain at least six rotational constants to determine its structure. Three of these are usually taken from the normal isotopic species, and three more could come from any isotopologue. That is, one could use another isotope of A or Y, or a second isotope of X in either position in the molecule.

Again, if the constants are known *precisely*, one can obtain the same structure, irrespective of which isotopologue is used in the fitting process. However, let us assume that the second molecule is that in which atom Y is isotopically substituted. In this case, differences in the moments of inertia of the normal and substituted species are going to be more sensitive to parameters such as the XY bond length and XXY bond angle than to parameters localized on the other end of the molecule like the AX bond length and the AXX bond angle. The fits of the geometrical parameters to the rotational constants are then going to be significantly more sensitive to the former class of parameters than to the latter. The best way to avoid this problem is to work with as many isotopologues as possible. If, for example, all 4 atomic positions are substituted with other isotopes, there will be 15 constants to fit to the 6 parameters, and the least-squares answer will not be systematically biased in the same way as those coming from the less-rigorous procedure.

3.2.2 REAL MOLECULES: VIBRATION-ROTATION INTERACTION CONSTANTS

When one moves from the idealized rigid rotator to real molecules, the rotational constants obtained from experiment are no longer exactly proportional to the reciprocal principal moments of inertia of the equilibrium structure, but rather are affected by the vibrations of the molecule [5–7]. The difference between the rigid-rotator-type constants associated with the equilibrium structure [see Equations 3.1 and 3.2, designated here as $A(B,C)_e$] and the measured ones is known as the *vibrational contribution*. Chapter 4 shows that the use of vibrational perturbation theory [11–13] leads to the power series expansion

$$A(B,C)^{[n]} = A(B,C)_e - \sum_i \alpha_i^{A(B,C)}\left(n_i + \frac{1}{2}\right)$$
$$+ \frac{1}{2}\sum_{ij} \gamma_{ij}^{A(B,C)}\left(n_i + \frac{1}{2}\right)\left(n_j + \frac{1}{2}\right) + \cdots \tag{3.5}$$

In this equation, $A(B,C)^{[n]}$ is the A (or B,C) rotational constant for the vibrational state described by the set of quantum numbers $\{n_1, n_2, \ldots\}$ and the coefficients of α_i and γ_{ij} are known as the first and second "vibration-rotation" interaction constants, respectively.

The algebraic form of the first-order vibration-rotation interaction constants can be derived in a tedious but straightforward way by applying perturbation theory to the vibration-rotation Hamiltonian for semirigid molecules written, for example, in the simplified form first derived by Watson [14,15]. The somewhat cumbersome result for the effect of normal mode k on constant A is [13,16,17]

$$\alpha_k^A = -2(A_e)^2 \left[\sum_\gamma \frac{3(a_k^{A\gamma})^2}{4\omega_k I_\gamma^e} + \sum_l \frac{(3\omega_k^2 + \omega_l^2)(\zeta_{kl}^A)^2}{\omega_k(\omega_k^2 - \omega_l^2)} + \pi \left(\frac{c}{h}\right)^{1/2} \sum_l \frac{a_k^{AA}\phi_{kll}}{\omega_k^{3/2}} \right] \quad (3.6)$$

with similar equations holding for the B and C constants. In this expression, ω_k is the harmonic frequency (in cm^{-1}) of normal mode k; $a_k^{A\gamma}$ is defined as the derivative of the (A, γ) element of the inertia tensor with respect to normal mode k, evaluated at the equilibrium geometry [i.e., $a_k^{A\gamma} \equiv (\partial I_{A\gamma}/\partial Q_k)_e$]; and ϕ_{kll} is a cubic force constant (in the dimensionless normal coordinate representation). In Equation 3.6, c and h are the speed of light and Planck's constant, respectively. Finally, ζ_{kl}^α is an element of the antisymmetric (and dimensionless) Coriolis ζ^σ matrix [18,19].

Some discussion of units is appropriate here. The vibration-rotation interaction constants have the same dimension as the rotational constants themselves. From Equation 3.6, one can see that the term in brackets must have dimensions of the reciprocal of the rotational constants. In the usual scenario favored by vibrational spectroscopists, where the rotational constants are in units of cm^{-1}, the quantity in braces must be in centimeters. The first term in brackets has the dimensions of $(I/\omega Q^2)$, that is, (mass)(length)^2/(length)$^{-1}$(mass)(length)^2, which, after cancellation, is the desired dimension of length. One can straightforwardly verify the same for the second term, and the third term can be analyzed as follows:

$$\begin{aligned}
\dim(c^{1/2}h^{-1/2}a\phi\omega^{-3/2}) &= l^{1/2}t^{-1/2} \; E^{-1/2}t^{-1/2} \; ml^2 m^{-1/2}l^{-1} \; l^{-1} \; l^{3/2} \\
&= l^2 t^{-1} m^{1/2} E^{-1/2} \\
&= l^2 t^{-1} m^{1/2} m^{-1/2}l^{-1}t \\
&= l
\end{aligned} \quad (3.7)$$

Here, the shorthand notation has been adopted in which m, l, and t stand for the *base physical quantities* mass, length, and time, respectively, and E for energy [20]. Thus, Equation 3.6 is appropriate for the case in which rotational constants are in units of cm^{-1}; multiplication of the term in brackets by the speed of light (c) in $cm \cdot s^{-1}$ gives the first vibration-rotation interaction constants in units of hertz.

EXERCISE 3.2

Suppose you have a diatomic molecule with equilibrium distance r_e, harmonic frequency ω_e (in cm^{-1}) and reduced mass μ. Also, assume that the molecule is governed by the Morse potential as follows:

$$V(q) = D_e[\exp(-aq) - 1]^2$$

with respect to the dimensionless coordinate q, where D_e is the dissociation energy. First reduce Equation 3.6 to the form appropriate to a diatomic molecule, and then evaluate the vibrational correction to the rotational constant for the aforementioned model system.

EXERCISE 3.3

Is the form of the central term in Equation 3.6 wrong? In particular, since the sum is unrestricted, is the term $k = l$ problematic? (Hint: Think about the properties of the Coriolis ζ^σ matrix mentioned in this chapter.)

3.2.3 PRACTICAL CONSIDERATION

3.2.3.1 Vibrational Corrections and "Alphas": Not the Same Thing

Equation 3.5 in the previous section gives a theoretical formula for rotational constants in an arbitrary vibrational state $\{n\}$. Although this is an important result by itself and provides predictions that can be used to find rotational lines of an excited vibrational state in a warm sample, it is actually not precisely the equation that is useful for the extraction of equilibrium rotational constants. Rather, what is needed is the vibrational contribution to any one rotational constant, and *not* the full set of excited-state rotational constants that are given, in principle, by Equation 3.6. If one rearranges this equation, the equilibrium rotational constants can be written in terms of the rotational constants determined for *any* vibrational state $\{n\}$, namely,

$$A(B,C)_e = A(B,C)^{\{n\}} + \sum_i \alpha_i^{A(B,C)}\left(n_i + \frac{1}{2}\right) - \cdots \tag{3.8}$$

One can, of course, choose this arbitrary vibrational state as the vibrational ground state, which yields

$$A(B,C)_e = A(B,C)^0 + \frac{1}{2}\sum_i \alpha_i^{A(B,C)} - \cdots \tag{3.9}$$

That is, the sum of the vibration–rotation interaction constants is needed to determine the equilibrium rotational constants and *not* the constants themselves. Of course, from the perspective of an experimentalist, there is no fundamental difference— clearly all the constants are needed in order to evaluate their sum; but to a theorist, this is a different task in principle. This is an important result that is not widely appreciated among theoreticians (see, however, Section 4.6.3); its consequences are discussed here.

First, let us consider the difference between the equation that is operative for the individual vibration–rotation interaction constants (Equation 3.6) and that gives directly the vibrational contribution to the rotational constants. The two equations shown together are

$$\alpha_k^A = -2(A_e)^2 \left\{ \sum_\gamma \frac{3(a_k^{A\gamma})^2}{4\omega_k I_\gamma^e} + \sum_l \frac{(3\omega_k^2 + \omega_l^2)(\zeta_{kl}^A)^2}{\omega_k(\omega_k^2 - \omega_l^2)} + \pi\left(\frac{c}{h}\right)^{1/2} \sum_l \frac{a_k^{AA}\phi_{kll}}{\omega_k^{3/2}} \right\} \tag{3.10}$$

and (using Equation 3.9)

$$\Delta A \equiv A_e - A_0 = -(A_e)^2 \left\{ \sum_{k\gamma} \frac{3(a_k^{A\gamma})^2}{4\omega_k I_\gamma^e} + \sum_{kl} \frac{(3\omega_k^2 + \omega_l^2)(\zeta_{kl}^A)^2}{\omega_k(\omega_k^2 - \omega_l^2)} \right. $$
$$\left. + \pi \left(\frac{c}{h}\right)^{1/2} \sum_{kl} \frac{a_k^{AA}\phi_{kll}}{\omega_k^{3/2}} \right\} \tag{3.11}$$

in which the right-hand sides at this point differ from one another in only two ways: a factor of 2 is present only in the top equation, and the summations extend over normal mode k in the lower equation. Anyone familiar with perturbation theory, however, recognizes something rather strange about the second equation. While the presence of an (potentially small) energy denominator in the top equation is not unexpected since this represents the property of an (potentially near-degenerate) excited state, the vibrational correction to the ground vibrational state is purely a property of a nondegenerate state. Therefore, if there are two vibrational modes with very similar frequencies (and connected by the Coriolis interaction), a very large contribution might occur in the second term in braces. This is a strange feature. However, this feature can be removed by the following straightforward algebraic procedure [21,22]. Consider the second term

$$\sum_{kl} \frac{(3\omega_k^2 + \omega_l^2)(\zeta_{kl}^A)^2}{\omega_k(\omega_k^2 - \omega_l^2)} \tag{3.12}$$

Using (twice) the antisymmetric property of the Coriolis ζ^σ matrix [18,19], the following steps lead to the removal of this problematic denominator:

$$\frac{1}{2} \sum_{kl} (\zeta_{kl}^A)^2 \left[\frac{(3\omega_k^2 + \omega_l^2)}{\omega_k(\omega_k^2 - \omega_l^2)} + \frac{(3\omega_l^2 + \omega_k^2)}{\omega_l(\omega_l^2 - \omega_k^2)} \right]$$
$$= \frac{1}{2} \sum_{kl} (\zeta_{kl}^A)^2 \left[\frac{(3\omega_k^2 + \omega_l^2)}{\omega_k(\omega_k^2 - \omega_l^2)} - \frac{(3\omega_l^2 + \omega_k^2)}{\omega_l(\omega_k^2 - \omega_l^2)} \right]$$
$$= \frac{1}{2} \sum_{kl} (\zeta_{kl}^A)^2 \left[\frac{(3\omega_k^2 + \omega_l^2)\omega_l - (3\omega_l^2 + \omega_k^2)\omega_k}{\omega_k\omega_l(\omega_k^2 - \omega_l^2)} \right]$$
$$= \frac{1}{2} \sum_{kl} (\zeta_{kl}^A)^2 \left[\frac{\omega_l^3 + 3\omega_k^2\omega_l - 3\omega_l^2\omega_k - \omega_k^3}{\omega_k\omega_l(\omega_k^2 - \omega_l^2)} \right] \tag{3.13}$$
$$= \frac{1}{2} \sum_{kl} (\zeta_{kl}^A)^2 \left[\frac{(\omega_l - \omega_k)^3}{\omega_k\omega_l(\omega_k - \omega_l)(\omega_k + \omega_l)} \right]$$
$$= -\frac{1}{2} \sum_{kl} (\zeta_{kl}^A)^2 \left[\frac{(\omega_l - \omega_k)^2}{\omega_k\omega_l(\omega_k + \omega_l)} \right]$$
$$= -\sum_{k<l} (\zeta_{kl}^A)^2 \left[\frac{(\omega_l - \omega_k)^2}{\omega_k\omega_l(\omega_k + \omega_l)} \right]$$

Thus, the *sum* of the first-order vibration-rotation interaction constants is not affected by small denominators as are the constants themselves, just as one expects since the sum of the constants corresponds to a nondegenerate ground-state property, while the constants themselves are associated with the property of a (potentially quasidegenerate) excited state. The truly important consequence of this feature is the following: while it is possible, in principle, to obtain the difference between the measured ground-state constants and the corresponding idealized equilibrium values experimentally (the determination of the entire set of first-order vibration-rotation interaction constants via rotationally resolved vibrational spectroscopy), such an effort might be impeded by the so-called vibrational Coriolis resonances [16]. However, one can certainly *calculate* the sum of the constants directly, blissfully ignoring those cases in which a Coriolis resonance is present. Stated alternatively, calculation of the vibrational correction is a more robust numerical procedure than that of the vibration-rotation interaction constants, or "alphas"; they are in fact fundamentally different calculations. Indeed, one cannot make the argument that a computational method unable to achieve agreement for the experiment for the alphas is therefore unable to accurately calculate the vibrational correction. This, instead, reflects a fundamentally flawed view of the situation.

One can take an entirely different approach and derive, directly, the difference between ground-state and equilibrium rotational constants *without even considering* the general vibrational state dependence of the rotational constants. In such a derivation, the so-called alphas are never encountered and one obtains Equation 3.11 directly [21,22]. A parallel exists in the computation of the *zero-point vibrational energy*, where the standard expression (the power series in $[n + 1/2]$) for vibrational term energies derived by perturbation theory—which includes potentially problematic denominators—can be resumed to eliminate the denominators [23], while precisely the same equation can be obtained directly using Rayleigh–Schrödinger perturbation theory.

To conclude this diversion, equilibrium rotational constants corresponding to the idealized rigid equilibrium structure of a molecule can be estimated by perturbation theory, provided one knows the ground-state rotational constants. The lowest-order correction can be calculated directly and is formally equivalent to one-half the sum of the first-order vibration-rotation interaction constants. Thus, if these are known, accurately, from experiment, the equilibrium rotational constants can be determined from laboratory data alone. However, the difficulties involved in extracting the interaction constants from experimental data extend beyond the already formidable task of obtaining the rotational constants for all single quantum-excited vibrational states of the molecule; they further include complications arising from the phenomenon of Coriolis resonance. The theoretical determination of the difference, however, is rigorously decoupled from any Coriolis resonances and is a numerically stable procedure.

3.2.3.2 Centrifugal Distortion

The rotational constants obtained from experimental studies of vibrational ground states differ in small details from those designated as $A(B,C)_0$ in this chapter. Typically, the rotational lines are fit to a so-called *S*- or *A*-reduced Hamiltonian (see Chapter 4), and the constants so obtained are properly designated as $A(B,C)_0^{(S)}$ or $A(B,C)_0^{(A)}$ accordingly. These differ from the ground-state vibrational constants of Equation 3.5 by centrifugal distortion corrections. In addition, there are also small electronic contributions to the moment of inertia. This and the next sections deal briefly with these arcane subjects and present an encyclopedic discussion of how the centrifugal distortion and electronic corrections are defined and how the various parameters (which depend on the Hamiltonian reduction approach) can be related to molecular parameters that can be found in specialized monographs [24]. The same topics will be addressed, from an experimental point of view, in Chapter 4.

For polyatomic molecules, there are quartic, sextic, etc., corrections due to centrifugal distortion (see Chapter 4), but only the lowest-order quartic terms are usually considered. They formally depend only on the harmonic force field and are in fact sometimes used to estimate very small vibrational frequencies in molecules [25].

Centrifugal distortion corrections to rotational constants tend to be very small in most cases. Moreover, in much of the *quantum chemical* literature, these effects are ignored in the determination of what we will define in the next section as a semiexperimental equilibrium structure—one that is obtained from *measured* rotational constants $A(B,C)_0$ augmented with *calculated* vibrational corrections. This is somewhat unfortunate, since one should ideally correct the measured constants in as rigorous a manner as possible. At the same time, it is a valid argument that the magnitudes of these centrifugal distortion corrections are probably, in most cases, smaller than the intrinsic error associated with the truncation of Equation 3.5 after the term linear in $[n + 1/2]$.

In any event, the inclusion or exclusion of this contribution to the calculations has a usually negligible effect on derived molecular structures. Computational chemistry programs such as CFOUR [26], SPECTRO [27], and GAUSSIAN [28] nevertheless provide quartic centrifugal distortion constants in both the A- and S-reduced representations of the Hamiltonian, as well as the correction to rotational constants that arise from these small effects. All in all, however, the correction to rotational constants due to centrifugal distortion effects is at least two orders of magnitude smaller than the vibrational corrections that are central to this chapter (see Table 4.4). The interested reader can find these effects discussed in some detail in complementary literature [21,22,24] and in Chapter 4.

3.2.3.3 Electronic Contribution to Rotational Constants

Another small effect that can be considered in very-high-accuracy studies is that ultimately the electrons themselves make contributions to the moments of inertia. There would be, in principle, two ways to address this effect: (1) One could use the nuclear masses of all the atoms in the molecule and treat the electronic contribution from that starting point. (2) Alternatively, one can use the atomic masses and make corrections from this perspective (in which the electronic masses are subsumed into the nuclear masses). A very detailed exposition on this point was given long ago by Flygare [29], and the preferred method is to use atomic masses, evaluate the full rotational **g**-tensor, and use this value to augment the rotational constants via the relationship,

$$\Delta A(B,C) = \frac{m_e}{m_p} \mathbf{g}^{A(B,C),A(B,C)} A(B,C)_e \qquad (3.14)$$

where m_e and m_p are the electron and proton masses, respectively. Elements of the dimensionless rotational $\mathbf{g}$-tensor comprise a (trivial) nuclear and a (nontrivial) electronic contribution, given by

$$\mathbf{g} = \mathbf{g}_e + \mathbf{g}_n \qquad (3.15)$$

$$\mathbf{g}_e = -\frac{h}{\mu_N} \left(\frac{\partial^2 E}{\partial \mathbf{B} \partial \mathbf{J}} \right)_e \qquad (3.16)$$

$$\mathbf{g}_n = \frac{1}{\mu_N} \sum_\gamma Z_\gamma (\mathbf{r}_\gamma \cdot \mathbf{r}_\gamma \mathbf{1} - \mathbf{r}_\gamma \mathbf{r}_\gamma) \mathbf{I}_e^{-1} \qquad (3.17)$$

In the aforementioned equations, μ_N is the nuclear magneton, $\mathbf{B}$ and $\mathbf{J}$ are the magnetic and the rotational angular momenta in the principal axis system, and $\mathbf{r}_\gamma$ are the coordinates of nucleus γ (with charge Z_γ in the center-of-mass frame).

The electronic part of the $\mathbf{g}$-tensor can be calculated from analytic second derivative methods [30,31], and its evaluation is greatly facilitated by the use of London (gauge-including) atomic orbitals [30]. The importance of these electronic contributions to the rotational constants has been studied systematically in the quantum chemical literature only recently [32]. While the contributions to the rotational constants can be large enough to significantly affect the inertia defect of small molecules with π-electron networks [33], the effect on equilibrium structures is generally negligible.

EXERCISE 3.4

The rotational $\mathbf{g}$-tensor for the hydrogen fluoride has been measured experimentally and is equal to 0.75449. Calculate the corresponding contribution to the rotational constant of HF, given that the equilibrium bond length of this molecule is 0.9169 Å. You will need to look up the atomic masses of ^{1}H and ^{19}F for this problem.

3.3　SEMIEXPERIMENTAL EQUILIBRIUM STRUCTURES

In Section 3.2, it was shown how the equilibrium structure defined ultimately through the Born–Oppenheimer approximation can be determined, even though the potential energy surface on which its definition is based does not truly exist. By means of a perturbation expansion, the rotational constants obtained in an experiment can be related to the idealized rigid-rotator "equilibrium" constants that are inversely proportional to the principal moments of inertia of the idealized equilibrium structure, but are not, in themselves, experimental observables.

The process by which one determines an equilibrium structure in this manner may take a number of forms. First, and most essential, is the experimental determination of the rotational constants of the ground vibrational state. If the number of constants is less than the number of degrees of geometrical freedom in the molecule (see Section 3.2.1), similar data must be obtained for a sufficient number of isotopic species. Of course, this step is required for any structural investigation involving microwave (MW) spectroscopy. The procedure for getting a structure from rotational constants was roughly sketched in Section 3.2.3, and involves least-squares fitting of structural parameters to rotational constants, *under the assumption* that the constants are related to the structure via Equation 3.1. When the rotational constants used are the ground-state $A(B,C)_0$ constants, this is known as the *effective* structure (see Chapters 4 and 8). When one uses the equilibrium constants, this is the *equilibrium* structure.

A fully experimental approach to the determination of an equilibrium structure might proceed as follows. Once the ground-state constants are available, the rotational levels need to be determined in vibrationally excited states, which necessitates the recording of rotationally resolved infrared or Raman spectra. The dependence of the rotational constants on the specific vibrational state is given, in principle, by Equation 3.5, but a number of assumptions would typically be made in this process for pragmatic reasons. First, it is really only reasonable to expect that one might obtain the excited-state constants for the modes that have significant infrared absorption, specifically the fundamental vibrations. For a molecule like F_2O of C_{2v} point-group symmetry, for example, these would be the constants $A(B,C)^{100}$, $A(B,C)^{010}$, and $A(B,C)^{001}$, respectively. From these and the assumption that higher than first-order terms do not contribute significantly to $A(B,C)^{[n]}$ (Equation 3.5), that is, assuming they are three or more orders of magnitude smaller than the rotational constants, one can obtain the first-order vibration-rotation interaction constants $\alpha_1^{A(B,C)}$, $\alpha_2^{A(B,C)}$, and $\alpha_3^{A(B,C)}$ and therefore

$$A_e = A_0 + \frac{1}{2}[\alpha_1^A + \alpha_2^A + \alpha_3^A]$$

$$B_e = B_0 + \frac{1}{2}[\alpha_1^B + \alpha_2^B + \alpha_3^B] \qquad (3.18)$$

$$C_e = C_0 + \frac{1}{2}[\alpha_1^C + \alpha_2^C + \alpha_3^C]$$

or, in terms of the measured rotational constants,

$$A_e = A_0 + \frac{1}{2}[3A_0 - A^{100} - A^{010} - A^{001}]$$

$$B_e = B_0 + \frac{1}{2}[3B_0 - B^{100} - B^{010} - B^{001}] \qquad (3.19)$$

$$C_e = C_0 + \frac{1}{2}[3C_0 - C^{100} - C^{010} - C^{001}]$$

This, of course, requires a serious laboratory effort and might not be feasible even for small molecules. Moreover, if there is a case of Coriolis resonance (see the discussion of denominators in Section 3.2.3.1), the perturbation expansion of Equation 3.5 will converge slowly and the use of the (linear) lowest-order expansion becomes invalid. Getting around this problem is difficult, because fitting both the alphas and the gammas (see Equation 3.5) to data would require the rotationally resolved study of all two-quantum transitions in addition to the fundamentals, which is an intractable task.

Alternatively, one can choose to *compute* the vibrational correction using the methods of electronic structure theory. In this case, the structure ultimately derived is often called a semiexperimental equilibrium structure. This term reflects the fact that the data going into this structure is both experimental (the rotational constants) and theoretical (the vibrational corrections). Such structures are neither truly theoretical in pedigree nor truly experimental.

In the opinion of the authors, this is the best method for determining equilibrium structures of larger molecules for the following reasons. The entirely experimental approach is fraught with difficulties in terms of the amount of data that needs to be obtained and issues such as Coriolis resonances. The intrinsic problem with the theoretical approach is slightly more difficult to grasp, but can be stated roughly as follows: the vibrational corrections to rotational constants are quite small, typically two orders of magnitude smaller than the constants themselves. Thus, if one uses theory and the calculated vibrational corrections are accurate to only 80%, one still obtains a reasonably good estimation of the equilibrium rotational constants $A(B,C)_e$. However, when the equilibrium structure is calculated by brute force using theoretical means, a method giving a 20% error in the vibrational corrections is unlikely to give a structure that is competitive with the semiexperimental equilibrium structure. The most accurate ab initio calculations of equilibrium structures are expensive affairs that make use of various extrapolation procedures (see Chapter 1 and references [32,34]).

3.3.1 Calculation of Vibrational Contributions

From Equation 3.6, one can see that the following four quantities are needed in order to obtain estimates of the equilibrium rotational constants from the corresponding ground-state constants:

1. The harmonic frequencies of the system or, equivalently, the quadratic force constants in the dimensionless normal coordinate representation.
2. The derivatives of the principal axis inertia tensor elements with respect to the normal coordinates Q_k.
3. The Coriolis ζ^σ matrices for the three principal axis directions.
4. The diagonal (ϕ_{iii}) and semidiagonal (ϕ_{ijj}) cubic force constants. Again, these are required in the dimensionless normal coordinate representation.

The first three quantities are all dependent on the quadratic force field of the molecule, the first of these being trivial. The derivatives of the inertia tensor can be determined analytically [14] or numerically, the simplest being the straightforward central-difference approach:

$$a_k^{\alpha\beta} \sim \frac{I^{\alpha\beta}(Q_k = \Delta) - I^{\alpha\beta}(Q_k = -\Delta)}{2\Delta} \tag{3.20}$$

for small values of displacement Δ. The Coriolis constants are given by the equation [16]

$$\zeta_{ij}^{\alpha} = -\zeta_{ji}^{\alpha} = \sum_k (l_{k\beta,i} l_{k\gamma,j} - l_{k\gamma,i} l_{k\beta,j}), \quad \alpha, \beta, \gamma = x, y, z \tag{3.21}$$

where the sum extends over all the atoms k, and $l_{k\beta,i}$ is one of the elements of the transformation matrix between the mass-weighted Cartesian coordinates d_k^{β} of atom k ($\beta = x, y, z$) and the normal coordinate Q_i. Thus, a straightforward calculation of the quadratic force field (and consequently, the normal coordinates) is sufficient to determine three of the four quantities involved in estimating the correction that must be applied to the ground-state rotational constants to estimate the equilibrium values.

The cubic constants require, by far, the largest computational effort in the calculation of vibrational corrections. Several different strategies might be envisioned for their calculation (see also Chapter 1). The simplest of such strategies might be the evaluation of the energy on a grid of points in dimensionless normal coordinate space, followed by a fit to a polynomial. Alternatively, one can take advantage of analytic and/or numerical derivative techniques that are available for virtually all quantum chemical methods. Although analytic third derivatives have been implemented only for the simplest Hartree–Fock (HF) approach [35] and are not to be found in any currently distributed quantum chemistry program package, a mixed use of analytic and numerical procedures is possible [36]. Given the availability of analytic first derivatives for many methods ranging from the very simple to the relatively complex, a procedure could be followed in which analytic first derivatives evaluated at points displaced from the equilibrium structure are used to calculate the cubic constants by central difference formula. Specifically, the required diagonal and semidiagonal constants are given by

$$\phi_{iij} = \frac{\left(\frac{\partial V}{\partial q_i}\right)_{q_j=\Delta}^{\text{analytic}} + \left(\frac{\partial V}{\partial q_i}\right)_{q_j=-\Delta}^{\text{analytic}} - 2\left(\frac{\partial V}{\partial q_i}\right)_0^{\text{analytic}}}{\Delta^2} \tag{3.22}$$

This approach would require $2N$ analytic gradient evaluations, where N is the number of normal coordinates. Note the distinction between the normal coordinates of $dim(Q) = m^{1/2}l$ and the dimensionless normal coordinates q (see Table 1.3).

Another approach is to use analytic second derivatives (which are available at the Hartree–Fock (HF) level in most program packages, and at the post-HF level in some others) and differentiate them once to obtain the desired cubic constants. That is,

$$\phi_{ijj} = \frac{\left(\frac{\partial^2 V}{\partial q_j^2}\right)^{\text{analytic}}_{q_i=\Delta} - \left(\frac{\partial^2 V}{\partial q_j^2}\right)^{\text{analytic}}_{q_i=-\Delta}}{2\Delta} \tag{3.23}$$

The advantage here, in addition to the increased numerical accuracy from numerical first, as opposed to second derivatives, is that only $2N_s$ second derivative calculations are needed, where N_s is the number of totally symmetric internal coordinates. Moreover, the full set of cubic constants ϕ_{ijk} (where i is a totally symmetric coordinate) can be obtained with no additional computational overhead, which is useful when vibrational corrections are needed for other isotopologues with the same symmetry.

Generally, one has to work with a number of isotopically substituted species when endeavoring to obtain a semiexperimental equilibrium structure. In such a case, the vibrational corrections must be evaluated for each isotopomer. Although the Born–Oppenheimer potential energy surfaces are identical in the geometric representation, they differ in the normal coordinate representation. Nevertheless, only a single calculation of the force field is needed.

The normal coordinates (and the quantities mentioned in points 1–3 of the list) can be obtained from the second derivatives of the energy *with respect to the Cartesian displacements of the nuclei* for any isotopic species. Appropriate mass-weighting of this matrix of second derivatives (the Hessian) followed by diagonalization provides the harmonic frequencies (which are the quadratic constants in the dimensionless normal coordinate representation), with the normal coordinates (needed for quantities discussed in 2 and 3) given by the eigenvectors of the mass-weighted Hessian [37,38]. If the full set of cubic constants is calculated, these can be transformed (linearly) into the geometric representation and then transformed back to the dimensionless normal coordinates associated with the isotopomer under study. There is a subtlety here, however. Because of the nonlinear transformation between internal coordinates and normal coordinates, the cubic (and all higher-order) normal coordinate force constants do not necessarily vanish when one of the coordinates represents a rotation of the molecule [39]. Therefore, in order to carry out the transformation, it is necessary to work with the *full* set of cubic force constants, specifically those that include the rotational coordinates obtained in the mass-weighted Hessian diagonalization. If the isotopic substitution does not break the framework symmetry of the unsubstituted species, the full set of force constants can be obtained from the set ϕ_{ijk}, where i is a totally symmetric coordinate (including rotation) of the unsubstituted species. If the analytic second derivative approach is followed, it is necessary to calculate only analytic second derivatives at displacements in the q coordinates that do not disturb the framework molecular symmetry, which is computationally expedient.

Finally, some remarks on the choice of the quantum chemical method are appropriate here (see also Chapter 1), and we must deal with the usual issues of *accuracy*, *precision*, and *cost*. To optimize the former, it is clearly necessary to use a quantum chemical approach that comes as close as possible to a solution of the electronic Schrödinger equation involving large basis sets and elaborate treatments of correlation, perhaps in conjunction with extrapolation techniques [32,34]. Precision is an issue only because some of the calculations outlined here have to be done numerically. Specifically, if one has access only to analytic first-derivative techniques, then both the harmonic force field and (especially) the cubic force constants are subject to numerical errors that are therefore quite sensitive to issues such as convergence criteria in solving the quantum chemical equations (the solutions are, in themselves, numerical).

With analytic second derivatives, judicious selection of convergence criteria for quantum chemical methods renders numerical "slop" in the class 1–3 parameters listed at the beginning of this section negligible, but one then has to make sure that the cubic force field is precisely calculated. This issue is ultimately governed by the degree of anharmonicity in the potential energy surface and the choice of step size (Δ in Equations 3.22 and 3.23). The CFOUR program system, in which the calculation of vibrational corrections is highly automated, performs consistency checks to ensure that the relevant parameters are numerically precise. Accuracy, of course, depends on the quantum chemical method chosen. Fully analytic third derivatives, of course, are the ultimate in precision. But, as mentioned earlier, these are available only at the HF-SCF level of theory [35], which then compromises their accuracy.

Finally, we must discuss the third issue associated with choosing a computational method—cost. And here, pragmatic considerations are operative, and different research objectives will result in different choices. Clearly, accuracy and precision issues might be optimized by using analytic second derivatives in conjunction with a large basis set and a highly correlated method such as CCSDTQ* [40–42], the coupled cluster method with full inclusion of up to quadruple excitations. This has become possible in recent years with the advent of analytic second derivatives for the entire coupled-cluster hierarchy [43]; but such calculations would be prohibitively expensive and probably run for months, even in a highly parallel computational environment. It would certainly be much faster to calculate vibrational corrections using analytic second derivatives (and therefore, more precise) using SCF, density functional theory (DFT) [44], or second-order perturbation theory (MP2) [45] methods. Some accuracy is lost, of course, but the effect on structures might be small enough that a user is satisfied. And indeed, if the structures derived from the SCF and MP2 vibrational corrections are not significantly different, then it is unlikely that the exact equilibrium structure will differ significantly from either.

3.3.2 Outline of Procedure for Obtaining Semiexperimental Equilibrium Structures

1. Collect experimental data on ground-state rotational constants of various isotopic species. Test accuracy of data and fidelity of transcription of the same by determining the r_0 structure based on the data.

2. Carry out a quantum chemical calculation of the quadratic and cubic force fields for one of the isotopic species. This is typically the so-called normal species that comprises the most common isotopes of all atoms.

3. Transform the quadratic and cubic force fields into those for the other, relevant isotopic species. Quantum chemical programs such as CFOUR do this automatically.

---

* CCSDTQ stands for coupled cluster with single (S), double (D), triple (T) and quadruple (Q) excitations.

4. Use Equation 3.11 to calculate the estimated equilibrium rotational constants for each isotopic species.

5. Carry out precisely the same least-squares calculation employed in step 1 of this list, but now using the semiexperimental $A(B,C)_e$ constants instead of the ground-state $A(B,C)_0$ constants. The resulting structure is the so-called semiexperimental equilibrium structure and should be an excellent approximation to the r_e structure that represents the corresponding minimum on the potential energy surface.

6. Depending on the accuracy level desired, carry out the aforementioned procedure using a few different levels of electronic structure theory and basis sets. If the derived structural parameters do not vary greatly when the basis set is enlarged or the treatment of electron correlation is improved, then they can be considered quite reliable and can provide some basis for assigning the associated uncertainties.

7. Consider a few caveats: the presence of very low frequency and/or highly anharmonic large-amplitude vibrations is generally associated with failures of low-order perturbation theory. That is, a molecule like methane or acetylene is a more suitable candidate for this approach than would be hydrogen peroxide or molecules with a freely rotating methyl group.

3.4 REPRESENTATIVE EXAMPLES

In this section, several representative examples are given for semiexperimental equilibrium structures. No effort is made to present an exhaustive review of the literature, but rather, emphasis is placed on work that was done by our group and some of our close collaborators after the seminal work on methane [2]. Specifically included is the pioneering research that Allen et al. started in the field nearly 20 years ago. With the advent of computer programs that are well-suited for calculating vibrational contributions to rotational constants, the number of semiexperimental equilibrium structures available today is quite extensive, and the number of such structures is expected to grow significantly in the coming years.

3.4.1 STUDIES OF ISOCYANIC ACID, KETENE, AND GLYCINE

Allen et al. used the semiexperimental procedure for the first time in the study of the equilibrium structure of isocyanic acid (HNCO) [21]. They employed experimental effective rotational constants ($A_0^{(S)}$, $B_0^{(S)}$ and $C_0^{(S)}$) of five isotopologues, (HNCO, DNCO, H^{15}NCO, NH^{13}CO and HNC^{18}O) corrected for centrifugal distortion and zero-point vibrational effects. The resulting equilibrium rotational constants were used to determined the r_e structure via the Kraitchman equations [24,47]. The vibrational corrections were estimated from pure ab initio quadratic and cubic force fields calculated at the restricted Hartree–Fock (RHF) level [46] (structure I, in reference [21]) and also from a harmonic experimentally scaled quantum mechanical force field (SQM) and cubic force constants computed at the RHF level and using the equilibrium structure (I). In the latter case the resulting equilibrium parameters were referred as structure (II). The peculiarity of this example, which often occurs in

TABLE 3.1
Semiexperimental Equilibrium Parameters of Isocyanic Acid (HNCO)[a]

	r(C–N)	r(N–C)	r(C–O)	θ(H–N–C)	θ(N–C–O)
(A_e, B_e)	1.00389	1.21450	1.16342	123.325	172.236
(A_e, C_e)	1.00333	1.21450	1.16344	123.289	172.236
(B_e, C_e)	1.00243	1.21457	1.16341	123.366	172.209
Final[b]	1.0030(20)	1.2145(6)	1.1634(4)	123.34(20)	172.32(20)

Source: From East, A. L. L., C. S. Johnson, and W. D. Allen. 1993. *J Chem Phys* 98:1299.

Note: Bond distances are in Å and bond angles in °.

[a] Structure derived using vibrational corrections calculated from SQM(CCSD)//expt. harmonic force field and RHF//expt cubic force constants. See [21] for more details.

[b] The structural parameters were weighted averages from the pairs of rotational constants (A_e, B_e), (A_e, C_e) and (B_e, C_e) with weighting facts chosen as 1/4, 1/4, and 1/2, respectively.

structural determination, was that the inertial defect among the corrected rotational constants did not vanish-in contrast to what theoretically should be the case for a planar molecule-. As a result of that, nuclear coordinates obtained from different pairs of rotational constants were slightly different. An arbitrary choice applied in this study was average the results obtained using the three different pairs of equilibrium rotational constants (see Table 3.1). The final r_e structure was, to the best of our knowledge, the first semiexperimental equilibrium structure determination made for this molecule.

Later, these authors presented an interesting idea during the process of studying the anharmononic force field and equilibrium structure of ketene (H_2CCO) [22]. They built a refinement procedure based on two types of cycles, that is, Cycles A, or Microiterations, and Cycles B, or Macroiterations. With this refinement procedure the final converged output consists not only of equilibrium structural parameters but also of the force field, first vibrational anharmonicity constants, rotation-vibration interaction constants, centrifugal distortion and Coriolis constants.

The A-cycles start keeping the reference geometry unchanged and submitting the quadratic force field to a self-consistent refinement with the empirical harmonic frequencies, which are obtained by doing several iterations with the vibrational anharmonicities derived after deperturbing resonances in the observed spectra. The B-cycles are initiated using as input vibration-rotation interaction constants (α_i^β), resulting from the harmonic frequencies (ω_i) and the cubic force field (φ_{ijk}) previously converged by doing A-cycles. These α_i^β constants are then employed to calculate vibrational corrections to rotational constants at the vibrational ground state (B_0^β), $\left(\dfrac{1}{2} \sum_i \alpha_i^\beta \right)$. These vibrational contributions, together with centrifugal distortion and electronic effects, are then used to correct sets of experimental rotational constants, B_0^β, from which an r_e structure is inferred by a least-squares refinement process and by doing B-cycles. The idea behind this double fit is to shift the reference geometry close to the true equilibrium configuration and refine it by carrying

TABLE 3.2
Semiexperimental Equilibrium Parameters of Ketene (H_2CCO)[a]

	$r(C=O)$	$r(C=C)$	$r(C-H)$	$\theta(H-C-H)$
r_e(I: ls-ABC)	1.16030(29)	1.31212(30)	1.07576(7)	121.781(12)
r_e(II: ls-BC)	1.16031(20)	1.31212(21)	1.07584(6)	121.789(9)
r_e(III: Kr-BC,D_2CCO)	1.16037	1.31222	1.07611	121.828
r_e(IV: Kr-BC,HDCCO)	1.16036	1.31223	1.07626	121.833

Source: East, A. L. L., W. D. Allen, and S. J. Klippenstein. 1995. *J Chem Phys* 102:8506.

Note: Bond distances are in Å and bond angles in °.

[a] ls-ABC and ls-BC stand for least-squares fits to istopomeric sets of (I_e^A, I_e^B, I_e^C) and (I_e^B, I_e^C) moment of inertia, respectively. Kr-BC refers to Kraitchman analysis applied to (I_e^B, I_e^C) sets using data from D_2CCO and HDCOO, respectively. The recommended equilibrium structure is (I; ls-ABC).

out B-cycles. In a subsequent step, the anharmonic force field at that new configuration is determined and, by repeating the A-cycles, this results in a new set of ω_i and φ_{ijk} parameters, which will generate a new set of α_i^β constants used to refine a new semiexperimental equilibrium structure by, again, doing B-cycles.

Table 3.2 is a summary of the semiexperimental equilibrium structures derived by Allen et al. using four different strategies (I, II, III and IV in Table 3.2) and data of six different isotopologues (H_2CCO, $H_2C^{13}CO$, $H_2C^{13}CO$, $H_2CC^{18}O$, HDCCO, and D_2CCO). In the approaches I and II, a least-squares fit to the moments of inertia (I_e^A, I_e^B, I_e^C) and (I_e^B, I_e^C), respectively, of the various isotopomers was carried out excluding I_e^A($H_2CC^{18}O$) in the former case. In methods III and IV, r_e was derived from the nuclear positions (a_i, b_i) as obtained by applying, once more, the Kraitchman equations [24,47]. The values obtained by the four approaches vary by 0.00007 Å, 0.00011 Å and 0.00050 Å for the CO, CC and CH distances, respectively and of 0.052° for the HCH angle. An curious structural aspect in ketene is the bond angle for the methylene skeleton, which is larger than 120°. This extension of the HCH angle is not found in related molecules as formaldehyde, ethylene or allene but, by contrast, is even more pronounced in diazomethane.

More recently, these authors have extended the semiexperimental determination of equilibrium structures to larger molecules and in particular to an interesting and important group of biomolecules, i.e, amino acids. For example, they have studied the two lowest-energy conformers of proline ($H_9C_5NO_2$) [48] and glycine ($H_5C_2NO_2$) (Gly-Ip and Gly-IIn)* [49]. For the latter geometries were optimized at the CCSD(T)/cc-pVTZ level† of theory and vibrational corrections to rotational constants were computed from MP2/6-31G harmonic and cubic force fields. Using

---

* I and II refer to the first and second, most stable conformers of glycine, respectively. "p" and "n" stands for conformations having C_s and C_1 point-group symmetries, respectively.

† CCSD(T) stands for the coupled-cluster method based on singles and doubles approximation augmented by a quasi-perturbative estimate of triple excitation effects; see reference [50]. cc-pVTZ stands for one of the correlation-consistent basis set of Dunning and coworkers ([51] and Chapter 1). (fc) indicates that core electrons were not correlated.

experimental effective rotational constants, B_0^β, for 5 isotopologues of Gly-Ip and 10 isotopologues of Gly-IIn, and carrying out a least-squares refinement of the structural parameters based on the empirically corrected B_e^β, improved semiexperimental equilibrium structures for both conformers were derived. The peculiarity of this example is that, due to the size of the molecules, some structural parameters were fixed to the ab initio values. For example, for the conformer Gly-Ip, there are fifteen structural degrees of freedom and 15 empirical rotational constants, which are insufficient to provide a well-defined structure without imposing constrains. Table 3.3 illustrates the results of fitting only the four heavy-atom distances and keeping the rest of the parameters fixed at the values computed at the CCSD(T)/cc-pVTZ level (Fit 1). In order to improve the semiexperimental structure a second fit was carried out (Fit 2) where more structural parameters were refined keeping only four structural constrains (see footnote b of Table 3.3). The deviation from the final refined structures (Fit 2) and the one calculated at the CCSD(T)/cc-pVTZ level of theory are reasonably small; the largest differences are 0.006 Å, for the N–H distance, and 0.5°, for the C–C–O angle. These results also show that the r_e structure predicted by the high level electronic theory method is an acceptable prediction.

TABLE 3.3
Ab Initio and Semiexperimental Equilibrium Parameters of Planar Neutral Glycine ($H_5C_2NO_2$)

	CCSD(T)/ cc-pVTZ	r_e(Fit 1)[a]	r_e(Fit 2)[b] Final
r(C–N)	1.446	1.448(4)	1.441(1)
r(C–C)	1.511	1.514(2)	1.511(1)
r(C=O)	1.204	1.203(1)	1.207(2)
r(C–O)	1.349	1.347(4)	1.353(1)
r(C–H)	1.088	–	1.0907(2)
r(N–H)	1.012	–	1.0065(2)
θ(C–C–O)	111.4	–	111.9(1)
θ(O–C–O)	123.1	–	123.2(1)
θ(C–C–N)	115.2	–	115.4(1)
θ(CH2 scissor)	105.9	–	105.95(3)
θ(CH2 wag)	5.1	–	5.4(1)

Source: Kasalová, V., W. D. Allen, H. F. Schaefer III, E. Czinki, and A. G. Császár. 2007. *J Comp Chem* 28:1373.

Note: Units are in Å for bond distances, in ° for bond angles, and in kHz for root mean square residual (rms). Standard errors are given in parentheses.

[a] Fit 1 releases r(C–N), r(C–C), r(C–O), and r(C=O) only.

[b] Fit 2 imposes r(O–H) = 0.9660 Å, θ(C–O–H) = 106.04°, θ(NH$_2$ scissor) = 104.98°, γ(NH$_2$ wag) = 57.67°. See reference [49] for details.

3.4.2 SELECTED OTHER STUDIES: SMALL- AND MEDIUM-SIZED MOLECULES CONTAINING FIRST-ROW ATOMS

The first work of Gauss et al. on the semiexperimental determination of r_e structures was the investigation of the equilibrium geometry of dioxirane (H_2COO) [52]. They exploited the use of analytical second derivatives for the numerical calculation of the third derivatives of the energy necessary in the framework of second-order vibrational perturbation theory [13]. They computed cubic force constants for three isotopologues of dioxirane ($^{13}CH_2O_2$, $CHDO_2$, and $CH_2^{18}O_2$) and with them and the harmonic force fields, they calculated vibration-rotation interaction constants, α_i^β, as well as quartic centrifugal distortion constants of the type τ_{ijkl}. With these parameters, they could conclusively determine equilibrium rotational constants, B_e^β, from the experimental B''^β constants based on the following expression:

$$B_e^\beta = B''^\beta + \frac{1}{2}\sum_r \alpha_r^\beta + \Delta B_\tau^\beta \qquad (3.24)$$

where ΔB_τ^β is the correction for centrifugal distortion effects and β refers to one of the principal axes of inertia.

What characterized this study was the use of two different semiexperimental procedures based on B_e^β values (calculated from the previous equation, and employing experimental B''^β in conjunction with theoretical vibrational corrections). The first approach determines the structure by a least-squares fitting to the isotopomeric B_e^β. The second approach derives the structure by a fit to isotopic shifts in the calculated B_e^β values rather than to the constants themselves. Structures inferred from the results of these two approaches and using B_e^β calculated at the CCSD(T)/cc-pVTZ level of theory are presented in Table 3.4. The final recommended equilibrium structure of dioxirane was as follows: $r_e(CO) = 1.3846 \pm 0.0005\,\text{Å}$, $r_e(OO) = 1.5133 \pm 0.0005\,\text{Å}$, $r_e(CH) = 1.0853 \pm 0.0015\,\text{Å}$, $\theta_e(HCH) = 117.03 \pm 0.20°$.

In propadienylidene (C_3H_2) [53], experimental ground-state rotational constants [54], B_0^β, were combined with vibrational corrections computed at the CCSD(T)/cc-pCVTZ level.* $r_e(C_1C_2) = 1.328 \pm 0.005\,\text{Å}$, $r_e(C_2C_3) = 1.287 \pm 0.001\,\text{Å}$, $r_e(CH) = 1.083 \pm 0.001\,\text{Å}$, $\theta_e(CCH) = 121.2 \pm 0.1°$ were recommended as semiexperimental equilibrium structures.

The equilibrium structure of methane was also reinvestigated under the semiexperimental approach [56]. In this case, two different procedures were followed. The first was based on CCSD(T) calculations with the cc-pCVXZ (X = D, T, Q, and 5) hierarchy of basis sets using an extrapolation scheme to account for the complete basis set (CBS) limit. The resulting equilibrium distance was $r_e = 1.08580$ Å. The second strategy was the experimental-theoretical approach that used experimental B_0^β constants for five isotopologues (CH_4, CH_3D, CH_2D_2, CHD_3, and CD_4) [57–61]

---

* CCSD(T) stands for the coupled-cluster method based on singles and doubles approximation augmented by a quasi-perturbative estimate of triple excitation effects; see reference [50]. cc-pCVXZ (X = D, T, Q, 5, etc.) are the core correlation-consistent basis set developed by Dunning and coworkers; see reference [55].

TABLE 3.4
Semiexperimental Equilibrium Parameters of Dioxirane (H_2COO)

Theoretical Method	r_e(CO)	r_e(OO)	r_e(CH)	θ_e(HCH)	Residual
		Least-squares fit to semiexperimental B_e^β values			
CCSD(T)/cc-pVTZ	1.3846(7)	1.5133(3)	1.0853(23)	117.06(35)	0.25
		Least-squares fit to semiexperimental B_e^β isotope shifts			
CCSD(T)/cc-pVTZ	1.3847(3)	1.5134(12)	1.0852(12)	117.00(16)	0.10

Source: Stanton, J. F., C. L. Lopreore, and J. Gauss. 1998. *J Chem Phys* 108:7190.
Note: Bond distances are in Å, bond angles in ° and root mean square resudual (rms) in kHz. Estimated uncertainties are given in parentheses.

and vibrational corrections calculated from quadratic and cubic force fields computed at the CCSD(T)/cc-pCVQZ level. The semiexperimental equilibrium distance was 1.0859 Å. Both structures agree almost perfectly and the final suggested r_e for CH_4 was $1.0859_5 \pm 0.0003$ Å.

Benzene, with twelve atoms and D_{6h} point-group symmetry, represented a new challenge and two approaches were followed to revise its r_e structure [62]. One was based on high-level quantum chemical calculations at the CCSD(T)/cc-pVQZ level. The other approach consisted of correcting experimental rotational constants (B_0^β) for four isotopologues (C_6H_6, $^{13}C_6H_6$, $^{13}C_6D_6$, and C_6D_6) using vibrational corrections based on harmonic and cubic force fields computed with the SDQ-MBPT(4) method [63] and the cc-pVTZ basis set.*

The structural parameters were derived by a least-squares fit to the empirically derived B_e^β. The agreement between the two procedures was nearly perfect (see Table 3.5), and the following equilibrium parameters were recommended: r_e(CC) = 1.3915 ± 0.0010 Å and r_e(CH) = 1.0800 ± 0.0020 Å. In the same study, mean internuclear distances (r_g) and distances between mean internuclear positions in the vibrational ground state of C_6H_6 (r_z) were also calculated and compared with experimental values [64] (see Chapter 8 for the definitions of the r_g and r_z structural parameters). The experimental and calculated r_g and r_z values agreed extremely well (see Table 3.5).

Because its equilibrium geometry was not well established, another prototype organic molecule of structural interest is cyclopropane (C_3H_6). Here, a comparison of the ab initio r_e structure calculated at the CCSD(T)/cc-pVQZ level of theory with that derived by least-squares fits of structural parameters to the vibrationally corrected B_e^β constants of C_3H_6 and $C_3H_4D_2$ was carried out [65]. Vibration-rotation constants were calculated from cubic and harmonic force fields computed at the

* cc-pVTZ stands for one of the correlation-consistent basis set of Dunning and coworkers [50]. (fc) indicates that core electrons were not correlated. MBPT(4) means many-body perturbation theory carried out to fourth order.

TABLE 3.5
Internuclear Distances of Benzene (C_6H_6)

	r_{CC}		r_{CH}	
	Reference [62]	Reference [104]	Reference [62]	Reference [104]
$r_e^{calculated}$	1.3911[a]		1.0800[a]	
$r_e^{semiexp}$	1.3914[b]	1.3902(2)	1.0802[b]	1.0862(15)

	Reference [62]	Reference [64]	Reference [62]	Reference [64]
r_g	1.3988[c]	1.399(1)	1.1005[c]	1.101(5)[d]
r_z	1.3964[c]	1.3976(15)	1.0846[c]	1.085(1)

Source: Gauss, J., and J. F. Stanton. 2000. *J Phys Chem A* 104:2865.

Note: Bond distances are in Å. Estimated uncertainties are given in parentheses. Rotational constants were taken from [105,106].

[a] Equilibrium distance computed at the CCSD(T)/cc-pVQZ level.
[b] Based on SDQ-MBPT(4)/cc-pVTZ cubic potential.
[c] Based on CCSD(T)/cc-pVQZ geometries using vibrational corrections calculated at the SDQ-MBPT(4)/cc-pVTZ level.

SDQ-MBPT(4)/cc-pVTZ level. The recommended parameters for cyclopropane were 1.5030 ± 0.0010 Å for $r_e(CC)$, 1.0786 ± 0.0010 Å for $r_e(CH)$, and $114.97 \pm 0.10°$ for $\alpha_e(HCH)$, which are in good agreement with an r_e geometry derived from gas-phase electron-diffraction information (see Tables 3.6 and 3.7). The agreement between the calculated and the experimental r_g and r_z parameters is also quite satisfactory.

Cis-hex-3-ene-1,5-diyne (C_6H_4) is another example of great interest because this molecule plays an important role in the Bergman cyclization and it is involved in the chemistry of the interestellar medium and pharmaceuticals against cancer. In the investigation of its equilibrium structure [66], effective rotational constants B_0^β were determined for five isotopologues and vibrational corrections were estimated by three approaches based on the use of different levels of theory for computing equilibrium geometries, quadratic and cubic force fields. In the first approach (I), quadratic and cubic force constants were calculated at the SCF/double-zeta plus polarization (DZP) level of theory. In the second approach (II), equilibrium structure and harmonic force constants were obtained at the SDQ-MBPT(4)/DZP level and cubic force field was computed at the SCF level using the DZP basis set [67]. The third approach (III) is identical to II, except that the DZP basis set is replaced by the TZ2P basis set [68].

The first molecule studied by the same group with atoms different from carbon, oxygen, and hydrogen was LiCCH [69]. Table 3.8 presents the semiexperimental equilibrium structure as well as the r_0 and r_s parameters [70] and the ab initio structure calculated at the CCSD(T)/(full WMR + diffuse) level of theory, where "full WMR" denotes the largest atomic natural orbital (ANO) basis of Widmark, Malmqvist, and Roos [71]. Results for the r_0 structure obtained by the least-squares fit to the experimental B_0^β are also included in Table 3.8. We can conclude that the vibrational effects are not entirely negligible. The r_0 distances for CC and CLi are

TABLE 3.6
Calculated and Experimental Internuclear Distances of Cyclopropane (C_3H_6)

	r(CC)			r(CH)		
	r_e	r_g	r_z	r_e	r_g	r_z
SDQ-MBPT(4)/ cc-pVTZ	1.4996	1.5098	1.5089	1.0753	1.0958	1.0811
CCSD(T)/ cc-pVQZ[a]	1.5019	1.5120	1.5111	1.0781	1.0986	1.0838
Experiment[b]	1.501(4)	1.5139(12)	1.5127(12)	1.083(5)	1.0991(20)	1.0840(20)
Experiment[c]	1.5101(23)	–	1.5157(69)	1.0742(29)	–	1.080(10)

Source: Gauss, J., D. Cremer, and J. F. Stanton. 2000. *J Phys Chem A* 104:1219.
Note: Bond distances in Å. Estimated uncertainties are given in parentheses.
[a] Vibrational corrections calculated at the SDQ-MBPT(4)/cc-pVTZ level.
[b] Obtained from gas-phase electron diffraction data [107].
[c] Obtained from microwave (MW) information [108].

TABLE 3.7
Equilibrium Structural Parameters r_e of Cyclopropane (C_3H_6) Obtained from Different Procedures and Sources

	MW[a]	GED[b]	This Work[c]	r_e^{BO} [d]
r(CC)	1.5101(23)	1.501(4)	1.5030(10)	1.5019
r(CH)	1.0742(29)	1.083(5)	1.0786(10)	1.0781
⟨(HCH)	115.85(33)	114.5(9)	114.97(10)	114.81

Source: Gauss, J., D. Cremer, and J. F. Stanton. 2000. *J Phys Chem A* 104:1219.
Note: Bond distances are in Å, bond angles in °. Estimated uncertainties are given in parentheses.
[a] From microwave (MW) data (see reference [108]).
[b] From gas-phase electron diffraction data (see reference [107]).
[c] Obtained from rotational constants determined in reference [108] and vibration-rotation interaction constants calculated at the SDQ-MBPT(4)/cc-pVTZ level of theory.
[d] Computed at the CCSD(T)/cc-pVQZ level.

slightly shorter than the apparent equilibrium internuclear separations. There is a remarkably good agreement between r_e parameters obtained in the fitting procedure and those obtained by direct quantum chemical calculations.

In a similar manner to the study on methane [56], the equilibrium structure of the ammonium radical (NH_4), one of the simplest Rydberg radicals, was also investigated [72]. A first approach consisted of the determination of an ab initio r_e structure using the CCSD [73,74] and CCSD(T) methods in conjunction with basis set

TABLE 3.8
Experimental, Semiexperimental, and Theoretical Structural Parameters (r_e) of LiCCH

Structure	Method	$r(CC)$	$r(CLi)$	$r(CH)$	References
r_0	Exp.	1.226	1.888	1.061	[69]
	Exp.	1.230	1.886	1.059	[68]
r_s	Exp.	1.226	1.891	1.066	[69]
r_e	Exp. + 6532/532[a]	1.2263	1.8901	1.0631	[68]
	Full WMR + diffuse[b]	1.2293	1.8919	1.0648	[68]

Source: Gauss, J., and J. F. Stanton. 2000. *Int J Quantum Chem* 77:405.

Note: Bond distances are in Å.

[a] From experimental rotational constants and vibrational corrections calculated with the CCSD(T) method and the ANO basis set of Widmark et al. [71].

[b] Computed using the CCSD(T) method and the full ANO basis set [71] adding diffuse basis functions.

extrapolation of the aug-cc-pVXZ (X = D, T, Q, and 5) hierarchy of basis sets [75,76]. The second treatment used experimental rotational constants [77], correcting them with vibration-rotation interaction constants calculated at the CCSD(T)/aug-cc-pVTZ level. The semiexperimental procedure leads to an r_e of 1.0363 Å versus the 1.0367 Å obtained by using the ab initio approach. The good agreement between the values resulting from these two methods is remarkable, and suggested that $r_e(NH)$ for the ammonium radical is 1.0365 Å with an uncertainty of 0.0005 Å.

In the same year, the r_e structure of allene [78] was studied on the basis of high-level quantum chemical calculations [CCSD(T)/cc-pVQZ and CCSD(T)/cc-pCVQZ] as well as by the analysis of experimental rotational constants for C_3H_4, C_3D_4, and $C_3H_2D_2$, with the latter vibrationally corrected using vibration-rotation interaction constants computed at the SDQ-MBPT(4)/cc-pVTZ level [63]. For the equilibrium geometry, a least-squares fit to the corrected rotational constants yielded $r(CH) = 1.081$ Å, $r(CC) = 1.307$ Å, and $\alpha(HCH) = 118.35°$. The results were in good agreement with the ab initio equilibrium structures as well as with a previous r_e structure based on gas-phase electron diffraction data (see Table 3.9). The final r_e-structure recommendation for allene was: $r_e(CC) = 1.307 \pm 0.001$ Å, $r_e(CH) = 1.081 \pm 0.002$ Å, $\theta_e(HCH) = 118.3 \pm 0.1°$. Table 3.9 also presents the r_g and r_z geometries. For vibrationally averaged distances, the agreement between computed and experimental results is satisfactory.

The r_e structures of two larger molecules such as (Z)-pent-2-en-4-ynenitrile [H_3C_5N, (A)] and maleonitrile [$H_2C_4N_2$, (B)] were also investigated [79]. The B_0^β constants of 10 isotopomers of (A) and 5 isotopomers of (B) were corrected using cubic force fields computed at the SCF/DZP level [46,67] in conjunction with geometries and quadratic potential calculated at the SCF/DZP and SDQ-MBPT(4) [63] levels of theory. The resulting semiexperimental equilibrium rotational constants, B_e^β, were used in a least-squares procedure to derive equilibrium parameters.

TABLE 3.9

Calculated and Semiexperimental Structural Parameters of Allene (C$_3$H$_4$)

		r(C–C)	r(C–H)	α(HCH)
r_0	Exp.[a]	1.3084(3)	1.0872(13)	118.2(2)
r_g	Exp.[b]	1.3129(9)	1.102(2)	
	Calc.[c]	1.3131	1.1001	
r_z	Exp.[d]	1.3091(4)	1.0862(10)	118.27(15)
	Exp.[b]	1.3093(7)	1.0865(19)	118.3(3)
	Calc.[c]	1.3095	1.0846	118.19
r_e	Exp.[b]	1.3082(10)	1.076(3)	118.2(5)
	Exp./theor.[e]	1.3069	1.0808	118.35
	Calc.[f]	1.3070	1.0800	118.16
	Calc.[g]	1.3074	1.0810	118.21

Source: Auer, A. A., and J. Gauss. 2001. *Phys Chem Chem Phys* 3:3001.

Note: Bond distances are in Å, bond angles in °. Estimated uncertainties are given in parentheses.

[a] From microwave (MW) data (see reference [109]).

[b] From gas-phase electron diffraction data (see reference [110]).

[c] From equilibrium geometry calculated at the CCSD(T)/cc-pVQZ level and vibrational corrections computed at the SDQ-MBPT(4)/cc-pVTZ level.

[d] Using rotational constants (see reference [111]).

[e] From experimental rotational constants [112–114] and vibrational corrections computed at the SDQ-MBPT(4)/cc-pVTZ level of theory.

[f] Optimized at the CCSD(T)/cc-pVQZ level.

[g] Optimized at the CCSD(T)/cc-pCVQZ level.

Two major studies on equilibrium structures were published in 2001 and 2002. The authors analyzed the accuracy of the equilibrium structures of a set of molecules containing first-row atoms [80,81]. These structures were obtained by using the semiexperimental approach in conjunction with high-level ab initio calculations. The conclusions were that except for diatomics, the r_e structures obtained by the semiexperimental procedure have higher quality, and their accuracy, with relative errors of 0.02%–0.06%, is superior to those experimentally determined. This is especially the case for larger molecules for which it is difficult to obtain experimental vibration-rotation interaction constants of a sufficient number of isotopologues. Among the constants calculated by a group of methods, the most accurate vibration-rotation interaction constants were those computed at the CCSD(T)/cc-pVQZ level. Regarding the ab initio structures, the CCSD(T) method is significantly more accurate than other models such as MP2 [45] and CCSD [73,74].

The investigation of individual molecules continued and a semiexperimental equilibrium structure for *trans*-glyoxal (CHO–CHO) was proposed a few years later [82] using the experimental ground-state rotational constants of four isotopic species (C$_2$H$_2$O$_2$, C$_2$HDO$_2$, C$_2$D$_2$O$_2$, and ^{13}C$_2$H$_2$O$_2$) and vibration-rotation interaction constants calculated at the CCSD(T)/cc-pVTZ level of theory. The least-squares fit yielded the following structural parameters: r_e^{SE}(C–C) = 1.51453 ± 0.00038 Å, r_e^{SE}

(C–H) = 1.10071 ± 0.00026 Å, r_e^{SE}(C=O) = 1.20450 ± 0.00027 Å, θ_e^{SE}(CCH) = $115.251 \pm 0.024°$, and θ_e^{SE}(HCO) = $123.472 \pm 0.019°$ (see Table 3.10 for more details).

Very recently, high-level quantum chemical calculations at the MP2 [45] and CCSD(T) levels in conjunction with Dunning's hierarchy of correlation-consistent basis sets (for example, cc-pVXZ and cc-pCVXZ), as well as the ANO2 basis set of Almlöf and Taylor [83], have been carried out for computing the equilibrium structure and harmonic and anharmonic force fields of diacetylene (H_2C_4) [84]. A semiexperimental equilibrium structure based on experimental rotational constants of 13 isotopologues and vibrational corrections estimated at different levels of theory was derived. The final recommended semiexperimental equilibrium structure was r_e^{SE}(CH) = 1.0615 Å, r_e^{SE}(C=C) = 1.2085 Å, and r_e^{SE}(C–C) = 1.3727 Å. These results agree well with the best ab initio r_e structure calculated at the CCSD(T)/cc-pCV5Z level of theory (see Table 3.11).

TABLE 3.10
Calculated and Semiexperimental Structural Parameters of *Trans*-Glyoxal (CHO–CHO)

	Semiexperimental Values	CCSD(T)/cc-pVQZ
r(C–C)	1.51453(38)[a]	1.51446
r(C–H)	1.10071(26)	1.09991
r(C–O)	1.20450(27)	1.20506
α(CCH)	115.251(24)	115.152
α(HCO)	123.472(19)	123.537

Source: Larsen, R. W., F. Pawlowski, F. Hegelund, P. Jørgensen, J. Gauss, and B. Nelander. 2003. *Phys Chem Chem Phys* 5:5031, Table 5.

Note: Bond distances are in Å and bond angles in °. Estimated uncertainties are given in parentheses.

[a] Fit of rotational constants B_e and C_e of four isotopologues of *trans*-glyoxal.

TABLE 3.11
Calculated and Semiexperimental Structural Parameters of Diacetylene (C_4H_2)

Method	r_{C-H}	$r_{C\equiv C}$	r_{C-C}
CCSD(T)/cc-pCV5Z	1.0617	1.2083	1.3737
r_0	1.0561	1.2079	1.3752
r_e^{SE} [(fc) CCSD(T)/cc-pVQZ]	1.0616	1.2084	1.3727
r_e^{SE} [CCSD(T)/cc-pCVQZ]	1.0615	1.2085	1.3727
r_e^{SE} [(fc) CCSD(T)/ANO2]	1.0614	1.2085	1.3726

Source: Thorwirth, S., M. E. Harding, D. Muders, and J. Gauss. 2008. *J Mol Spectrosc* 251:220.

Note: In the semiexperimental equilibrium geometries, the methods in parentheses refer to the level of theory used to calculate the quadratic and cubic force fields. Bond distances are in Å.

3.4.3 SELECTED OTHER STUDIES: SMALL AND MEDIUM MOLECULES CONTAINING SECOND-ROW ATOMS AND BEYOND

The first study of the equilibrium structure for molecules containing atoms from the second row of the periodic table was devoted to the two cyclic isomers of SiC_3 [33]. The best ab initio r_e parameters were calculated at the CCSD(T)/cc-pCVQZ level and vibration-rotation interaction constants were evaluated from quadratic and cubic force fields computed at the CCSD(T)/cc-pVTZ level of theory. The r_e structures were obtained by least-squares adjustment of atomic coordinates to moments of inertia based on the B_e and C_e constants for the six isotopomers of isomer I (isomer with a transannular CC-bond) and the three constants for the seven isotopomers of II (isomer with a transannular Si–C bond) [85–88]. Based on the agreement between the purely ab initio equilibrium structures and those obtained by least-squares fitting to moments of inertia derived from the semiexperimental equilibrium rotational constants, the following parameters were recommended.

For isomer I,

$$r_e(\text{Si–C*}) = 1.828 \pm 0.002\,\text{Å}$$

$$r_e(\text{C–C*}) = 1.433 \pm 0.002\,\text{Å}$$

$$r_e(\text{C–C}) = 1.483 \pm 0.002\,\text{Å}$$

For isomer II,

$$r_e(\text{Si–C*}) = 1.886 \pm 0.002\,\text{Å}$$

$$r_e(\text{C–C*}) = 1.343 \pm 0.002\,\text{Å}$$

$$r_e(\text{Si–C}) = 2.020 \pm 0.003\,\text{Å}$$

The study on the two cyclic SiC_3 isomers was followed by the investigation of one of the species involved in the chemical vapor deposition processes of chlorosilane [89–92], that is, monochlorosilylene (HSiCl) [93]. For this molecule, the best ab initio equilibrium structure was predicted at the CCSD(T)/cc-pCVQZ level of theory. A semiexperimental equilibrium structure was derived by least-squares fit to B_e^β constants obtained from experimental vibrational ground-state rotational constants of four isotopologues of HSiCl [94] and theoretical vibrational corrections calculated at the CCSD(T)/cc-pVTZ and CCSD(T)/cc-pCVTZ levels of theory. Considering both the purely ab initio and semiexperimental equilibrium structures, the recommended r_e parameters were:

$$r_e(\text{Si–H}) = 1.514 \pm 0.001\,\text{Å}$$

$$r_e(\text{Si–Cl}) = 2.071 \pm 0.001\,\text{Å}$$

$$\theta_e(\text{HSiCl}) = 95.0 \pm 0.5^\circ$$

In the case of diethynyl sulfide [S(CCH)$_2$] [95], an ab initio r_e structure was calculated at the CCSD(T) level using the cc-pwCVTZ basis set [96,97] for sulfur and the cc-pCVTZ basis set for carbon and hydrogen. The semiexperimental equilibrium structure was determined from the experimental rotational constants of four isotopic species [S(CCH)$_2$, S(CCH)$_2$, S(CCD)$_2$, and S(^{13}C^{13}CH)$_2$], and vibrational corrections to them were estimated from quadratic and cubic force fields calculated at the SCF/DZP level. The recommended r_e structure is an average of columns 3 and 4 of Table 3.12, namely, r(S–C) = 1.708 ± 0.020 Å, r(C≡C) = 1.211 ± 0.010 Å, r(C–H) = 1.061 ± 0.010 Å, θ(C–S–C) = 100.5 ± 1.0°, θ(S–C≡C$_{(outside angle)}$) = 174.4 ± 1.5°, and θ(C≡C–H$_{(inside angle)}$) = 177.0 ± 2.5°.

The structures of the all-sulfur molecules, thiozone (S$_3$) and tetrasulfur (S$_4$) were also examined [98]. Ab initio equilibrium geometries were obtained with the CCSD(T)/cc-pVTZ level of theory. Semiexperimental equilibrium structures were based on experimental B_0^β values of three isotopomers of S$_3$ and S$_4$ and vibrational corrections were calculated at the CCSD(T)/cc-pVTZ level. For S$_3$, the bond length and angle are nearly the same for the experimental (r_0) and the semiexperimental equilibrium (r_e^{SE})

TABLE 3.12
Structural Parameters of Diethynyl Sulfide S(C≡CH)$_2$

	Experimental		Ab Initio
	r_0^a	r_e^b	r_e^c
Bond lengths			
S–C	1.7230	1.7209	1.6947
C≡C	1.2066	1.2102	1.2109
r_{C-H}	1.0573	1.0596	1.0633
Bond angles			
C–S–C	100.24	100.21	100.83
S–C≡C$_{outer}^d$	173.30	173.56	175.31
C≡C–H$_{inner}^d$	174.80	174.99	179.10
Dihedral angles			
S–C≡C–H	180.00	180.00	180.00
C–S–C≡C	180.00	180.00	180.00

Source: Matzger, A. J., K. D. Lewis, C. E. Nathan, et al. 2002. *J Phys Chem* 106:12110.

Note: Bond distances are in Å and bond angles in °.

a Fit of the I_a, I_b ground-state moments of inertia for four isotopic species.
b Fit of the I_a, I_b moments of inertia corrected for vibration-rotation interaction effects.
c CCSD(T); basis sets were cc-pwCVTZ for sulfur and cc-pCVTZ for carbon and hydrogen.
d The "outer" implies that the C≡C bonds bend away from each other. The "inner" implies that the C–H bonds bend toward each other.

TABLE 3.13

Structural Parameters of S_3 and S_4

| Structure | S_3 | | S_4 | | |
	$r(S=S)$	$\alpha(SSS)$	$r(S=S)_{outer}$	$r(S-S)_{inner}$	$\alpha(SSS)$
r_0^a	1.917(1)	117.36(6)	1.899(7)	2.173(32)	103.9(3)
(r_e^{BO})	1.938	117.0	1.920	2.177	103.6
(r_e^{SE})	1.914(2)	117.33(5)	1.898(5)	2.155(10)	104.2(2)

Source: Thorwirth, S., M. C. McCarthy, C. A. Gottlieb, P. Thaddeus, H. Gupta, and J. F. Stanton. 2005. *J Chem Phys* 123:054326.

Note: Bond distances are in Å and bond angles in °. Theoretical equilibrium (r_e^{BO}) structures were calculated at the CCSD(T)/cc-pVTZ level of theory. The semiexperimental equilibrium (r_e^{SE}) structures were determined by the procedure described in the chapter.

a Experimental r_0 structures are from references [115,116].

structures (see Table 3.13) possibly due to very small vibrational anharmonicity. The bond length in the theoretical equilibrium (r_e^{BO}) structure calculated at the CCSD(T) level with the cc-pVTZ basis set is about 0.02 Å longer than the experimental value. This might be because expansion of the basis set tends to reduce internuclear distances while higher level electron correlation methods tend to increase them. In S_3, the CCSD(T) method provides an adequate treatment of correlation; the major shortcoming of the calculation is the cc-pVTZ basis set. A similar behavior is observed in S_4.

The equilibrium structures of the *cis* and *trans* isomers of 1-chloro-2-fluoroethylene have recently been studied [99,100] using experimental ground-state rotational constants of several isotopologues and vibrational corrections estimated from quadratic and cubic force fields calculated at the MP2/cc-pVTZ and CCSD(T)/cc-pVTZ level of theory. The resulting r_e structures are presented in Table 3.14. In this table, the semiexperimental equilibrium structures are compared with the best estimated ab initio equilibrium parameters obtained, this time taking into account the CBS limit and core-valence (CV) electron correlation effects.

Similarly, the equilibrium structure of bromofluoromethane has been established [101] based on new rotational constants for the mono- and bideuterated bromofluoromethane ($CDH^{79}BrF$, $CDH^{81}BrF$, $CD_2Br^{79}F$, and $CD_2Br^{81}F$) as well as on those for other isotopologues, and vibrational corrections have been calculated with MP2 and CCSD(T) methods and different basis sets. Equilibrium structures were obtained by a least-squares fit to the vibrationally corrected experimental equilibrium rotational constants. The semiexperimental equilibrium structures agree well with the theoretical predictions derived from CCSD(T) calculations after extrapolation to the CBS limit, inclusion of CV correlation corrections and, for the first time, inclusion of relativistic effects.

Very recently, silanethione ($H_2Si=S$) was for the first time experimentally characterized by rotational spectroscopy [102]. The equilibrium structure was evaluated

TABLE 3.14
Experimental Equilibrium Geometry of *Cis*- and *Trans*-1-Chloro-2-Fluoroethylene (CHF=CHCl): Semiexperimental Equilibrium Structure from Experimental Ground-State Rotational Constants and Theoretical Vibration-Rotation Interaction Constants

	Semiexperimental r_e			Best Theoretical[d]
	I[a]	II[b]	III[c]	
	Trans isomer			
C_1–Cl	1.7181(13)	1.7189(9)	1.7187(9)	1.7163
C_1–H	1.0772(4)	1.0772(1)	1.0772(1)	1.0775
C_1–C_2	1.324[e]	1.324[e]	1.324[e]	1.3240
C_2–F	1.3398(17)	1.3387(11)	1.3383(12)	1.3376
C_2–H	1.0789(3)	1.0784(1)	1.0784(1)	1.0785
$\theta(ClC_1C_2)$	120.51(2)	120.52(1)	120.54(1)	120.63
$\theta(HC_1C_2)$	123.01(27)	123.09(4)	123.09(5)	122.95
$\theta(FC_2C_1)$	120.17(4)	120.14(2)	120.15(2)	120.14
$\theta(HC_2C_1)$	125.67(9)	125.84(5)	125.80(6)	125.82
	Cis isomer			
C_1–Cl	–	1.7130(5)	1.7128(6)	1.7107
C_1–H	–	1.0776(4)	1.0776(4)	1.0764
C_1–C2	–	1.3242(12)	1.3240(14)	1.3249
C_2–F	–	1.3316(3)	1.3317(3)	1.3310
C_2–H	–	1.0800(5)	1.0802(6)	1.0787
$\theta(ClC_1C_2)$	–	123.07(1)	123.07(1)	123.10
$\theta(HC_1C_2)$	–	120.73(8)	120.74(9)	120.43
$\theta(FC_2C_1)$	–	122.59(5)	122.61(6)	122.53
$\theta(HC_2C_1)$	–	123.48(2)	123.50(2)	123.43

Source: Cazzoli, G., C. Puzzarini, A. Gambi, and J. Gauss. 2006. *J Chem Phys* 125:0544313; Puzzarini, C., G. Cazzoli, A. Gambi, and J. Gauss. 2006. *J Chem Phys* 125:054307.

Note: Bond distances are in Å and bond angles in °. Estimated uncertainties are given in parentheses.

[a] Experimental data: only deuterated isotopologues. Theoretical data: MP2/cc-pVTZ force fields.

[b] Experimental data: all isotopologues available. Theoretical data: MP2/cc-pVTZ force fields.

[c] Experimental data: all isotopologues available. Theoretical data: CCSD(T)/cc-pVTZ force fields.

[d] Best estimated pure ab initio structure (CBS + CV).

[e] Refer to fixed at the best estimate ab initio value.

through a combination of rotational constants from a total of 10 isotopologues and by using high-level quantum chemical methods. The ab initio equilibrium structure was calculated using the CCSD(T) method and taking into account CBS and CV corrections effects. The vibrational correction to rotational constants were estimated from quadratic and cubic force fields computed at the CCSD(T)/cc-pV(Q + d)Z(fc) level of theory. The resulting semiexperimental equilibrium structural parameters were r_e^{SE}(Si–H) = 1.4735 Å, r_e^{SE}(Si–S) = 1.9357 Å and α_e^{SE}(HSiH) = 110.33°. Meanwhile, the best theoretical estimate resulted in r_e^{BO} (Si–H) = 1.4739 Å, r_e^{SE}(Si–S) = 1.9362 Å and α_e^{BO}(HSiH) = 110.23°.

Finally, the semiexperimental equilibrium structures of silacyclopropenylidene (c-C_2H_2Si) and silapropa-dienylidene (H_2CCSi) have been recently determined [103] using experimental ground-state rotational constants of four isotopologues of both molecules and vibrational corrections calculated by an extensive set of levels of theory.

The examples sketched in the second part of this chapter represent only a small portion of the studies involving the determination of semiexperimental equilibrium structures. Nevertheless, the cases show the effectiveness of the approach and also underline the importance of accurate quantum-chemical models in this procedure. The evaluation of semiexperimental structures has expanded substantially starting from the earliest studies on small molecules (e.g., ketene) to larger systems such as benzene, as well as from molecules containing only first-row elements to those with one or more sulfur, silicon, or halogen atoms and even to structures harboring atoms from the third row of the periodic table. The increase in sophistication of the quantum-chemical methods used in the semiexperimental approach has been illustrated. The first computations employed the rather simple Hartree–Fock approximation and the most recent computations on bromofluoromethane use high level coupled-cluster methods in conjunction with basis set extrapolation techniques, and include relativistic effects. The continuous advances in quantum-chemical methods, computational resources, least-square refinements schemes, and experimental techniques have made the determination of semiexperimental equilibrium structures a consolidated strategy in the evaluation of molecular structures.

REFERENCES

1. Born, M., and J. R. Oppenheimer. 1927. *Ann Phys* 84:457.
2. Pulay, P., W. Meyer, and J. E. Boggs. 1978. *J Chem Phys* 68:5077.
3. Bartell, L. S., K. Kuchitsu, and R. J. Deneui. 1961. *J Chem Phys* 35:1211.
4. Kemble, E. C. 1958. *The Fundamental Principles of Quantum Mechanics with Elementary Applications.* New York: Dover Publications.
5. Wollrab, J. E. 1967. *Rotational Spectra and Molecular Structure.* New York: Academic Press.
6. Kroto, H. W. 1975. *Molecular Rotation Spectra.* Bristol: Wiley.
7. Townes, C. H., and A. L. Schawlow. 1955. *Microwave Spectroscopy.* New York: McGraw-Hill.
8. Ladd, M. 1998. *Symmetry and Group Theory in Chemistry.* Chichester, UK: Horwood.
9. Robinson, E. A. 1981. *Least Squares Regression Analysis in Terms of Linear Algebra.* Houston: Goose Pond Press.
10. Hollas, J. M. 1998. *High Resolution Spectroscopy.* Chichester, UK: Wiley.
11. Nielsen, H. H. 1951. *Rev Mol Phys* 23:90.
12. Nielsen, H. H. 1959. In *Handbuch der Physik,* ed. S. Flügge, Part. I, Vol. 37, 171. Berlin: Springer-Verlag.

13. Mills, I. M. 1972. In *Molecular Spectroscopy: Modern Research*, ed. K. N. Rao and C. W. Mathews. New York: Academic Press.
14. Watson, J. K. G. 1968. *Mol Phys* 15:479.
15. Watson, J. K. G. 1970. *Mol Phys* 19:465.
16. Papoušek, D., and M. R. Aliev. 1982. *Molecular Vibrational-Rotational Spectra*. New York: Elsevier Scientific.
17. Birss, F. W. 1976. *Mol Phys* 31:491.
18. Cyvin, S. J. 1968. *Molecular Vibrations and Mean-Square Amplitudes*. Oslo: Elsevier.
19. Allen, H. C., and P. C. Cross. 1963. *Molecular Vib-Rotors*. New York: Wiley.
20. Mills, I., T. Cvitaš, K. Homann, N. Kallay, and K. Kuchitsu. 1993. *Quantities, Units and Symbols in Physical Chemistry*. Oxford, Boston: Blackwell Scientific Publications.
21. East, A. L. L., C. S. Johnson, and W. D. Allen. 1993. *J Chem Phys* 98:1299.
22. East, A. L. L., W. D. Allen, and S. J. Klippenstein. 1995. *J Chem Phys* 102:8506.
23. Simmonett, A. C., F. A. Evangelista, W. D. Allen, and H. F. Schaefer III. 2007. *J Chem Phys* 127:014306.
24. Gordy, W., and R. L. Cook. 1984. *Microwave Molecular Spectra*. New York: Wiley.
25. Yamamoto, S., and S. Saito. 1994. *J Chem Phys* 101:5484.
26. CFOUR, a quantum chemical program package written by Stanton, J. F., J. Gauss, M. E. Harding, and P. G. Szalay with contributions from Auer, A. A., R. J. Bartlett, U. Benedikt, C. Berger, D. E. Bernholdt, Y. J. Bomble, O. Christiansen, M. Heckert, O. Heun, C. Huber, T. -C. Jagau, D. Jonsson, J. Jusélius, K. Klein, W. J. Lauderdale, D. A. Matthews, T. Metzroth, D. P. O'Neill, D. R. Price, E. Prochnow, K. Ruud, F. Schiffmann, S. Stopkowicz, J. Vázquez, F. Wang, and J. D. Watts and the integral packages MOLECULE (Almlöf, J., and P.R. Taylor), PROPS (Taylor, P.R.), ABACUS (Helgaker, T., H. J. Aa. Jensen, P. Jørgensen, and J. Olsen), and ECP routines by A. V. Mitin and C. van Wüllen. See http://www.cfour.de for the current version.
27. Gaw, J. F., A. Willetts, W. H. Green, and N. C. Handy. 1991. SPECTRO: A program for the derivation of spectroscopic constants from provided quartic force fields and cubic dipole fields. In *Advances in Molecular Vibrations and Collision Dynamics*, ed. J. M. Bowman and M. A. Ratner, Vol. 1B. London: Jai Press.
28. Frisch, M. J., G. W. Trucks, H. B. Schlegel, et al. 2003. *Gaussian 03, Revision B.03*. Pittsburgh, PA: Gaussian, Inc.
29. Flygare, W. 1974. *Chem Rev* 74:653.
30. Gauss, J., K. Ruud, and T. Helgaker. 1996. *J Chem Phys* 105:2804.
31. Gauss, J., K. Ruud, and M. Kállay. 2007. *J Chem Phys* 127:074101.
32. Puzzarini, C., M. Heckert, and J. Gauss. 2008. *J Chem Phys* 128:194108.
33. Stanton, J. F., J. Gauss, and O. Christiansen. 2001. *J Chem Phys* 114:2993.
34. Heckert, M., M. Kállay, D. P. Tew, W. Klopper, and J. Gauss. 2006. *J Chem Phys* 125:044108.
35. Maslen, P. E., N. C. Handy, R. D. Amos, and D. Jayatilaka. 1992. *J Chem Phys* 97:271.
36. Stanton, J. F., and J. Gauss. 1991. *Int Rev Phys Chem* 19:61.
37. Wilson Jr., E. B., J. C. Decius, and P. C. Cross. 1955. *Molecular Vibrations*. New York: Dover Publications.
38. Califano, S. 1976. *Vibrational States*. Bristol: Wiley.
39. Hoy, A. R., I. M. Mills, and G. Strey. 1972. *Mol Phys* 24:1265.
40. Kucharski, S. A., and R. J. Bartlett. 1991. *Theor Chim Acta* 80:387.
41. Kucharski, S. A., and R. J. Bartlett. 1992. *J Chem Phys* 97:4282.
42. Oliphant, N., and L. Adamawicz. 1991. *J Chem Phys* 94:1229.
43. Gauss, J., and J. F. Stanton. 1997. *Chem Phys Lett* 276:70. And references in there.
44. Parr, R. G., and W. Yang. 1989. *Density-Functional Theory of Atoms and Molecules*. New York: Oxford University Press.

45. Cremer, D. 1998. Møller-plesset perturbation theory. In *Encyclopedia of Computational Chemistry*, ed. P. von Ragué Schleyer. New York: Wiley.

46. Szabo, A., and N. S. Ostlund. 1989. *Modern Quantum Chemistry. Introduction to Advanced Electronic Structure Theory*. Mineola, NY: Dover Publications.

47. Kraitchman, J. 1953. *Am J Phys* 21:17.

48. Allen, W. D., E. Czinki, and A. G. Császár. 2004. *Chem Eur J* 10:4512.

49. Kasalová, V., W. D. Allen, H. F. Schaefer III, E. Czinki, and A. G. Császár. 2007. *J Comp Chem* 28:1373.

50. Dunning Jr., T. H. 1989. *J Chem Phys* 90:1008.

51. Raghavachari, K., G. W. Trucks, J. A. Pople, and M. Head-Gordon. 1989. *Chem Phys Lett* 157:479.

52. Stanton, J. F., C. L. Lopreore, and J. Gauss. 1998. *J Chem Phys* 108:7190.

53. Gauss, J., and J. F. Stanton. 1999. *J Mol Struct* 485:43.

54. Langer, W. D., T. Velusame, T. B. H. Kuiper, and R. Peng. 1997. *Astrophys J* 480:L63.

55. Woon, D. E., and T. H. Dunning. 1995. *J Chem Phys* 103:4572.

56. Stanton, J. F. 1999. *Mol Phys* 97:841.

57. Tarrago, G., M. Dang-Nhu, G. Poussigue, G. Guelachvili, and C. Amiot. 2975. *J Mol Spectrosc* 57:246.

58. Tarrago, G., M. Delaveau, L. Fusina, and G. Guelachvili. 1987. *J Mol Spectrosc* 126:149.

59. Deroche, J. C., and G. Guelachvili. 1975. *J Mol Spectrosc* 56:76.

60. Dupre-Macquaire, J., and G. Tarrago. 1982. *J Mol Spectrosc* 96:170.

61. Lolck, J.-E., G. Poussigue, E. Pascaud, and G. Guelachvili. 1985. *J Mol Spectrosc* 111:235.

62. Gauss, J., and J. F. Stanton. 2000. *J Phys Chem A* 104:2865.

63. Bartlett, R. J. 1981. *Ann Rev Phys Chem* 32:359.

64. Tamagawa, K., T. Iijima, and M. Kimura. 1976. *J Mol Struct* 30:243.

65. Gauss, J., D. Cremer, and J. F. Stanton. 2000. *J Phys Chem A* 104:1219.

66. McMahom, R. J., R. J. Halter, R. L. Fimmen, et al. 2000. *J Am Chem Soc* 122:939.

67. Redmon, L. T., G. D. Purvis, and R. J. Bartlett. 1979. *J Am Chem Soc* 101:2856.

68. Szalay, P. G., J. F. Stanton, and R. J. Bartlett. 1992. *Chem Phys Lett* 193:573.

69. Gauss, J., and J. F. Stanton. 2000. *Int J Quantum Chem* 77:405.

70. Grotjahn, D. B., A. J. Apponi, M. A. Brewster, J. Xin, and L. M. Ziurys. 1998. *Angew Chem* 110:2824.

71. Widmark, P. O., P. -Å. Malmqvist, and B. O. Roos. 1990. *Theor Chim Acta* 79:290.

72. Sattelmeyer, K. W., H. F. Schaefer III, and J. F. Stanton. 2001. *J Chem Phys* 114:9863.

73. Bartlett, R. J., and J. F. Stanton. 1994. *Rev Comput Chem* 5:65.

74. Lee, T. J., and G. E. Scuseria. 1995. In *Quantum Mechanical Electronic Structure Calculations with Chemical Accuracy*, ed. S. R. Langhoff, 47. Dordrecht: Kluwer Academic.

75. Dunning, T. H. 1992. *J Chem Phys* 90:1007.

76. Kendall, R. A., T. H. Dunning, and R. J. Harrison. 1992. *J Chem Phys* 96:6796.

77. Signorell, R., H. Palm, and F. Merkt. 1997. *J Chem Phys* 106:6523.

78. Auer, A. A., and J. Gauss. 2001. *Phys Chem Chem Phys* 3:3001.

79. Halter, R. J., R. L. Fimmen, R. J. McMahon, S. A. Peebles, R. L. Kuczkowski, and J. F. Stanton. 2001. *J Am Chem Soc* 123:12353.

80. Bak, K. L., J. Gauss, P. Jørgensen, J. Olsen, T. Helgaker, and J. F. Stanton. 2001. *J Chem Phys* 114:6548.

81. Pawlowski, F., P. Jørgensen, J. Olsen, et al. 2002. *J Chem Phys* 116:6482.

82. Larsen, R. W., F. Pawlowski, F. Hegelund, P. Jørgensen, J. Gauss, and B. Nelander. 2003. *Phys Chem Chem Phys* 5:5031.

83. Almlöf, J., and P. R. Taylor. 1987. *J Chem Phys* 86:4070.

84. Thorwirth, S., M. E. Harding, D. Muders, and J. Gauss. 2008. *J Mol Spectrosc* 251:220.

85. Alberts, I. L., R. S. Grey, and H. F. Schaefer. 1990. *J Chem Phys* 93:5046.

86. Rittby, C. M. L. 1992. *J Chem Phys* 95:5409.

87. Gomei, M., R. Kishi, A. Nakajima, S. Iwata, and K. Kaya. 1997. *J Chem Phys* 107:10051.
88. McCarthy, M. C., A. J. Apponi, and P. Thaddeus. 1999. *J Chem Phys* 111:7175.
89. Ban, V. S., and S. L. Gilbert. 1975. *J Electronchem Soc* 122:1382.
90. Ban, V. S., and S. L. Gilbert. 1975. *J Electronchem Soc* 122:1389.
91. Sedgwick, T. O., and J. E. Smith Jr. 1976. *J Electrochem Soc* 173:254.
92. Smith Jr., J. E., and T. O. Sedgwick. 1977. *Thin Solid Films* 40:1.
93. Vázquez, J., and J. F. Stanton. 2002. *J Phys Chem A* 106:4429.
94. Harper, W. W., and D. J. Clouthier. 1997. *J Chem Phys* 106:9461.
95. Matzger, A. J., K. D. Lewis, C. E. Nathan, S. A. Peebles, R. A. Peebles, R. L. Kuczkowski, J. F. Stanton and J. J. Oh. 2002. *J Phys Chem, A*, 106:12110.
96. Peterson, K. A., and T. H. Dunning Jr. 2002. *J Chem Phys* 117:10548.
97. Balabanov, N. B., and K. A. Peterson. 2005. *J Chem Phys* 123:064107.
98. Thorwirth, S., M. C. McCarthy, C. A. Gottlieb, P. Thaddeus, H. Gupta, and J. F. Stanton. 2005. *J Chem Phys* 123:054326.
99. Cazzoli, G., C. Puzzarini, A. Gambi, and J. Gauss. 2006. *J Chem Phys* 125:0544313.
100. Puzzarini, C., G. Cazzoli, A. Gambi, and J. Gauss. 2006. *J Chem Phys* 125:054307.
101. Puzzarini, C., G. Cazzoli, A. Baldacci, A. Baldan, C. Michauk, and J. Gauss. 2007. *J Chem Phys* 127:164302.
102. Thorwirth, S., J. Gauss, M. C. McCarthy, F. Shindo, and P. Thaddeus. 2008. *Chem Commun* 5292.
103. Thorwirth, S., and M. E. Harding. 2009. *J Chem Phys* 130:214303.
104. Pliva, J., J. W. C. Johns, and L. Goodman. 1991. *J Mol Spectrosc* 148:427.
105. Hollenstein, H., S. Piccirillo, M. Quack, and M. Snels. 1990. *Mol Phys* 71:759.
106. Pliva, J., J. W. C. Johns, and L. Goodman. 1990. *J Mol Spectrosc* 140:214.
107. Yamamoto, S., M. Nakata, T. Fukayama, and K. Kuchitsu. 1985. *J Phys Chem* 89:3298.
108. Endo, Y., M. C. Chang, and E. Hirota. 1987. *J Mol Spectrosc* 126:63.
109. Maki, A. G., and R. A. Toth. 1977. *J Mol Spectrosc* 65:366.
110. Ohshima, Y., S. Yamamoto, M. Nakata, and K. Kuchitsu. 1987. *J Phys Chem* 91:4696.
111. Hegelund, F., J. L. Duncan, and D. C. McKean. 1977. *J Mol Spectrosc* 65:366.
112. Anttila, R., M. Koivusaari, J. Kauppinen, and F. Hegelund. 1981. *J Mol Spectrosc* 87:393.
113. Hegelund, F., and J. Kauppinen. 1985. *J Mol Spectrosc* 110:106.
114. Hegelund, F., N. Andressen, and M. Koivusaari. 1987. *J Mol Spectrosc* 149:4696.
115. McCarthy, M. C., S. Thorwirth, C. A. Gottlieb, and P. Thaddeus. 2004. *J Am Chem Soc* 126:4096.
116. McCarthy, M. C., S. Thorwirth, C. A. Gottlieb, and P. Thaddeus. 2004. *J Chem Phys* 121:632.

4 Spectroscopy of Polyatomic Molecules
Determination of Rotational Constants

Agnès Perrin, Jean Demaison, Jean-Marie Flaud, Walter J. Lafferty, and Kamil Sarka

CONTENTS

4.1   INTRODUCTION

Chapter 3 explains how to determine equilibrium rotational constants using experimental ground state constants and rovibrational corrections calculated with the aid of an ab initio anharmonic force field. This chapter demonstrates that in some cases it is possible to obtain these equilibrium constants using only spectroscopic information alone. As usual in molecular spectroscopy, the operators are written in boldface. In rovibrational spectroscopy, the experimental data are the frequencies of the transitions between the energy levels and one of the main tasks of the spectroscopist is to describe and explain the observed spectrum and to determine the molecular parameters. This chapter is not intended to be a review of molecular spectroscopy. Only the notions essential to understand how accurate (equilibrium) rotational constants are obtained will be given. The reader is directed to several reference books, starting from the historical [1–8], to the most recent [9–16]. Furthermore, a glossary as well as the definition of the quantum numbers, operators, and so on is given in Appendix IV.1.

The molecular parameters are usually determined by the least-squares method (Chapter 2) using as input data the observed frequencies and as model a Hamiltonian briefly described in Section 4.2. However, it is not as easy as it may seem because, quite often, severe correlation problems appear (see Sections 2.2.2 and 2.2.3). One obvious case, which is easy to trace, is when the available data do not permit the determination of some parameters (too little data). However, there is a much more complicated kind of difficulty when some parameters are highly correlated. Another difficulty is the correct treatment of the interactions between the different rovibrational levels, which is one of the most difficult problems in experimental rovibrational spectroscopy. Several typical examples will be described in detail. Their main purpose is to show that it is often not straightforward to determine accurate rotational constants.

4.2   THE ROVIBRATIONAL HAMILTONIAN

4.2.1   CONSTRUCTION OF THE HAMILTONIAN

Molecular vibration-rotation spectra are arguably the most important source of information on the structures of molecules. The relation between the spectra and the structure comes from the two fundamental equations of the quantum mechanics. The first, given by Bohr [17], says that if the electromagnetic radiation is passing through the medium, part of the radiation is absorbed or emitted if and only if the frequency of the absorbed radiation f_{mn} satisfies the Bohr's equation

$$hf_{mn} = E_n - E_m \qquad (4.1)$$

where E_n and E_m are two energy levels of the molecule absorbing or emitting the radiation and h is the Planck's constant.

In spectroscopy, we are able to measure the frequencies f_{mn} (usually in MHz units) or equivalent wave numbers $\omega_{mn} = f_{mn}/c$ (usually in cm^{-1} units), where c is the speed of light in vacuum ($c = 299{,}792{,}458$ ms^{-1}). Actually, Equation 4.1 is the basis for all spectroscopic methods because if we let the electromagnetic radiation of varying frequency f pass through the studied sample, we can observe the frequencies at which the radiation is absorbed. Such record is called a spectrum and the spectra provide us with information on the energy levels of the studied molecule. The Raman spectroscopy data which are sometimes used in the vibration-rotation analysis are also processed according to Equation 4.1.

The second equation that provides the link to the structure of the molecules is the Schrödinger equation, which is the cornerstone equation of quantum mechanics and says that if the quantum mechanical operator $\mathbf{H}$ operates on its eigenfunction Ψ_i describing the state i of the molecule with energy E_i then

$$\mathbf{H}\Psi_i = E_i\Psi_i \qquad (4.2)$$

The information on the structure of a molecule must be therefore contained in the Hamiltonian $\mathbf{H}$, and we shall now concentrate on this operator.

In the classical mechanics, the Hamiltonian $\mathbf{H}$ is the operator describing the total energy of the system. The molecule as a whole can fly through the space and this motion, called *translation*, can be described by the position $\mathbf{R}_0$ and velocity $\mathbf{v}$ of its center of mass. The energy E_{TRANSL} corresponding to the translational motion of a molecule with total mass M is

$$E_{\text{TRANSL}} = \frac{1}{2}M\mathbf{v}^2 \qquad (4.3)$$

The molecule is also rotating; this rotation can be described by the three angles, two of them, say θ and φ, specifying the orientation in space of the axis of rotation, and the third, say χ, specifying how much has the molecule rotated about this axis. For symmetric-top molecules, these three angles coincide with the Euler angles, describing the relative orientation of the spaces-fixed and molecule-fixed axes. For the definition of the Eulerian angles, see reference [6]. The free molecule can rotate about the three principal axes of rotation $\alpha = x, y, z$ and the energy of the rotational motion E_{ROT} can be expressed as

$$E_{\text{ROT}} = \sum_{\alpha=x,y,z} \frac{J_\alpha^2}{2I_{\alpha\alpha}} \qquad (4.4)$$

where

$$I_{\alpha\alpha} = \sum_i m_i \delta_i^2 \qquad (4.5)$$

is the moment of inertia with respect to the rotation axis α, δ_i is the perpendicular distance of the ith atom of mass m_i from the axis of rotation, and J_α is the component of the angular momentum

$$\mathbf{J} = \sum_i m_i \mathbf{R}_i \times \mathbf{v}_i \tag{4.6}$$

where $\mathbf{R}_i = (X_i, Y_i, Z_i)$ is the position vector of atom i in the space-fixed system of the Cartesian coordinates. Finally, the atoms in the molecule vibrate about their equilibrium positions, and the vibrations are characterized by vibrational displacement coordinates with respect to equilibrium $d_{i\alpha}$, which are assumed to take small values compared to the interatomic distances. The vibrations are characterized by their frequencies f. Again, if, for example, a diatomic molecule is vibrating, the energy E_{VIB} is given by the expression

$$E_{\text{VIB}} = \frac{1}{2} k \Delta_{\text{max}}^2 \tag{4.7}$$

where the amplitude Δ_{max} is the maximum displacement of the interatomic distance from its equilibrium length and k is the force constant, which we can imagine as the strength of a spring holding the two atoms together.

Equations 4.4 through 4.7 give us also the first insight into the information on the structure that can be derived from them. The translational motion is not particularly interesting because the only information it could provide would be on the molecular mass M. On the contrary, rotational motion depends on the moments of inertia $I_{\alpha\alpha}$, which are clearly determined by the geometry of the molecule, and therefore rotation energies will provide information on the molecular geometry. For example, for a molecule XY_2 the moment of inertia about the axis passing through the atom X and dissecting in half the angle $\alpha = \angle(\text{YXY})$ is given by the expression $I = 2m_Y r_{X-Y}^2 \sin^2(\alpha/2)$ (see Table V.1 in Appendix V). Finally, the vibration energies will provide us with information on the force constants and the forces keeping the atoms in the molecule together.

The Hamiltonian in Equation 4.2, and in particular its kinetic energy part, takes the simplest form in the space fixed system of the Cartesian coordinates. However, these coordinates are not very suitable for subsequent work because they mix the three simple motions described above and instead of simple trajectories, we obtain a very complicated pattern. For this reason, we use the transformation from the Cartesian coordinates to the following coordinates, which are "natural" for translation, rotation, and vibration:

$$\mathbf{R}_i = \mathbf{R}_0 + \mathbf{S}^{-1}(\theta, \varphi, \chi)(\mathbf{a}_i + \mathbf{d}_i) \tag{4.8}$$

where $\mathbf{R}_0$ is the position vector of the center of mass of the molecule, the elements of the (3×3) matrix of direction cosines $\mathbf{S}^{-1}(\theta, \varphi, \chi)$, transforming the vectors from the molecule-fixed to the space-fixed system of coordinates, are functions of the Euler angles. Finally, the vectors $\mathbf{a}_i$ and $\mathbf{d}_i$ describe the equilibrium position of atom i ($\mathbf{a}_i$) and the displacement from the equilibrium position ($\mathbf{d}_i$) in the system of coordinates fixed in the molecule. Note that the vectors $\mathbf{a}_i$ are the constant vectors describing the equilibrium geometrical structure of the molecule in the system of axes rotating with the molecule, whereas the vectors $\mathbf{d}_i$, the displacements from the equilibrium position, are variables describing the vibrations of the molecule.

There is also a mathematical reason for the transformation to the new coordinates. The time-independent Schrödinger equation (Equation 4.2) is a partial differential equation with many variables. The standard way of solving such equations is by separation of variables leading to the several equations with lower number of variables in each of them that replace the original one. We will see in Equation 4.20 that at least in a good approximation, the Schrödinger equation can indeed be separated into simple equations by the transformation to the new variables.

We can now proceed and construct the new Hamiltonian using the new system of coordinates. This is, however, a rather involved and lengthy process, and the reader is referred to the several excellent presentations in the literature [6,8,15].

One form of the resulting rovibrational quantum mechanical Hamiltonian can be written in a Watson's form [18]

$$\mathbf{H} = \frac{\hbar^2}{2} \sum_{\alpha,\beta=x,y,z} \mu_{\alpha\beta}(J_\alpha - \pi_\alpha)(J_\beta - \pi_\beta) + \frac{1}{2} \sum_{k=1}^{3N-6} P_k^2 - \frac{\hbar^2}{8} \sum_{\alpha=x,y,z} \mu_{\alpha\alpha} + V \qquad (4.9)$$

where the operator J_α is the αth component of the dimensionless total angular momentum of the molecule in the molecule-fixed axis system, π_α is a component of the vibrational angular momentum defined below in Equation 4.13, μ is the 3×3 inverse tensor to inertia tensor I, and P_k is the momentum conjugated to the normal coordinate Q_k ($P_k = -i\hbar\,\partial/\partial Q_k$). The *normal coordinates* Q_k are introduced by transforming the displacements of atoms from equilibrium $d_{i\alpha}$.

$$m_i^{1/2} d_{i\alpha} = \sum_{k=1}^{3N-6} l_{i\alpha,k} Q_k \qquad (4.10)$$

$\mathbf{l}$ is a matrix that permits us to transform the $3N$ vibrational variables (N being the number of atoms) $d_{i\alpha}$ into the normal coordinates Q_k. There are two reasons for this transformation. First, we now have the correct number of coordinates (for a nonlinear molecule: $3N = 3$ for translation + 3 Euler angles for rotation + $3N - 6$ normal coordinates for vibrations). The second reason is linked with the expression for the potential energy V. We choose the transformation (Equation 4.10) in such a way that all the cross terms in the quadratic parts of the potential energy expansion vanish and we have (see Section 1.4)

$$V = \frac{1}{2}\sum_r \lambda_r Q_r^2 + \frac{1}{6}\sum_{rst} \Phi_{rst} Q_r Q_s Q_t + \cdots \qquad (4.11)$$

where the summations are over all normal coordinates and

$$\lambda_r = \left(\frac{\partial^2 V}{\partial Q_r^2}\right)_e, \quad \Phi_{rst} = \left(\frac{\partial^3 V}{\partial Q_r \partial Q_s \partial Q_t}\right)_e \tag{4.12}$$

In addition, π_α is usually called vibrational angular momentum (although it would be more accurate to call it Coriolis coupling operator) and is defined as

$$\hbar \pi_\alpha = \sum_{l,k} \zeta_{lk}^\alpha Q_l P_k \tag{4.13}$$

with the Coriolis coupling constant ζ_{lk}^α defined as

$$\zeta_{lk}^\alpha = -\zeta_{kl}^\alpha = \sum_{\beta,\gamma} e_{\alpha\beta\gamma} \sum_i l_{i\beta,l} l_{i\gamma,k} \tag{4.14}$$

$e_{\alpha\beta\gamma}$ is the Levi-Civita symbol, equal to +1 if α, β, γ is a cyclic permutation, equal to −1 if α, β, γ is anticyclic, and equal to zero otherwise. The (3×3) generalized inverse inertia tensor $\boldsymbol{\mu}$ can be decomposed as follows [19]:

$$\boldsymbol{\mu} = (\mathbf{I''})^{-1} \mathbf{I}^e (\mathbf{I''})^{-1} \tag{4.15}$$

where $\mathbf{I}^e$ is the diagonal inertia tensor in the reference configuration with the components

$$I_{\alpha\alpha}^e = \sum_i m_i (a_{i\beta}^2 + a_{i\gamma}^2) \quad (\alpha,\beta,\gamma \text{ all different}) \tag{4.16}$$

while

$$I_{\alpha\beta}'' = I_{\beta\alpha}'' = \delta_{\alpha\beta} I_{\alpha\alpha}^e + \frac{1}{2} \sum_k a_k^{\alpha\beta} Q_k \tag{4.17}$$

where

$$a_k^{\alpha\beta} = a_k^{\beta\alpha} = 2 \sum_i m_i^{1/2} \left\{ \delta_{\alpha\beta} \sum_\gamma a_{i\gamma} l_{i\gamma,k} - a_{i\alpha,k} l_{i\beta,k} \right\} \tag{4.18}$$

$\delta_{\alpha\beta}$ is the Konecker-delta symbol and $a_{i\alpha}$, $a_{i\beta}$, and $a_{i\gamma}$ are the Cartesian coordinates of atom i in the principal axis system (PAS; for which the inertia tensor is diagonal, see Section 5.2.2). The Hamiltonian in Equation 4.9 is not particularly suitable for direct calculation, and one of the reasons is the fact that $\boldsymbol{\mu}$ is the inverse matrix and therefore in the next step we expand it around the equilibrium position using the fact that the vibrational displacement coordinates are much smaller than the interatomic distances. The expanded Hamiltonian $\mathbf{H}_{EXP}$ can be written as a sum of the terms

$$\mathbf{H}_{EXP} = \sum_m \sum_{n=1,2} \mathbf{H}_{mn} \qquad (4.19)$$

where indices m, n denote the respective powers of vibrational and rotational operators. The most important part of the Hamiltonian, the *zero-order approximation* Hamiltonian $\mathbf{H}^0$ consists of the following terms:

$$\mathbf{H}^0 = \mathbf{H}_{20} + \mathbf{H}_{02} = \frac{1}{2}\sum_k (P_k^2 + \lambda_k Q_k^2) + \frac{\hbar^2}{2} \sum_{\alpha=x,y,z} \frac{J_\alpha^2}{I_{\alpha\alpha}^e} \qquad (4.20)$$

The Hamiltonian $\mathbf{H}^0$ is also called the rigid rotor-harmonic (RRHO) oscillator Hamiltonian, because we obtain Equation 4.20 from Equation 4.9 by neglecting all π_α, which is tantamount to neglecting Coriolis interaction between the vibrations and rotations, and by neglecting Q_k-dependent part of inertia tensor $I''_{\alpha\beta}$ in Equation 4.17 so that the molecule rotates like a rigid body. We neglect the cubic and higher-order terms in the expansion of the potential energy V in Equation 4.11, obtaining separate harmonic oscillators. Finally, Watson has shown [19] that the term with $\sum \mu_{\alpha\alpha}$ in Equation 4.9 can be neglected in the $\mathbf{H}^0$. The last term in Equation 4.20 is usually abbreviated to $\sum_{\alpha=x,y,z} B^\alpha J_\alpha^2$, and B^α are the *rotational constants*. Notice that we have achieved in Equation 4.20 the desired separation of rotation from vibration (and also the mutual separation of vibrations), and we can write

$$\psi^0 = \psi^0_{VIB}\psi^0_{ROT}, \quad E^0 = E^0_{VIB} + E^0_{ROT} \qquad (4.21)$$

Expressions for ψ^0_{VIB}, E^0_{VIB}, ψ^0_{ROT}, and E^0_{ROT} can be found in references [6,15,20].

It is also important that in Equations 4.9 and 4.20 we have been able to express the rotational Hamiltonian in terms of the components of the total angular momentum. The total angular momentum has a special position in quantum mechanics, and a very powerful apparatus has been developed for this operator. In particular, for the operators $\mathbf{J}^2 = J_x^2 + J_y^2 + J_z^2$, J_z we know their eigenfunction and eigenvalues to be

$$\mathbf{J}^2|J,k\rangle = J(J+1)|J,k\rangle$$
$$J_z|J,k\rangle = k|J,k\rangle \qquad (4.22)$$

It is customary to use instead of the operators J_x, J_y the "ladder" operators $J_\pm$

$$J_\pm|J,k\rangle = (J_x \pm iJ_y)|J,k\rangle = \sqrt{J(J+1)-k(k\pm1)}|J,k\pm1\rangle \qquad (4.23)$$

In Equations 4.22 and 4.23, J,k are the rotational quantum numbers characterizing the rotational states and $k = -J, -J+1, \ldots J-1, J$, and $|J,k\rangle$ are the eigenfunctions. When the operators $J_\pm$ are acting on the eigenfunctions, they decrease or increase the quantum number k by 1 like the steps on the ladder and that is the reason why they are called "ladder" operators. Equations 4.22 and 4.23 enable us to express the matrix elements of all rotational operators in a closed form.

4.2.2 TRANSFORMATIONS OF THE HAMILTONIAN

The rovibrational molecular Hamiltonian in Equation 4.9 is not easily suitable for calculating the energy levels (however, variational computations employ this operator), and in the next step, it is expanded using the fact that the vibrational displacements are much smaller than are the interatomic distances in molecules. The expanded Hamiltonian $\mathbf{H}_{EXP}$ can be written as a sum of the terms (see Equation 4.19):

$$\mathbf{H}_{EXP} = \sum_m \sum_{n=1,2} \mathbf{H}_{mn} \tag{4.24}$$

This Hamiltonian expressed as a power series is much more amenable to treatment, but its principal disadvantage is that the individual terms connect different vibrational states in the matrix form, so that the matrix is "full."

$$\mathbf{H}_{EXP} = \begin{pmatrix}
H_{v_1 v_1} & H_{v_1 v_2} & H_{v_1 v_3} & H_{v_1 v_4} & \cdots & H_{v_1 vk} \\
H_{v_2 v_1} & H_{v_2 v_2} & H_{v_2 v_3} & H_{v_2 v_4} & \cdots & H_{v_2 vk} \\
H_{v_3 v_1} & H_{v_3 v_2} & H_{v_3 v_3} & H_{v_3 v_4} & \cdots & H_{v_3 vk} \\
H_{v_4 v_1} & H_{v_4 v_2} & H_{v_4 v_3} & H_{v_4 v_4} & \cdots & H_{v_4 vk} \\
\vdots & \vdots & \vdots & \vdots & \ddots & \vdots \\
H_{v_k v_1} & H_{v_k v_2} & H_{v_k v_3} & H_{v_k v_4} & \cdots & H_{v_k v_k}
\end{pmatrix} \tag{4.25}$$

In the next step, the matrix is block-diagonalized by the contact transformation with the operator $\mathbf{S}$ [18]

$$\mathbf{H}_{EFF} = \tilde{\mathbf{H}} = \exp(i\mathbf{S})\mathbf{H}_{EXP}\exp(-i\mathbf{S}) = \begin{pmatrix}
\tilde{H}_{v_1} & 0 & 0 & 0 & \cdots & 0 \\
0 & \tilde{H}_{v_2} & 0 & 0 & \cdots & 0 \\
0 & 0 & \tilde{H}_{v_3} & 0 & \cdots & 0 \\
0 & 0 & 0 & \tilde{H}_{v_4} & \cdots & 0 \\
\vdots & \vdots & \vdots & \vdots & \ddots & \vdots \\
0 & 0 & 0 & 0 & \cdots & \tilde{H}_{v_k}
\end{pmatrix} \tag{4.26}$$

and the Hamiltonians $\tilde{H}_{v_i}$ are then the effective Hamiltonians for the vibrational states v_i. The actual transformation is carried out by expanding the exponentials and can be expressed in a commutator form as follows [18,20]:

$$\mathbf{H}_{EFF} = \exp(i\mathbf{S})\mathbf{H}_{EXP}\exp(-i\mathbf{S})$$

$$= \mathbf{H}_{EXP} + \left[i\mathbf{S}, \mathbf{H}_{EXP}\right] + \frac{1}{2}\left[i\mathbf{S}, \left[i\mathbf{S}, \mathbf{H}_{EXP}\right]\right] + \frac{1}{6}\left[i\mathbf{S}, \left[i\mathbf{S}, \left[i\mathbf{S}, \mathbf{H}_{EXP}\right]\right]\right] + \cdots \tag{4.27}$$

The results of the transformation are twofold:

1. The Hamiltonian is block-diagonalized, and we can treat each vibrational state (or manifold of states) as completely isolated from other vibrational states.
2. The transformation contributes to the individual terms in the effective Hamiltonians H_{v_i} and produces spectroscopic constants for each vibrational state such as the rotational constants, the centrifugal distortion constants, and so on.

The transformation operator $\mathbf{S}$ is a sum of the operators $\mathbf{S}_i$, $\mathbf{S} = \sum_i \mathbf{S}_i$, and the individual operators $\mathbf{S}_i$ are selected in such a way that they remove from the transformed Hamiltonian the terms responsible for the block nondiagonal matrix elements such as $H_{v_i v_k}$. The process of constructing the operators $\mathbf{S}_i$ is described in [18,20]. In certain cases in which the vibrational energies of the interacting states are too close, the process fails because the transformation operators are too large.

In such cases, we do not try to remove the interaction operators in the Hamiltonian, and, for example, if the close-lying vibrational states were v_2 and v_3, the block-diagonal Hamiltonian would take the form

$$\mathbf{H}_{\mathrm{EFF}} = \tilde{\mathbf{H}} = \exp(i\mathbf{S})\mathbf{H}_{\mathrm{EXP}}\exp(-i\mathbf{S}) = \begin{pmatrix} \tilde{H}_{v_1} & 0 & 0 & 0 & \cdots & 0 \\ 0 & \tilde{H}_{v_2 v_2} & \tilde{H}_{v_2 v_3} & 0 & \cdots & 0 \\ 0 & \tilde{H}_{v_3 v_2} & \tilde{H}_{v_3 v_3} & 0 & \cdots & 0 \\ 0 & 0 & 0 & \tilde{H}_{v_4} & \cdots & 0 \\ \vdots & \vdots & \vdots & \vdots & \ddots & \vdots \\ 0 & 0 & 0 & 0 & \cdots & \tilde{H}_{v_k} \end{pmatrix} \quad (4.28)$$

That is, we shall have the effective Hamiltonians for the isolated states and also the effective Hamiltonian for the polyad $v_2 v_3$. As an example, Table IV.6 in Appendix IV presents the block-diagonalized Hamiltonian matrix for the SO_2 molecule. Thus, the diagonal blocks—the effective Hamiltonians—can be of three types:

1. If the state i is an isolated nondegenerate vibrational state, such as ($v_n = 1, n \in A_1$, $v_k = 0$ for all $k \neq n$ for a C_{3v} symmetric top), the corresponding block is vibrationally one-dimensional.
2. If the separation between two or more vibrational levels is not large compared to the matrix elements connecting them (we say that the states are in resonance), the interaction blocks between them cannot be removed by the contact transformation because of resonance denominators causing a singularity or very slow convergence of perturbation treatment and they must be treated in the same diagonal block. We can distinguish two cases.
   a. Because of molecular symmetry, the harmonic vibrational frequencies of two vibrations (E vibrations for symmetric-top and spherical-top

molecules or Π vibrations for linear molecules) or three vibrations (F vibrations for spherical-top molecules) become exactly degenerate. Typical example would be $v_t = 1, l_t = 1, t \in E$ states of molecules of C_{3v} point-group symmetry. Such resonances are called *essential resonances*.

b. A near-degeneracy that does not result from symmetry is called an *accidental degeneracy* and the off-diagonal matrix elements produce *accidental resonances*.

In cases 2a and 2b, the vibrational dimension of the effective Hamiltonian is two or more, depending on the actual vibrational states.

We concentrate now on the effect of reductions. There are both practical and theoretical aspects of the reduction procedures. On a practical level, once we have an effective Hamiltonian for a given vibrational state (or for a polyad of such states, for the definition of a polyad, see Appendix IV.1) with spectroscopic constants, we can attempt to determine their values by fitting them to the experimental frequencies of transitions between the rotation-vibration states. Such fitting means that we try to obtain the values of the spectroscopic constants that give theoretical predictions, which provide the best agreement with the experimental data. In the fitting process, often problems are encountered that may be alternatively denoted as *correlations* or *indeterminacies* or *collinearities* (see Section 2.2.2.2). No matter what name is given to the problem, in practice, it means that the matrix of normal equations or *design* matrix is nearly singular, and the fitting procedure diverges or converges poorly and some of the determined spectroscopic constants are highly correlated.

Historically, the problem had been recognized initially when it became obvious that for the asymmetric rotors, only five out of the six "independent" quartic centrifugal distortion constants can be determined and the sixth one has to be constrained. The same problem was later encountered for many effective Hamiltonians for various states [21,22]. The problem is usually treated by constraining as many spectroscopic constants as is necessary to predetermined values (usually zero). Obviously, the values of the remaining fitted constants are affected by this treatment. The following practical questions emerge: How many and which constants should be constrained? What is exactly the effect of the constraints on the remaining parameters? The correct answers to such questions are becoming more and more important with the fast development of ab initio computations that are able to provide independently the values of the spectroscopic constants. Fortunately, the theory of reductions can provide the answers to these questions (see Chapters 1 and 3).

The feature of the reduction theory that may cause confusion is that it uses formally exactly the same equation as does the contact transformation block-diagonalizing the Hamiltonian (compare Equations 4.27 and 4.29).

$$\begin{aligned}
\mathbf{H}_{RED} &= \exp(i\mathbf{S}^R)\mathbf{H}_{EFF}\exp(-i\mathbf{S}^R) \\
&= \mathbf{H}_{EFF} + \left[i\mathbf{S}^R, \mathbf{H}_{EFF}\right] + \frac{1}{2}\left[i\mathbf{S}^R, \left[i\mathbf{S}^R, \mathbf{H}_{EFF}\right]\right] \\
&\quad + \frac{1}{6}\left[i\mathbf{S}^R, \left[i\mathbf{S}^R, \left[i\mathbf{S}^R, \mathbf{H}_{EFF}\right]\right]\right] + \cdots
\end{aligned} \qquad (4.29)$$

where both operators S and S^R must satisfy the same criteria of being Hermitian, odd to time reversal, and invariant to all operations of molecular symmetry group. It is therefore important to point out the differences between the two transformations.

First, the transformation operator S^R in reductions is different from the operator S in block diagonalization. Whereas the operator S^R is vibrationally block-diagonal, the operator S is block-nondiagonal, which can become apparent from the following matrices showing the matrix elements of both operators in the basis of harmonic oscillator-rigid rotor wave functions:

$$\mathbf{S^R} = \begin{pmatrix} S^R_{v_1} & 0 & 0 & 0 & \cdots & 0 \\ 0 & S^R_{v_2} & S^R_{v_2 v_3} & 0 & \cdots & 0 \\ 0 & S^R_{v_3 v_2} & S^R_{v_3} & 0 & \cdots & 0 \\ 0 & 0 & 0 & S^R_{v_4} & \cdots & 0 \\ \vdots & \vdots & \vdots & \vdots & \ddots & \vdots \\ 0 & 0 & 0 & 0 & \cdots & S^R_{v_k} \end{pmatrix} \tag{4.30}$$

$$\mathbf{S} = \begin{pmatrix} 0 & S_{v_1 v_2} & S_{v_1 v_3} & S_{v_1 v_4} & \cdots & S_{v_1 v_k} \\ S_{v_2 v_1} & 0 & 0 & S_{v_2 v_4} & \cdots & S_{v_2 v_k} \\ S_{v_3 v_1} & 0 & 0 & S_{v_3 v_4} & \cdots & S_{v_3 v_k} \\ S_{v_4 v_1} & S_{v_4 v_2} & S_{v_4 v_3} & 0 & \cdots & S_{v_4 v_k} \\ \vdots & \vdots & \vdots & \vdots & \ddots & \vdots \\ S_{v_k v_1} & S_{v_k v_2} & S_{v_k v_3} & S_{v_k v_4} & \cdots & 0 \end{pmatrix} \tag{4.31}$$

Perhaps an even more important distinction between the two transformations lies in the reasons we apply them. The transformation (Equation 4.27) has been chosen by us for the very positive effect of decoupling the Hamiltonian $\mathbf{H_{EXP}}$ into a set of effective Hamiltonians, whereas the unitary transformation (Equation 4.29) has been, so to speak, forced on us unwillingly; its effect is negative, it is the source of indeterminacies, and we must apply it to find the correct constraints and their effects on the remaining fitted parameters. This is the practice-related explanation.

The deeper reasons follow from the principles of quantum mechanics. For example, Mekhtiev and Hougen [23] showed that the problem with indeterminacies would not exist if the quantization of the operator $\mathbf{J}$ would allow more than $2J + 1$ energy samples for a given J.

The result of the transformation (Equation 4.29) is that the spectroscopic constants after the transformation $\tilde{X}$ are expressed as the function of the original spectroscopic constants X and the parameters s_i of the reduction operator as follows:

$$\tilde{X}_j = X_j + \Delta X_j(X_i, s_1, s_2, \dots s_n) \tag{4.32}$$

4.2.3 THE ASYMMETRIC TOP

4.2.3.1 Reduction of the Hamiltonian

The Hamiltonian has the form

$$\mathbf{H}_{EFF} = E_v + \mathbf{H}_{rot} \tag{4.33}$$

where E_v is the purely vibrational energy and the rotational Hamiltonian may be written in cylindrical tensor form

$$\mathbf{H}_{rot} = B_{200}\mathbf{J}^2 + B_{020}J_z^2 + T_{400}(\mathbf{J}^2)^2 + T_{220}\mathbf{J}^2 J_z^2 + T_{040}J_z^4$$

$$+ \frac{1}{2}\left[B_{002} + T_{202}\mathbf{J}^2 + T_{022}J_z^2, J_+^2 + J_-^2 \right]_+ \tag{4.34}$$

$$+ \frac{1}{2}\left[T_{004}, J_+^4 + J_-^4 \right]_+ + \cdots$$

in which $[A, B]_+$ represents the anticommutator $AB + BA$, $J_\pm$ is the ladder operator, B_{ijk} and T_{ijk} are the cylindrical components of rotational and centrifugal constants.

Watson [21,24] has shown that of the six quartic centrifugal distortion constants T, only five combinations are experimentally determinable, and from Equation 4.32

$$\tilde{T}_{400} = -\tilde{D}_J = T_{400} - \frac{1}{2}s_{111}(B_x - B_y) \tag{4.35a}$$

$$\tilde{T}_{220} = -\tilde{D}_{JK} = T_{220} + 3s_{111}(B_x - B_y) \tag{4.35b}$$

$$\tilde{T}_{040} = -\tilde{D}_K = T_{040} - \frac{5}{2}s_{111}(B_x - B_y) \tag{4.35c}$$

$$\tilde{T}_{022} = 2\tilde{R}_5 = T_{022} - s_{111}(2B_z - B_x - B_y) \tag{4.35d}$$

$$\tilde{T}_{004} = \tilde{R}_6 = T_{004} + \frac{1}{4}s_{111}(B_x - B_y) \tag{4.35e}$$

$$\tilde{T}_{202} = -\tilde{\delta}_J = T_{202} \tag{4.35f}$$

where the reduction coefficient s_{111} is the coefficient of the cubic transformation operator $\mathbf{S}_3$ and the subscripts 1,1,1 in the coefficient s_{111} correspond to the powers of the operators J_x, J_y, J_z. In principle, s_{111} can take any value, but care must be taken that it does not exceed the order of magnitude limit in order to ensure a fast convergence of the Hamiltonian as follows:

$$s_{111} \approx \kappa^4 \qquad (4.36)$$

where the smallness parameter κ is best defined by the relation

$$T \approx \kappa^4 B \qquad (4.37)$$

and T and B are typical values of quartic centrifugal distortion constants and rotational constants, respectively, for the molecule under study.

The most often used constraint, called A-reduction, is

$$s_{111} = -\frac{4T_{004}}{B_x - B_y} \qquad (4.38)$$

However, this reduction blows up if the molecule is close to the symmetric-top limit. In this case, the usual constraint, called S-reduction, is

$$s_{111} = \frac{T_{022}}{2B_z - B_x - B_y} \qquad (4.39)$$

The rotational constants are also affected by the reduction, as in Equation 4.32. This is discussed in Section 4.3.3.

4.2.3.2 Choice of Reduction and Representation

The A-reduction is used for asymmetric tops and the S-reduction is preferred for molecules close to the symmetric top. Furthermore, there are six different ways to identify the (x, y, z) reference system with the (a, b, c) PAS. In practice, two different representations are used: I^r where $x = b$, $y = c$, $z = a$ and which is best for prolate molecules [asymmetry parameter $\kappa = \dfrac{(2B - A - C)}{(A - C)} < 0$] and III^r where $x = a$, $y = b$, $z = c$ and which is thought to be better for oblate molecules ($\kappa > 0$). However, the choice of the right reduction and of the right representation is not always obvious. Three criteria may be used for this choice: s, the standard deviation of the fit, s_{111}, the reduction coefficient and κ, the condition number (not to be confused with the asymmetry parameter; see Equation 2.16). All diagnostics should be as small as possible.

There are very few studies of the choice of the best reduction and representation. However, a comparative study was run on the oblate molecule carbonyl fluoride, OCF_2 (asymmetry parameter, $\kappa = +0.980$). Cohen and Lewis-Bevan [25] could demonstrate that, for the ground state, any of the combinations (S, III^r), (A, I^r), or (S, I^r) gives excellent fits of the combined experimental data (microwave measurements and infrared data). On the other hand, the combination (A, III^r) is clearly inappropriate since the fit of the experimental data is about 100 times worse than for the other choices. Similar results were found for dimethylsulfoxide $(CH_3)_2SO$ (see Table 4.1).

TABLE 4.1

Centrifugal Distortion Analysis of $(CH_3)_2SO$ (Asymmetry Parameter $\kappa = +0.886$)

	p^a	s^b	s_{111}^c	κ^d
S-I^r	18	16	7.6×10^{-8}	327
S-III^r	18	19	3.5×10^{-8}	663
A-I^r	18	16	4.8×10^{-7}	225
A-III^r	18	51	8.2×10^{-6}	10778

Source: Margulès, L., R. A. Motiyenko, E. A. Alekseev, and J. Demaison. 2010. *J Mol Spectrosc* 260:23–9.

[a] Number of parameters in the fit.
[b] Standard deviation of the fit in kilohertz.
[c] Reduction parameter.
[d] Condition number of the fit (see Equation 2.16).

4.2.4 THE LINEAR MOLECULE

In linear molecules, all atoms lie on the symmetry axis. These molecules have either $D_{\infty h}$ or $C_{\infty v}$ symmetry. Diatomic molecules, a special case of linear molecules, are considered in Chapter 6.

For linear molecules with more than two atoms, bending modes in which the atoms move perpendicular to the symmetry axis are degenerate and introduce another quantum number, ℓ, with $\ell = v_i, v_i - 2, \ldots -v_i$ and $J = |\ell|, |\ell| +1, |\ell|+2, \ldots$, where ℓ indicates the units of quantum momentum about the internuclear axis. States with $\ell = 0$ are labeled Σ states. States with $\ell = \pm 1, \pm 2, = \pm 3$ are called Π, Δ, Φ states, and so on. Only one rotational constant B_v has to be considered, and the rotational energy levels are represented as

$$E_r = B_v \left[J(J+1) \right] - D_v \left[J(J+1) \right]^2 + \cdots \qquad \text{for } \Sigma \text{ states } (\ell = 0) \qquad (4.40a)$$

and, for Π states with $|\ell| = 1$,

$$E_r = B_t \left[J(J+1) - \ell^2 \right] - D_t \left[J(J+1) - \ell^2 \right]^2 \pm \frac{1}{4} q_t (v_t + 1) J(J+1) + \cdots \qquad (4.40b)$$

In the last term of this expression, v_t is the vibrational quantum number of the tth degenerate bending mode and q_t, called the ℓ-type doubling constant, is the coupling constant of the vibration and rotation for this mode. For higher states with $|\ell| > 1$, Equation 4.40b can also be used in the first approximation, but without the ℓ-type doubling term.

4.2.5 THE SYMMETRIC-TOP MOLECULE

In symmetric tops, two rotational constants are identical. Two classes of symmetric tops are possible, oblate tops with $A = B > C$, like phosphine, PH_3, and prolate tops with $A > B = C$, like chloromethane, $CH_3{}^{35}Cl$ (see also Section 5.2.4). For a nondegenerate vibrational state of a prolate top, the rotational energy levels are

$$E_r = B_v J(J+1) + (A_v - B_v)K^2 - D_J^v \left[J(J+1) \right]^2$$
$$- D_{JK}^v J(J+1)K^2 - D_K^v K^4 + \cdots \tag{4.41}$$

For an oblate top, A is replaced by C. In this expression, K is the absolute value of the projection of J along the axis of the symmetry of the molecule.

For a degenerate vibration, a Coriolis interaction between vibration and rotation arises and the term $-2(A_v\zeta)K\ell$ should be added to Equation 4.41 for a prolate top (for an oblate top, A is replaced by C). The definition of ℓ is the same as for linear molecules (see Section 4.2.4).

The rotational constants are slightly affected by the reduction of the Hamiltonian, Equation 4.32, but the correction is usually negligible [22].

4.2.6 THE SPHERICAL-TOP MOLECULE

A spherical molecule is a special case of a symmetric top in which the three rotational constants are equal (Section 5.2.4). In this case, the apparatus of irreducible tensors is applied and the rotational Hamiltonian may be written as follows:

$$H_r = B_v \mathbf{J}^2 - D_v (\mathbf{J}^2)^2 + D_v^{4t} \Omega_4 + \cdots \tag{4.42}$$

where Ω_4 is a fourth-rank tensor operator whose explicit expression is given by Watson [21].

A sophisticated tensorial formalism developed in Dijon is well-adapted to spherical tops [67].

4.3 DETERMINATION OF EQUILIBRIUM ROTATIONAL CONSTANTS

4.3.1 VIBRATIONAL CORRECTION

The analysis of the spectra gives the rotational constants B_v^ξ for a given vibrational state v. In a perturbational treatment, the rotational constant B_v^ξ is given by (see also Equation 3.8)

$$B_v^\xi = B_e^\xi - \sum_i \alpha_i^\xi \left(v_i + \frac{d_i}{2} \right) + \sum_{i \geq j} \gamma_{ij}^\xi \left(v_i + \frac{d_i}{2} \right) \left(v_j + \frac{d_j}{2} \right) + \sum_{i \geq j} \gamma_{ij}^\xi l_i l_j + \cdots \tag{4.43}$$

where $\xi = a, b, c$. The summations are over all vibrational states, each characterized by a quantum number v_i and a degeneracy d_i. The parameters α_i^ξ and γ_i^ξ are called vibration-rotation interaction constants of different order. B_e^ξ is the equilibrium rotational constant. The convergence of the series expansion is usually fast, α_i^ξ being about two orders of magnitude smaller than B_e^ξ and γ_{ij}^ξ two orders of magnitude

TABLE 4.2

Vibrational Contribution to the Rotational Constants (in Megahertz)

Molecule	B_e	$C = \sum \dfrac{\alpha_i d_i}{2}$	C/B (%)	$D = \sum \dfrac{\gamma_{ij} d_i d_j}{4}$	D/C (%)
HCN	44511.620	198.137	0.45	2.395	1.21
FCN	10586.782	32.604	0.31	−0.248	−0.76
ClCN	5982.8975	12.0644	0.20	−0.0207	−0.17
BrCN	4126.5059	6.2838	0.15	0.0596	0.95
ICN	3329.0568	3.5084	0.11	−0.0049	−0.14

Source: Demaison, J. 2007. *Mol Phys* 105:3109–38. With permission.

smaller than α_i^ξ (see Table 4.2). Such results are presented for the bent asymmetric-top SO_2 in Table IV.8 of Appendix IV. The calculation of the α-constants from the cubic force field is explained in Section 3.2.2.

To determine an equilibrium structure, we should know the equilibrium rotational constants, which are obtained from the ground-state rotational constants and the rotational constants of all fundamentally excited vibrational states (see Equation 4.43). In principle, microwave as well as infrared spectroscopies may be used to get the rotational constants B_v^ξ of the excited states. Given its accuracy, microwave spectroscopy should be preferred; however, the intensity of a rotational transition is proportional to the population of its lower state, which is given by the Boltzmann law exp $(-E''/kT)$. If the energy of the excited vibrational state is high, above 1000–2000 cm^{-1}, the population of its rotational levels will be small and the rotational transitions between them will be too weak to be observed. In this rather common case, only infrared spectroscopy may be used to determine the α constants. It is thus obvious that the combination of both microwave and infrared spectroscopy is well suited to obtain the equilibrium rotational constants (see Sections 4.5 and 4.6). Another point worth noting is that different techniques have different selection rules and are, thus, more or less adapted to determine some parameters (see Appendix IV.2).

4.3.2 ELECTRONIC CORRECTION

Since the electrons tend to follow the motion of the nuclei, the bulk of the electronic contribution to the rotational constants can be taken into account by employing atomic rather than nuclear masses (see Ref. 3 and Section 3.2.3.3). This is a very good approximation for most molecules, and it is about the only feasible one for most polyatomic molecules. However, a small correction for unequal sharing of the electrons by the atoms and for nonspherical distribution of the electronic clouds around the atoms is sometimes nonnegligible and has to be taken into account.

The total angular momentum **J** of a molecule may be written as the sum of **N**, the angular momentum due to the rotation of the nuclei, and **L**, the angular momentum of the electrons. The rotational Hamiltonian for the nuclear system plus the Hamiltonian for the unperturbed electronic energies may be written as

$$
\mathbf{H} = \frac{1}{2}\sum_{\xi}\frac{\mathbf{N}_{\xi}^2}{I_{\xi}} + \mathbf{H}_e = \frac{1}{2}\sum_{\xi}\frac{(\mathbf{J}_{\xi}-\mathbf{L}_{\xi})^2}{I_{\xi}} + \mathbf{H}_e
$$

$$
= \underbrace{\frac{1}{2}\sum_{\xi}\frac{\mathbf{J}_{\xi}^2}{I_{\xi}} + \mathbf{H}_e}_{\mathbf{H}^0 = \mathbf{H}_R + \mathbf{H}_e} \underbrace{-\sum_{\xi}\frac{\mathbf{J}_{\xi}\mathbf{L}_{\xi}}{I_{\xi}} + \frac{1}{2}\sum_{\xi}\frac{\mathbf{L}_{\xi}^2}{I_{\xi}}}_{\mathbf{H}'}
\tag{4.44}
$$

Since $\mathbf{L}_{\xi}$ is very small, the last term can be neglected and $\mathbf{H}'$ can be treated as a perturbation of $\mathbf{H}^0$. We now assume that the molecule is not in a pure $^1\Sigma$ state $\psi_0^{(0)}$ ($L=0$) but in a perturbed state $\psi_0^{(1)}$, which has some electronic momentum. The correct effective rotational Hamiltonian is then

$$
\mathbf{H}_{\text{eff}} = \left\langle \psi_0^{(1)} \middle| \mathbf{H}_R + \mathbf{H}' \middle| \psi_0^{(1)} \right\rangle
\tag{4.45}
$$

A simple perturbation calculation up to second order gives

$$
\mathbf{H}_{\text{eff}} = \frac{1}{2}\sum_{\xi}\mathbf{J}_{\xi}^2\left(\frac{1}{I_{\xi}} - \frac{2}{I_{\xi}^2}\sum_{n\neq0}\frac{\left|\langle n|\mathbf{L}_{\xi}|0\rangle\right|^2}{E_n - E_0}\right)
\tag{4.46}
$$

This is equivalent to the definition of an effective moment of inertia $(I_{\xi})_{\text{eff}}$ by

$$
\frac{1}{(I_{\xi})_{\text{eff}}} = \frac{1}{I_{\xi}} - \frac{2}{I_{\xi}^2}\sum_{n\neq0}\frac{\left|\langle n|\mathbf{L}_{\xi}|0\rangle\right|^2}{E_n - E_0}
\tag{4.47}
$$

where I_{ξ} on the right is calculated using the nuclear masses. This effective moment of inertia can be expressed as a function of the molecular rotational g factor in the PAS, whose definition is

$$
g_{xx} = \frac{M_p}{I_x}\sum_i Z_i(y_i^2 + z_i^2) - \frac{2M_p}{mI_x}\sum_{n\neq0}\frac{\left|\langle n|\mathbf{L}_x|0\rangle\right|^2}{E_n - E_0}
\tag{4.48}
$$

and g_{yy} and g_{zz} are obtained by cyclic permutation. In this equation, M_p is the mass of the proton and m the mass of the electron.

The effective rotational constant B_{eff} (obtained from the analysis of the rotational spectrum) is therefore

$$
(B^{\xi})_{\text{eff}} = B^{\xi} + \frac{m}{M_p}g_{\xi\xi}(B^{\xi})_n
\tag{4.49}
$$

where $\xi = a, b, c$, B^{ξ} is the rotational constant calculated with atomic masses and $(B^{\xi})_n$ the rotational constant calculated with nuclear masses.

TABLE 4.3

Electronic Contribution to the Rotational Constants

(B_0 and ΔB in Megahertz)

Molecule	B_0	g	ΔB
LiH	229965.07	−0.6584	82.467
CO	57635.966	−0.2691	8.448
CO_2	11698.4721	−0.05508	0.351
OCS	6081.4921	−0.028839	0.096
OCSe	4017.6537	−0.01952	0.043
CS_2	3271.4882	−0.02274	0.041
HCN	44315.9757	−0.098	2.365
$FC^{15}N$	10186.2903	−0.0504	0.280
$ClC^{15}N$	5748.0527	−0.0385	0.121
$BrC^{15}N$	3944.8441	−0.0385	0.083
$IC^{15}N$	3225.5485	−0.0325	0.057
HC≡C–CN	4549.067	−0.0213	0.053
O_3, A	105536.235	0.642	−36.903
O_3, B	13349.2547	−0.119	0.865
O_3, C	11834.3614	−0.061	0.393

Source: Demaison, J. 2007. *Mol Phys* 105:3109–38. With permission.

The *g* factor can be obtained experimentally from the analysis of the Zeeman effect on the rotational spectrum [3,26] but it is much easier to calculate it ab initio [27] (see Section 3.2.3.3). A few typical results are given in Table 4.3. As expected, the correction is the largest for very light molecules (such as LiH), and it rapidly decreases when the mass of the molecule increases. There are, however, a few exceptions. As the expression of *g* shows, in Equation 4.48, *g* may become large when an electronic excited state is close to the ground state (because the denominator $E_n - E_0$ becomes small). This is the case for ozone (O_3), where the electronic correction is extremely large: $g_{aa} = -2.9877(9)$ [28] leads to a huge electronic correction of −173 MHz.

4.3.3 CENTRIFUGAL DISTORTION CORRECTION

The rotational constants are slightly affected by the reduction as in Equation 4.32. Watson [21] has shown that only the following linear combinations can be determined from the analysis of the spectra:

$$B_z = B_z^{(A)} + 2\Delta_J = B_z^{(S)} + 2D_J + 6d_2 \tag{4.50a}$$

$$B_x = B_x^{(A)} + 2\Delta_J + \Delta_{JK} - 2\delta_J - 2\delta_K = B_x^{(S)} + 2D_J + D_{JK} + 2d_1 + 4d_2 \tag{4.50b}$$

$$B_y = B_y^{(A)} + 2\Delta_J + \Delta_{JK} + 2\delta_J + 2\delta_K = B_y^{(S)} + 2D_J + D_{JK} - 2d_1 + 4d_2 \tag{4.50c}$$

$B_\xi^{(A)}$ are the experimental constants in the so-called A-reduction, $B_\xi^{(S)}$ are the experimental constants in the so-called S-reduction, and B_ξ are the determinable constants (where $\xi = x, y, z$). However, these latter constants are still contaminated by the centrifugal distortion. As shown by Kivelson and Wilson [29], the rigid rotor constants B_ξ' are given by

$$B_x' = B_x + \frac{1}{2}(\tau_{yyzz} + \tau_{xyxy} + \tau_{xzxz}) + \frac{1}{4}\tau_{yzyz} \tag{4.51}$$

where B_y' and B_z' are obtained by cyclic permutation of $x, y,$ and z. (To avoid confusion, B' is sometimes called B^{unpert} and B, B^{det}). The problem is that the τ constants are experimentally determinable only for a planar molecule by the planarity relations of Dowling [30] (see Appendix IV.3). For a nonplanar molecule, τ constants can be calculated from the harmonic force field, obtained either experimentally or ab initio. Compared to the other corrections, the centrifugal distortion correction is generally quite small except for very light molecules (see Table 4.4). Furthermore, it is different from zero only for asymmetric-top molecules and, in this case, it generally remains much larger than the experimental accuracy. The easiest way to estimate the centrifugal distortion correction is to calculate it ab initio (see Section 3.2.3.2).

TABLE 4.4

Centrifugal Contribution to the Rotational Constants of NH$_2$, HCOOH, and H$_2$C=C=O (in Megahertz)

	Exp.	$B_\xi^{det} - B_\xi^{exp}$ [a]	$B_\xi^{unpert} - B_\xi^{det}$ [b]
		NH$_2$ [c]	
A	712634.609(5)	63.329	−71.312
B	388213.098(9)	−145.871	−96.806
C	244843.726(9)	22.438	201.733
		HCOOH [d]	
A	77512.2354(11)	0.0200	−0.0643
B	12055.1065(2)	−0.1556	0.0154
C	10416.1151(2)	0.0231	0.1375
		H$_2$C=C=O [e]	
A	282101.19(41)	0.0062	−0.5589
B	10293.3212(8)	0.4852	−0.5177
C	9915.9055(8)	0.4858	−0.1768

[a] Equations 4.50, $\xi = a, b, c$, exp $= A$ or S.
[b] Equation 4.51.
[c] Reference [62].
[d] Reference [52].
[e] Reference [63].

4.4　RESONANCES

4.4.1　Coriolis Interaction

Coriolis interactions are caused by the coupling of the total angular momentum J_ξ and the vibrational angular momentum π_ξ (see Equation 4.9) [6]. The Coriolis interactions are implicitly taken into account in the calculation of the α constants (see second term of Equation 3.6) by a second-order perturbation treatment,

$$-\alpha_k^\xi(\text{Cor.}) = \frac{2(B_e^\xi)^2}{\omega_k} \sum_{l \neq k} (\zeta_{kl}^\xi)^2 \frac{3\omega_k^2 + \omega_l^2}{\omega_k^2 - \omega_l^2} \tag{4.52}$$

where ζ_{kl}^ξ is the Coriolis zeta constant (see Equation 4.14), which couples Q_k and Q_l through rotation about the ξ axis. It is different of zero if the direct product of the symmetry species of the vibrational states k and l contains the species of the rotation R_ξ [31].

EXAMPLE 4.1: OZONE, O_3

Ozone has three vibrational modes, v_2 (OOO bending mode) and v_1 and v_3 (symmetric and antisymmetric O–O stretching modes) [32]. For $^{16}O_3$, the three harmonic frequencies, $\omega_2(A_1) = 714.967$ cm^{-1}, $\omega_1(A_1) = 1133.01$ cm^{-1}, and $\omega_3(B_2) = 1087.277$ cm^{-1}, are in the approximate relation $\omega_1 \approx \omega_3$. The three rotations R_x, R_y, and R_z under the C_{2v} group belong to the species B_1, B_2, and A_2, respectively. Since $A_1 \times B_2 = B_2$, ζ_{13}^y is nonvanishing (y is perpendicular to the molecular plane).

When $\omega_k \approx \omega_l$ and $\zeta_{kl}^\xi \neq 0$, the perturbation calculation breaks down and the resonance itself has to be treated by the construction and diagonalization of a matrix of the rotational states of the two coupled vibrations, the main term being

$$\langle v_k, v_l | \mathbf{H} | v_k + 1, v_l - 1 \rangle = 2iB_e^\xi \zeta_{kl}^\xi \left[\sqrt{\frac{\omega_l}{\omega_k}} + \sqrt{\frac{\omega_k}{\omega_l}} \right] \sqrt{\frac{(v_k + 1)v_l}{4}} J_\xi = h_{vv'}^{1\xi} J_\xi \tag{4.53}$$

During the analysis of the spectra, it often happens that higher-order terms are necessary. For example, for $\xi = a$, the next term is $h_{vv'}^{2a} \left[J_b, iJ_c \right]_+$, and for $\xi = b$, or c, the second-order terms can be found by permutation. The analysis of Coriolis resonance is often difficult because the least-squares problem is strongly nonlinear. Furthermore, the Coriolis interaction parameter is highly correlated with the rotational constants and with the energy difference between the two interacting vibrational states. For this reason, the second-order term, $h_{vv'}^{2a}$, may be more significant than the first-order one, $h_{vv'}^{1\xi}$, as discussed by Perevalov and Tyuterev [33] (see Section 4.4.3).

However, when the α constants are summed (to determine the equilibrium rotational constants), the resonance contribution disappears, as follows (see also Section 3.2.3.1):

$$-\frac{1}{2} \sum_k \alpha_k^\xi = -\sum_{l>k} \frac{(B_e^\xi \zeta_{kl}^\xi)^2 (\omega_k - \omega_l)^2}{\omega_k \omega_l (\omega_k + \omega_l)} \tag{4.54}$$

In other words, even if the Coriolis interaction is handled only approximately (which is often the case), this is not important for determining the equilibrium rotational constants because there is a compensation of errors. However, Equation 4.54 is only a first-order approximation because it neglects off-diagonal terms in the Coriolis interaction (for a more exact formulation, see Appendix IV.4).

EXAMPLE 4.2: STRONG CORIOLIS INTERACTION—*TRANS*-FORMIC ACID, HCOOH, V_7 AND V_9 DYAD

Formic acid possesses nine nondegenerate vibrational modes. The two lowest fundamental states 7^1 (v_7 is the OCO scissor mode) and 9^1 (v_9 is the COH torsion mode) corresponding to infrared bands located at 626.166 cm^{-1} and 640.725 cm^{-1}, respectively, are far from the other fundamental or combination states. These two states are coupled through strong $7^1 \Leftrightarrow 9^1$ A- and B-type Coriolis resonances because the energy separation between the resonating states is small (only 14 cm^{-1}) and because the values of the first-order coupling parameters involved in the expansion of the Coriolis resonances are large ($h_{7,9}^{1A} = 31514.52$ MHz and $h_{7,9}^{1B} = 7874.21$ MHz, to be compared with $A_0 \approx 77512$ MHz and $B_0 \approx 12055$ MHz. It is interesting to compare the results of the three most recent experimental studies [34–36]. The quality of the experimental data used for these analyses has been significantly improved starting from the study performed in 1979 (infrared analysis at medium resolution ~0.02 cm^{-1}) up to the analyses performed in 2002 and 2006, where infrared data obtained from high-resolution Fourier transform spectra were combined with measurements of rotational transitions within the 7^1 and 9^1 excited states.

The theoretical models used for these three studies are basically identical in their form: The v-diagonal parts of the Hamiltonian model consist of an A-reduced Watson-type Hamiltonian written in the I^r representation, and the off-diagonal parts account both for A- and B-type Coriolis interactions. Using this form of the Hamiltonian matrix, the available experimental data for the 7^1 and 9^1 interacting energy levels were introduced in a least-squares fit leading to the determination of a set of band centers and rotational and Coriolis constants. In this way, the experimental data could be reproduced within their experimental accuracy. However, the three different sets of parameters are very different for the A, B, and C rotational constants, and the disagreements significantly exceed the uncertainty stated for these constants. The α_7^ξ and α_9^ξ constants for the 7^1 and 9^1 are collected in Table 4.5. This table gives also the sums $\alpha_7^\xi + \alpha_9^\xi$. These values are also compared to their ab initio counterparts. As anticipated, the sums $\alpha_7^\xi + \alpha_9^\xi$ are in excellent agreement for all experimental or ab initio studies. These sums may be used for a determination of structure because the perturbations mostly cancel out even if the $7^1 \Leftrightarrow 9^1$ resonances are not accounted for correctly (see Equation 4.54).

On the other hand, for the individual α_7^ξ and α_9^ξ the disagreements are important. This is because the expansion of the off-diagonal $7^1 \Leftrightarrow 9^1$ A- and B-type Coriolis operators was pursued up to different orders. Table 4.6 gives a short description of the rotational operators, which are involved in the expansions of the A- and B-type Coriolis operators. At a higher order, the Coriolis operators can be written as an expansion of $_k^n[A,B]_+$ rotational operators of degree n in J involving symmetrized tensorial products of $J_z, J^2, (J_- + J_+)$ and/or $(J_- - J_+)$ rotational operators. The values obtained for the first order A- and B-type Coriolis

TABLE 4.5

α_v^A, α_v^B and α_v^B for the 7^1 and 9^1 Excited States of *Trans*-Formic Acid, *Trans*-HCOOH

State	α_v^A	α_v^B	α_v^C	Reference
7^1	−31.7(14)	14.22(13)	21.04(5)	34
	−243.1(9)	20.37(6)	21.142(4)	35
	−416.943(4)	6.6193(2)	21.366(2)	36
9^1	−388.0(13)	54.49(13)	10.97(8)	34
	−172.8(9)	48.21(6)	10.834(4)	35
	0.670(4)	61.8860(2)	11.050(1)	36
$7^1 + 9^1$	−419.7	68.7	32.0	34
	−416.3	68.5	32.4	35
	−416.3	68.5	32.4	36
	−432.66	64.95	31.07	51[a]

[a] Calculated from the ab initio anharmonic force field.

TABLE 4.6

Description of the A- and B-Type Coriolis Resonances Coupling the $7^1 \Leftrightarrow 9^1$ Interacting States of *Trans*-HCOOH

Reference	[34]	[35]	[36]
A-type Coriolis $h_{7,9}^{1A}$ (MHz)	29,993.20	31,514.52	32,695.284
Number of $h_{7,9}^{kA}$ parameters	2	3	6
Maximum order in J^n in the ${}_k^n[A,B]_+$ rotational operator	2	3	7
B-type Coriolis $h_{7,9}^{1B}$ (MHz)	8,044.91	7,874.21	8,246.18
Number of $h_{7,9}^{kB}$ parameters	1	3	6
Maximum order in J^n in the ${}_k^n[A,B]_+$ rotational operator	1	3	7

parameters $h_{7,9}^{1A}$ and $h_{7,9}^{1B}$ are also given in Table 4.6. When going from the first [34] to the last study [36], the number (k_{max}) of $h_{7,9}^{kX}$ parameters and the maximum order (n_{max}) in J^n of the ${}_k^n[A,B]_+$ rotational operators involved in the $7^1 \Leftrightarrow 9^1$ off-diagonal operators increases dramatically (from $k_{max} = 3$ in 1979 up to $k_{max} = 12$ in 2006, and from $n_{max} = 3$ in 1979 up to $n_{max} = 12$ in 2006, respectively. On the other hand, the rotational expansion performed in the v-diagonal operators stops at $n = 8$. Therefore, the parameters are "effective" and do not have a clear physical meaning.

Coriolis resonances are also present in linear and symmetric-top molecules. An example of Coriolis resonance for linear molecule, HCO$^+$, is discussed in Appendix IV.5.

4.4.2 ANHARMONIC RESONANCES

The vibrational potential energy is usually expanded in terms of dimensionless normal coordinates as

$$V = \frac{1}{2} \sum_r \omega_r q_r^2 + \frac{1}{6} \sum_{rst} \phi_{rst} q_r q_s q_t + \frac{1}{24} \sum_{rstu} \phi_{rstu} q_r q_s q_t q_u + \cdots \qquad (4.55)$$

where ϕ_{rst} and ϕ_{rstu} are the cubic and quartic force constants, respectively. They are different from zero only if the product of the symmetry species of the vibrational states (r, s, t and r, s, t, u, respectively) is totally symmetric.

When two states of the same symmetry have nearly the same energy, a resonance is present and the second-order perturbation calculation of the anharmonic constants x_{rs} breaks down (for the definition of x, see Appendix V.1). The simplest and most common of these resonances is the Fermi resonance, which occurs whenever $2\omega_r \approx \omega_s$ or $\omega_r + \omega_s \approx \omega_t$. In this case, the resonance itself has to be treated by the construction and diagonalization of a matrix of the two coupled vibrations, the main term being

$$\langle v_r, v_s, v_t | \mathbf{H}^{\mathrm{Fermi}} | v_r + 1, v_s + 1, v_t - 1 \rangle = \phi_{rst} \left(\frac{v_t(v_r + 1)(v_s + 1)}{8} \right)^{1/2} \qquad (4.56)$$

For example, $v_r, v_s, v_t = 0, 0, 1$, or

$$\langle v_r, v_s | \mathbf{H}^{\mathrm{Fermi}} | v_r + 2, v_s - 1 \rangle = \frac{\phi_{rrs}}{2} \left(\frac{(v_r + 1)(v_r + 2)v_s}{2} \right)^{1/2} \qquad (4.57)$$

For example, $v_r, v_s = 0, 1$. Higher-order resonances are possible, for instance, the quartic Darling–Dennison resonance between two overtones states where $2\omega_r \approx 2\omega_s$.

As for the Coriolis resonances, it often happens that higher-order terms are necessary and the analysis is often difficult because the least-squares problem is strongly nonlinear. Furthermore, the interaction parameters are highly correlated with the rotational constants and with the energy difference between the two interacting vibrational states. The operator for anharmonic resonances may be written as follows [37]:

$$h_{v,v'}^{\mathrm{Anh}} = h_{v,v'}^{0\,\mathrm{Anh}} + h_{v,v'}^{1\,\mathrm{Anh}}\, J_{xy}^2 + h_{v,v'}^{2\,\mathrm{Anh}}\, \mathbf{J}^2 + h_{v,v'}^{3\,\mathrm{Anh}}\, J_z^2 + \cdots \qquad (4.58)$$

with $J_{xy}^2 = J_x^2 - J_y^2$ and where the first term $h_{v,v'}^{0\,\mathrm{Anh}}$ is given in Equations 4.56, or 4.57 in the case of a Fermi resonance (with Anh = Fermi). It is important to emphasize that h^0 is difficult to determine from the experimental spectra because this parameter is highly correlated with the band centers E_v and $E_{v'}$ of the resonating states (see Section 2 2.2). An example of such a difficulty is presented in Section 4.6.3.1

The individual α constants calculated from an anharmonic force field are free of contribution from anharmonic resonances. On the other hand, the experimental α constants are affected and have to be carefully corrected.

EXAMPLE 4.3: OXYGEN DIFLUORIDE, F$_2$O

The three fundamental bands are centered at $v_1 = 928$ cm^{-1} (OF stretching mode) $v_2 = 461$ cm^{-1} (FOF bending mode), and $v_3 = 832$ cm^{-1} [38]. There is a strong Fermi resonance coupling the energy levels belonging to the $1^1 \Leftrightarrow 2^2$ interacting states. It is clear that accounting or neglecting this resonance has a strong influence on the α_i^g. The Hamiltonian matrix to be diagonalized takes the following form:

$$
\begin{array}{c|cc}
 & |100> & |020> \\
\hline
|100> & \mathbf{H}_{100} & h_{100,020}^{Fermi} \\
|020> & h_{100,020}^{Fermi} & \mathbf{H}_{020}
\end{array}
\tag{4.59}
$$

The vibration-rotation eigenstates are $|001, R>$ and $|020, R>$, (resp. $|001, R>_F$ and $|020, R>_F$), where R are short-hand notation for the rotational quantum numbers neglecting (resp. accounting for) the Fermi resonance and the first three digits are the values of the vibrational quantum numbers v_1, v_2, v_3.

The matrix given in Equation 4.59 is diagonalized by the following unitary transformation:

$$|100, R>_F = a|100, R> + b|020, R> \tag{4.60a}$$

$$|020, R>_F = -b|100, R> + a|020, R> \tag{4.60b}$$

with the normalizing condition $a^2 + b^2 = 1$.

The rotational energy $|020, R>_F = -b|100,$, taking into account the Fermi resonance is given by taking an average of the rotational Hamiltonian over the wave function (Equation 4.60).

$$
\begin{aligned}
E(^F A_{100}, {}^F B_{100}, {}^F C_{100}) = & \, a^2 \left\langle 100, R | \mathbf{H}_{100} | 100, R \right\rangle \\
& + b^2 \left\langle 020, R | \mathbf{H}_{020} | 020, R \right\rangle
\end{aligned}
\tag{4.61}
$$

$$
\begin{aligned}
E(^F A_{100}, {}^F B_{100}, {}^F C_{100}) = & \, E(A_{100}, B_{100}, C_{100}) \\
& + b^2 \left[E(A_{020}, B_{020}, C_{020}) - E(A_{100}, B_{100}, C_{100}) \right]
\end{aligned}
\tag{4.62}
$$

A similar equation is obtained for the levels $|020, R>$.

Now we can estimate the effect of this resonance on the rotational constants, assuming that the resonance is not too strong and that the rotational energy is a linear function of the rotational constants (which is true only to the first order for an asymmetric top).

TABLE 4.7

Experimental and Unperturbed α_{100}^{ξ} Constants (in Megahertz) for the (100) State of F_2O

	Uncorrected[a]	Corrected[b]
α_{100}^{A}	−439.35	38.55
α_{100}^{B}	72.08	65.77
α_{100}^{C}	39.92	7.18

Source: Morino, Y., and S. Saito. 1966. *J Mol Spectrosc* 19:435–53.

[a] Experimental uncorrected value.

[b] Unperturbed value corrected for the effect of the Fermi resonance.

$$^{F}B_{100}^{\xi} = B_{100}^{\xi} + b^2 (B_{020}^{\xi} - B_{100}^{\xi}) \tag{4.63a}$$

$$^{F}B_{020}^{\xi} = B_{020}^{\xi} - b^2 (B_{020}^{\xi} - B_{100}^{\xi}) \tag{4.63b}$$

with $\xi = a$, b, and c.

One has to keep in mind that the following approximate relation holds quite well:

$$^{F}B_{100}^{\xi} + {}^{F}B_{020}^{\xi} = B_{100}^{\xi} + B_{020}^{\xi} \tag{4.64}$$

The effect of the Fermi resonance is particularly strong for the constants α_{100}^{ξ} (see Table 4.7).

4.4.3 REDUCTION OF THE HAMILTONIAN IN CASE OF RESONANCES

The interaction operators for the Coriolis resonances and anharmonic resonances have to be reduced in the same way as the centrifugal distorted Hamiltonian [33] (see Section 4.2.2). The consequence is that some interaction parameters cannot be uniquely determined from the experimental data. This is particularly important when one tries to compare experimental interaction parameters with their values computed from a force field.

4.5 DETERMINATION OF THE ROTATIONAL CONSTANTS OF AN ISOLATED VIBRATIONAL STATE

4.5.1 GROUND-STATE ROTATIONAL CONSTANTS

If a molecule possesses a permanent dipole moment, the best method to determine rotational constants is rotational spectroscopy. However, if the molecule has no dipole moment, there is no observable pure rotational spectrum (although perturbation-allowed transitions may sometimes be observed, see Appendix IV.2 and IV.6.1). Furthermore, if a molecule is light or near-symmetric, the information furnished by rotational spectroscopy may be insufficient. In these cases, infrared or Raman spectroscopies are useful. It is indeed possible to record the rotational spectrum of a molecule without dipole moment by Raman spectroscopy.

Moreover, the method of ground-state combination differences (GSCDs) of rovibrational transitions is also quite useful. This method is based on the Ritz principle [39]. When two rovibrational transitions originating from the ground state arrive at the same upper level, the frequency difference of these two transitions only depends on the ground-state rotational constants. One drawback is that infrared measurements are significantly less accurate than microwave or millimeter measurements (~3–30 MHz to be compared with ~1–30 kHz in the microwave). For this reason, GSCDs are useful for molecules without dipole moment or to obtain levels with quantum numbers inaccessible to rotational spectroscopy. The particular case of the determination of the axial rotational constant of a symmetric-top molecule is discussed in Appendix IV.6.

4.5.2 EXCITED STATE ROTATIONAL CONSTANTS

Rotational spectroscopy is well-suited to determining the rotational constants of low-lying vibrational states, but the most general method is high-resolution infrared spectroscopy. The assignment of an infrared spectrum usually is done iteratively. An example is described in Appendix IV.7. The assignment of an isolated band is straightforward because, the predictions are reliable since, in the absence of perturbations, the upper-state energy levels are modeled accurately using a standard rotational Hamiltonian. However, assignments may be sometimes difficult, for example, in congested Q branches. This is also discussed in Appendix IV.7.

4.6 DETERMINATION OF THE ROTATIONAL CONSTANTS OF A PERTURBED VIBRATIONAL STATE

4.6.1 CASE OF WEAKLY PERTURBED STATES

A typical example where weakly perturbed states are encountered is sulfur dioxide, SO_2. It is a particularly interesting molecule because its structure can be obtained using only two rotational constants of one single isotopologue. Furthermore, it is one of the most studied molecules by high-resolution spectroscopy. Its equilibrium structure was first determined by Morino et al. [40]. The α-constants were derived from the analysis of the rotational spectra of all fundamental vibrations. Later, Saito [41] extended the analysis of the rotational spectra to overtone and combination states. He was thus able to determine all higher-order γ-constants and to show that their effect on the structure is quite small. He also pointed out the presence of weak Fermi resonances.

Subsequently, Morino et al. [42] recalculated the equilibrium structure using the recent spectroscopic data. They used a full set of γ-constants and they eliminated the small contributions of Fermi and Darling–Dennison resonances. This new structure

---

TABLE 4.8

Experimental Equilibrium Structures of SO$_2$

r_e(SO)	∠(OSO)	Reference
1.4308(2)	119.32(3)	40[a]
1.43080(1)	119.329(2)	42
1.430782(15)	119.3297(30)	44[b]
1.4307932(40)	119.328985(24)	45

Note: Distance is in Å, angle in °.

[a] $A_e = 60485.32$ MHz, $B_e = 10359.51$ MHz, $C_e = 8845.82$ MHz.

[b] $A_e = 60502.21(26)$ MHz, $B_e = 10359.234(38)$ MHz, $C_e = 8845.105(40)$ MHz.

---

was in excellent agreement with the previous one, but was one order of magnitude more precise (see Table 4.8). The accuracy of this structure was confirmed by high-level ab initio calculations [43]. Particularly, the calculated α-constants were found in good agreement with the experimental ones. In references [44,45] the infrared spectra of $^{32}S^{16}O_2$ and $^{34}S^{16}O_2$ were reanalyzed and a new set of α- and γ-constants were determined taking into account the resonances coupling the 1^1, 2^2 and 3^1 interacting energy levels (see Appendix IV.8 for more details). They were thus able to obtain a refined structure of both isotopologues.

It is striking that all structures reported in Table 4.8 are in excellent agreement, the only improvement from 1964 to 2009 being a significant increase in precision. The great stability of the structure is mainly due to the following two factors:

1. The γ-constants and the anharmonic resonances are weak. Table 4.8 shows that the equilibrium rotational constants do not vary much from 1964 to 1993.
2. The structure is determined using only the rotational constants of one isotopologue, the system of equations is thus well conditioned (see Sections 5.3.4 and 5.5.4.3).

4.6.2 CASE OF SIGNIFICANTLY PERTURBED STATES

Carbonyl sulfide, OCS, is a linear molecule whose structure is completely defined by two parameters as SO$_2$. There are, however, two important differences:

1. The molecule is linear; thus, the rotational constants of two isotopologues are necessary to define the structure.
2. There is a Fermi resonance significantly affecting the value of the equilibrium rotational constant.

The equilibrium structure was first determined by Morino and Matsumura [46]. They measured the rotational spectra of various vibrationally excited states of OC^{32}S and OC^{34}S, including all fundamental states. They found at least two nonnegligible

Fermi resonances, one involving $v_1 = 1$ ($v_1 = 2062$ cm^{-1}) and $v_2 = 4^0$ ($v_2 = 520$ cm^{-1}) and another one between $v_3 = 1$ ($v_3 = 859$ cm^{-1}) and $v_2 = 2^0$. However, as the molecule is linear, the rotational energy linearly depends on the rotational constant and the sum rule (i.e., the trace of the energy matrix) holds between perturbed (experimental) and unperturbed rotational constants (with superscript 0), as follows:

$$B_{100} + B_{001} + B_{02^00} = B_{100}^0 + B_{001}^0 + B_{02^00}^0 \tag{4.65}$$

Thus, the equilibrium rotational constant may be simply determined as follows:

$$\begin{aligned} B_e &= B_0 + \frac{1}{2}\left[(B_0 - B_{100}) + (B_0 - B_{001}) + (B_0 - B_{02^00})\right] \\ &= \frac{1}{2}\left[5B_0 - B_{100} - B_{001} - B_{02^00}\right] \end{aligned} \tag{4.66}$$

Using the equilibrium rotational constants of OC^{32}S and OC^{34}S, Morino and Matsumura were able to obtain a rather accurate equilibrium structure (see Table 4.9). On the other hand, if the Fermi resonances are neglected and the equilibrium rotational constants are calculated from the α-constants derived from the three fundamental states,

$$\begin{aligned} B_e &= B_0 + \frac{1}{2}\left[(B_0 - B_{100}) + (B_0 - B_{001}) + 2(B_0 - B_{01^10})\right] \\ &= B_0 + \frac{1}{2}\left[\alpha_{100} + \alpha_{001} + 2\alpha_{010}\right] \end{aligned} \tag{4.67}$$

the equilibrium structure is significantly different; $r_e(OC) = 1.161$ Å and $r_e(CS) = 1.558$ Å (see Table 4.9).

TABLE 4.9

Experimental Equilibrium Structures of OCS

$r_e(OC)$	$r_e(CS)$	$r_e(OS)$	Reference
1.1572(20)	1.5606(20)	2.7178(5)	46[a]
1.1545(2)	1.5630(2)		47
1.15431(24)	1.56283(37)	2.71714(13)	48
1.155386(21)	1.562021(17)		64
1.156064(15)	1.561475	2.717539	65
1.15617(14)	1.56140(14)	2.717560(29)	49[b]
0.0029	0.0024	0.0007	Range

Note: Distance is in Å.

[a] $B_e(OCS) = 6099.225$ MHz or 6098.167 MHz if the Fermi resonances are neglected.

[b] $B_e(OCS) = 6099.496(7)$ MHz.

Morino and Nakagawa [47] determined an experimental anharmonic force field using the available spectroscopic data. They were thus able to correct the ground-state rotational constants of eight isotopologues and to derive a new equilibrium structure, which was found in rather good agreement with the previous one. However, the data (eight data for two parameters) are not fully compatible indicating an important role of higher-order effects.

Later, Maki and Johnson [48] determined the experimental equilibrium rotational constants of $O^{13}CS$ and ^{18}OCS by rotational spectroscopy using the same method as Morino and Matsumura [46]. They confirmed the existence of discrepancies due to the omission of weak resonances and the γ-constants. The problem was solved by Lahaye et al. [49], who performed a global analysis (i.e., taking into account all possible resonances) and determined all γ-constants for four isotopologues. These authors were thus able to obtain a very precise equilibrium structure. It is interesting to note that in all cases, $r_e(OS) = r_e(OC) + r_e(CS)$ is accurately determined whereas the errors on $r_e(OC)$ and $r_e(CS)$ nearly compensate each other, which is an indication of strong correlation (see Section 2.2.2.2). Indeed, Morino and Nakagawa [47] found a correlation coefficient of -0.9998 between $r_e(OC)$ and $r_e(CS)$ (with a large condition number, $\kappa = 204$). The ill-conditioning problem is confirmed by the fact that the values of the equilibrium rotational constants do not vary much from 1967 to 1987. An example of another "difficult" linear molecule, HCO^+, is illustrated in Appendix IV.5, and the case of HCP is discussed in Appendix IV.9.

4.6.3 Case of Strongly Perturbed States

4.6.3.1 $v_6 = 1$ State of *Trans*-HCOOH

When the energy levels of a fundamental state f are resonating with those from at least one (usually dark) overtone or combination state labeled as d, it is clear that the values obtained for the A_f, B_f, and C_f rotational constants of the fundamental state f are at least partly correlated with those $(A_d, B_d,$ and $C_d)$ of the resonating state d, depending on the quality of the theoretical model used to account for these resonances. In this case, the α_f^ξ are perturbed and the effects of the different resonances are no more canceled in the sum $\sum_i \alpha_i^\xi$ and the derived α_f^ξ constants cannot be used to conclusively determine the equilibrium rotational constants, although these constants permit the reproduction of the rovibrational data.

The resonances affecting the strong v_6 band (C–O stretch) located at 1104.9 cm^{-1} of *trans*-HCOOH is a good example of this problem. The v_8 weak band (out of plane CH bend) is located at 1033 cm^{-1}, and there exist weak A- and B-type resonances coupling the energy levels of 6^1 with those of the 8^1 vibrational state. Accounting for these resonances, most (but not all) of the existing experimental data for the 6^1 and 8^1 energy levels were reproduced within their experimental accuracy [50], and in this way, the sums $\alpha_6^\xi + \alpha_8^\xi = 446.835$, 105.387, and 60.567 MHz for $\xi = A, B,$ and C, respectively, are in good agreement with the ab initio predictions, which are 404.30, 99.48, and 57.32 MHz, respectively [51].

However, some series of levels of the 6^1 state are not reproduced correctly by the calculation. For this reason, Baskakov et al. [52] performed a new and more

TABLE 4.10

$h_{v,v'}^{0\,\text{Fermi}}$ or $h_{v,v'}^{0\,\text{Darling–Dennison}}$ **Parameters and Energy Differences ($\Delta\omega$) Leading to Fermi and Darling–Dennison Resonances for *Trans*-Formic Acid**

	$h_{v,v'}^{0\,\text{Fermi}}$ or $h_{v,v'}^{0\,\text{Darling Dennison}}$		
	Experimental[a]	Ab initio[b]	$\Delta\omega$
$5^1 \Leftrightarrow 9^2$	41.53	31.77	−32.12
$4^1 \Leftrightarrow 9^2$	27.90	−23.52	66.23
$6^1 \Leftrightarrow 9^2$	0	−7.74	−218.80
$4^1 \Leftrightarrow 7^2$	0	−1.58	160.93
$6^1 \Leftrightarrow 7^2$	−25.04	1.46	−124.10
$5^1 \Leftrightarrow 7^2$	−2.41	−0.03	62.58
$7^2 \Leftrightarrow 9^2$	0	−8.36	−94.70

a   Reference [52].
b   Reference [51].

extended analysis in which the 6^1 state is involved in a complicated scheme of resonances with six other states, some of them dark (see Appendix IV.1). Although this new analysis is much more sophisticated, some of the derived parameters are unsatisfactory. Indeed, according to the values of the interacting parameters, several resonances that are considered to be very weak (resp. very strong) by the experimental study are predicted to be very strong (resp. very weak) by the ab initio calculations. This is shown in Table 4.10, which compares the very different values obtained for the first-order parameters ($h_{v,v'}^{0,\text{Anh}}$, see Equation 4.58) involved in the expansion of the Fermi type or Darling–Dennison resonance operators. For example, the experimental and ab initio values of $h_{6,77}^{0\,\text{Fermi}}$ are −25.04 and 1.46 cm^{-1}, respectively.

According to the detailed line-intensity study performed by Vander Auwera et al. [53] in the 10–μm spectral range, which covers the v_6 and several interacting bands, the $2v_7$ band is extremely weak, indicating that the mixing between the $6^1 \Leftrightarrow 7^2$ wave functions, and therefore the $6^1 \Leftrightarrow 7^2$ Fermi resonance is extremely weak, in agreement with the ab initio prediction. The consequence is that the new experimental $\alpha_6^\xi + \alpha_8^\xi$ values are not compatible with the ab initio studies. For example, Baskakov [52] gives $\alpha_6^A + \alpha_8^A = -823.7$ MHz instead of +446.835 MHz as calculated ab initio. This conclusion is strengthened by the fact that several centrifugal distortion constants do not have the right order of magnitude. This example is typical of the problems that are encountered when a fundamental state is involved in many resonances. The available experimental data are not sufficient to reliably determine all the interacting parameters and the ab initio predictions are not accurate enough to fix them.

TABLE 4.11
$\sum_i \alpha_i^\xi / 2$ for H_2CO (in Megahertz)

ξ	Experimental[a]	Ab initio[b]
A	3757.8	3244.2
B	99.2	157.6
C	274.1	301.8

[a] Reference [54,55].
[b] Reference [66].

4.6.3.2 Equilibrium Rotational Constants of Formaldehyde

The situation described in Section 4.6.3.1 is not exceptional even for very simple molecules. For instance, a full set of experimental α-constants has been recently determined for H_2CO [54,55] but, as the spectra are extremely perturbed, the experimental $\sum_i \alpha_i^\xi$ do not appear as accurate when they are compared with the values calculated from a very accurate ab initio anharmonic force field (see Table 4.11).

Furthermore, the states $v_1 = 1$ and $v_5 = 1$ of D_2CO are in interaction with many dark states. For this reason, the unperturbed rotational constants of these two states are not yet known with an acceptable accuracy. In conclusion, there is not yet enough experimental information to determine an experimental equilibrium structure for formaldehyde.

4.6.3.3 Equilibrium Structure of Fluoromethane, CH_3F

The molecule CH_3F of C_{3v} symmetry has six fundamental vibrations, three nondegenerate of symmetry A_1: $v_1 = 2919.6$ cm^{-1}, $v_2 = 1459.4$ cm^{-1}, and $v_3 = 1048.6$ cm^{-1} and three doubly degenerate of symmetry E: $v_4 = 2999.0$ cm^{-1}, $v_5 = 1467.8$ cm^{-1}, and $v_6 = 1182.7$ cm^{-1}. From the energy difference between the corresponding states, these vibrations are usually arranged in three independent dyads (sets of 2 interacting states, see Appendix IV.1): v_1/v_4, v_2/v_5, and v_3/v_6 with a Coriolis interaction inside each dyad. This allows one to obtain satisfactory fits provided that some transitions are excluded (which is a common procedure in infrared spectroscopy). However, a careful analysis leads to the conclusion that the two dyads v_2/v_5 and v_3/v_6 are not isolated from each other but have to be analyzed as a tetrad [56]. The problem is that the two methods give very different α-constants (see Table 4.12). High-level ab initio calculations [57] confirm that the tetrad analysis gives rotational constants that are more accurate. This example shows that it is sometimes extremely difficult to determine whether the analysis of a spectrum and the derived constants are reliable.

4.7 NONORTHORHOMBIC ROTATIONAL HAMILTONIAN

According to symmetry considerations, the form of the rotational operator given in Equation 4.34 is, in principle, suitable only for orthorhombic molecules. As compared to Equation 4.34, the rotational Hamiltonian for a C_s-type molecule with a

TABLE 4.12
α-Constants for the Tetrad $\nu_2/\nu_5/\nu_3/\nu_6$ of CH_3F (in Megahertz)

ξ		α_2^ξ	α_3^ξ	α_5^ξ	α_6^ξ	S^a
Dyad analysis	a	−688.3	294.2	−1336.8	−477.1	−2010.9
	b	64.4	338.6	57.1	116.6	375.3
Tetrad analysis	a	−693.1	291.1	1574.5	−728.0	645.5
	b	124.1	261.3	−11.3	85.7	267.1

Source: Papousek, D., P. Pracna, M. Winnewisser, S. Klee, and J. Demaison. 1999. *J Mol Spectrosc* 196:319–23.

[a] $S = \dfrac{(\alpha_2^\xi + \alpha_3^\xi)}{2 + \alpha_5^\xi + \alpha_6^\xi}$.

(x, z) planar structure should include additional nonorthorhombic terms (one quadratic and three quartic terms) as follows:

$$H_v^{NO} = h_v^{1NO}\left[J_x, J_z\right]_+ + h_v^{2NO}\mathbf{J}^2\left[J_x, J_z\right]_+ + h_v^{3NO}\left[J_z^2,\left[J_x, J_z\right]_+\right]_+ \\ + h_v^{4NO}\left[J_{xy}^2,\left[J_x, J_z\right]_+\right]_+$$

(4.68)

NO meaning nonorthorhombic. However, Watson [21] has shown that these nonorthorhombic terms can, in principle, also be removed by suitable contact transformations, but it is sometimes convenient to retain these nonorthorhombic terms, that is, principally to use an axis system that is not the PAS. Indeed, when there is a Coriolis-like interaction, there is a set of axes, different from the PAS, which minimizes this interaction [58].

In the presence of a strong interaction, the standard Watson Hamiltonian in its PAS form may not be able to reproduce the energy levels without using an unreasonably number of parameters. In this case, the parameters may have a limited physical meaning. Such a situation mainly happens in the following two cases:

1. When there is a strong Coriolis interaction as for $7^1 \Leftrightarrow 9^1$ interacting states of *trans*-formic acid (see Section 4.4.1)
2. When there is a large amplitude motion

Obviously, the determined rotational constants have to be transformed into the PAS before using them for a structure determination. A typical example is the $\nu_9 = 1$ state (torsion) of nitric acid, HNO_3. This state is well isolated, but the rotational spectrum is nevertheless perturbed by the large-amplitude motion of the OH group. Petkie et al. [59] have analyzed the rotational spectrum of this state using a standard Watson Hamiltonian as well as a nonorthorhombic Hamiltonian. The results

TABLE 4.13
Analysis of the Rotational Spectrum of the $v_9 = 1$ State of HNO_3 (in Megahertz)

	Watson	Nonorthorhombic	Converted[a]
A	12998.94120(100)	12961.8487(1907)	12998.9991
B	12015.16454(85)	12052.2569(1907)	12015.1067
C	6255.23197(62)	6255.23036(29)	6255.2304
D_{ab}[b]		−187.3874(4628)	
s[c]	0.115	0.068	

Source: Petkie, D. T., P. Helminger, R. A. H. Butler, S. Albert, and F. C. De Lucia. 2003. *J Mol Spectrosc* 218:127–30.

[a] Nonorthorhombic rotational constants converted in the principal axis system. It is not possible to give precise values of the standard deviation because the correlation matrix is not given in the original work.

[b] D_{ab} is the usual notation for the nondiagonal term $H_v^{1\ NO}$.

[c] Standard deviation of the fit (in MHz).

are summarized in Table 4.13. The nonorthorhombic Hamiltonian gives a better fit as anticipated, but the derived rotational constants are less precise by one order of magnitude.

4.8 CONCLUSIONS

Rovibrational spectroscopy is by far the best method to determine very accurate ground-state rotational constants. It can also obtain equilibrium rotational constants, but, in this latter case, an extremely thorough analysis is required to avoid contamination of these constants by resonances. Such an analysis is often difficult and time-consuming and, in the end, one is never sure that all resonances have been correctly taken into account, except perhaps when the analysis is assisted by ab initio calculations. This is the main weakness of the method, because to derive the equilibrium structure of a polyatomic molecule, very accurate equilibrium rotational constants of several isotopologues are required, at least in most cases (the main exception is the XY_n molecules for which it is possible to determine the structure using the rotational constants of one single isotopologue). Unfortunately, an isotopic substitution leads generally to only a small variation of the rotational constants, the consequence being that the system of normal equations is ill-conditioned (see Chapter 2). Another weak point worth noting is that the effect of the neglected (or mishandled) resonances is not taken into account in the calculation of the uncertainties. For this reason, the precision of the structural parameters is often too optimistic. However, a thorough analysis of all significant resonances has been conducted for the isotopologues of some simple molecules (as OCS, see Section 4.6.2) and, in this case, the derived experimental equilibrium structure is accurate.

REFERENCES

1. Amat, G., H. H. Nielsen, and G. Tarrago. 1971. *Rotation–Vibration of Polyatomic Molecules*. New York: Marcel Dekker.
2. Allen, H. C. J., and P. C. Cross. 1963. *Molecular Vib-Rotors*. New York: Wiley.
3. Gordy, W., and R. L. Cook. 1984. *Microwave Molecular Spectra*. New York: Wiley.
4. Herzberg, G. 1991. *Molecular Spectra and Molecular Structure. II: Infrared and Raman Spectra*. Malabar, FL: Krieger.
5. Herzberg, G. 1991. *Molecular Spectra and Molecular Structure. III: Electronic Spectra and Electronic Structure of Polyatomic Molecules*. Malabar, FL: Krieger.
6. Papousek, D., and M. R. Aliev. 1982. *Molecular Vibrational-Rotational Spectra*. Amsterdam: Elsevier.
7. Townes, C. H., and A. L. Schawlow. 1975. *Microwave Spectroscopy*. New York: Dover.
8. Wilson, E. B., J. C. Decius, and P. C. Cross. 1980. *Molecular Vibrations: The Theory of Infrared and Raman Vibrational Spectra*. New York: Dover.
9. Bernath, P. F. 2005. *Spectra of Atoms and Molecules*. Oxford: Oxford University Press.
10. Bunker, P. R., and P. Jensen. 1998. *Molecular Symmetry and Spectroscopy*. 2nd ed. Ottawa, ON: NRC Press.
11. Califano, S. 1976. *Vibrational States*. New York: Wiley.
12. Demtröder, W. 2006. *Molecular Physics: Theoretical Principles and Experimental Methods*. New York: Wiley.
13. Duxbury, G. 2000. *Infrared Vibration-Rotation Spectroscopy: From Free Radicals to the Infrared Sky*. New York: Wiley.
14. Hollas, J. M. 1998. *Modern Spectroscopy*. New York: Dover.
15. Kroto, H. W. 1992. *Molecular Rotation Spectra*. New York: Dover.
16. Lafferty, W. J. 1997. Some aspects of high resolution infrared spectroscopy. In *Lectures on Molecular Physics*, ed. J. L. Doménech and V. J. Herrero. Madrid: CSIC.
17. Atkins, P. W. 1978. *Physical Chemistry*. Oxford: Oxford University Press.
18. Aliev, M. R., and J. K. G. Watson. 1985. Higher order effects in the vibration rotation spectra of semi-rigid molecules. In *Molecular Spectroscopy; Modern Research*, ed. K. N. Rao, vol. III, 1–67. Orlando: Academic Press.
19. Watson, J. K. G. 1968. *Mol Phys* 19:479–90.
20. Sarka, K., and J. Demaison. 2000. Perturbation theory, effective Hamiltonians and force constants. In *Computational Molecular Spectroscopy*, ed. P. Jensen and P. R. Bunker, 255–303. New York: Wiley.
21. Watson, J. K. G. 1977. Aspects of quartic and sextic centrifugal effects on rotational energy levels. In *Vibrational Spectra and Structure*, ed. J. R. Durig, vol. 6, 1–89. Amsterdam: Elsevier.
22. Sarka, K., D. Papousek, J. Demaison, H. Mäder, and H. Harder. 1997. Rotational spectra of symmetric top molecules: Correlation-free reduced forms of Hamiltonians, advances in measuring techniques, and determination of molecular parameters from experimental data. In *Vibration-Rotational Spectroscopy and Molecular Dynamics*, ed. D. Papousek, 116–238. Singapore: World Scientific Publishing.
23. Mekhtiev, M. A., and J. T. Hougen. 2000. *J Mol Spectrosc* 199:284–301.
24. Watson, J. K. G. 1967. *J Chem Phys* 46:1935–9.
25. Cohen, E. A., and W. Lewis-Bevan. 1991. *J Mol Spectrosc* 148:378–84.
26. Flygare, W. H. 1974. *Chem Rev* 74:653–87.
27. Gauss, J., K. Ruud, and T. Helgaker. 1996. *J Chem Phys* 105:2804–12.
28. Meerts, W. L., S. Stolte, and A. Dymanus. 1977. *Chem Phys* 119:467–72.
29. Kivelson, D., and E. B. Wilson. 1952. *J Chem Phys* 20:1575–9.
30. Dowling, J. M. 1961. *J Mol Spectrosc* 6:550–3.
31. Jahn, H. A. 1939. *Phys Rev* 56:680–3.

32. Flaud, J. -M., and R. Bacis. 1998. *Spectrochim Acta* 54A:3–16.
33. Perevalov, V. I., and V. G. Tyuterev. 1982. *J Mol Spectrosc* 96:56–76.
34. Deroche, J. C., J. Kauppinen, and E. Kyrö. 1979. *J Mol Spectrosc* 73:379–94.
35. Perrin, A., J. -M. Flaud, B. Bakri, J. Demaison, O. L. Baskakov, S. V. Sirota, M. Herman, and J. Vander Auwera. 2002. *J Mol Spectrosc* 216:203–13.
36. Baskakov, O. I., I. A. Markov, E. A. Alekseev, R. A. Motiyenko, J. Lohilahti, V. -M. Horneman, B. P. Winnewisser, I. R. Medvedev, and F. C. De Lucia. 2006. *J Mol Struct* 795:54–77.
37. Flaud, J. -M., C. Camy-Peyret, R. A. Toth. 1981. *Water Vapour Line Parameters From Microwave to Medium Infrared*. Oxford: Pergamon Press.
38. Morino, Y., and S. Saito. 1966. *J Mol Spectrosc* 19:435–53.
39. Ritz, W. 1911. *Gesammelte Werke—Oeuvres*. Gauthier-Villars, Paris: Societe suisse de physique.
40. Morino, Y., Y. Kikuchi, S. Saito, and E. Hirota. 1964. *J Mol Spectrosc* 13:95–118.
41. Saito, S. 1969. *J Mol Spectrosc* 30:1–16.
42. Morino, Y., M. Tanimoto, and S. Saito. 1988. *Acta Chem Scand A* 42:346–51.
43. Martin, J. M. L. 1998. *J Chem Phys* 108:2791–800.
44. Flaud, J. -M., and W. J. Lafferty. 1993. *J Mol Spectrosc* 161:396–402.
45. Lafferty, W. J., J. -M. Flaud, E. H. Abib Ngom, and R. L. Sams. 2009. *J Mol Spectrosc* 253:51–4.
46. Morino, Y., and C. Matsumura. 1967. *Bull Chem Soc Jpn* 40:1095–100.
47. Morino, Y., and T. Nakagawa. 1968. *J Mol Spectrosc* 26:496–523.
48. Maki, A. G., D. R. Johnson. 1973. *J Mol Spectrosc* 47:226–33.
49. Lahaye, J. -G., R. Vandenhaute, and A. Fayt. 1987. *J Mol Spectrosc* 123:48–83.
50. Baskakov, O. I., and J. Demaison. 2002. *J Mol Spectrosc* 211:262–72.
51. Demaison, J., M. Herman, and J. Liévin. 2007. *J Chem Phys* 126:164305/1–9.
52. Baskakov, O. L., E. A. Alekseev, R. A. Motiyenko, J. Lohilahti, V. -M. Horneman, S. Alanko, B. P. Winnewisser, I. R. Medvedev, F. C. De Lucia. 2006. *J Mol Spectrosc* 240:188–201.
53. Vander Auwera, J., K. Didriche, A. Perrin, and F. Keller. 2007. *J Chem Phys* 126:124311/1–9.
54. Margulès, L., A. Perrin, R. Janecková, S. Bailleux, C. P. Endres, T. Giesen, and S. Schlemmer. 2009. *Can J Phys* 87:425–55.
55. Perrin, A., A. Valentin, and L. Daumont. 2006. *J Mol Struct* 780–1:28–44.
56. Papousek, D., P. Pracna, M. Winnewisser, S. Klee, and J. Demaison. 1999. *J Mol Spectrosc* 196:319–23.
57. Demaison, J., J. Breidung, W. Thiel, and D. Papousek. 1999. *Struct Chem* 10:129–33.
58. Pickett, H. M., E. A. Cohen, and J. S. Margolis. 1985. *J Mol Spectrosc* 110:186–214.
59. Petkie, D. T., P. Helminger, R. A. H. Butler, S. Albert, and F. C. De Lucia. 2003. *J Mol Spectrosc* 218:127–30.
60. Margulès, L., R. A. Motiyenko, E. A. Alekseev, and J. Demaison. 2010. *J Mol Spectrosc* 260:23–9.
61. Demaison, J. 2007. *Mol Phys* 105:3109–38.
62. Müller, H. S. P., H. Klein, S. P. Belov, G. Winnewisser, I. Morino, K. M. T. Yamada, and S. Saito. 1999. *J Mol Spectrosc* 195:177–84.
63. Johns, J. W. C., L. Nemes, K. M. T. Yamada, T. Y. Wang, J. L. Doménech, J. Santos, P. Cancio, D. Bermejo, J. Ortigoso, and R. Escribano. 1992. *J Mol Spectrosc* 156:501–3.
64. Foord, A., J. G. Smith, and D. H. Whiffen. 1975. *Mol Phys* 29:1685–704.
65. Whiffen, D. H. 1980. *Mol Phys* 39:391–405.
66. Puzzarini, C., M. Heckert, and J. Gauss. 2008. *J Chem Phys* 128:194108/1–7.
67. http://icb.u-bourgogne.fr/OMR/SMA/SHTDS/

CONTENTS OF APPENDIX IV (ON CD-ROM)

5 Determination of the Structural Parameters from the Inertial Moments

Heinz Dieter Rudolph and Jean Demaison

CONTENTS

5.1 INTRODUCTION

Earlier reviews on the state of experiment and theory for the determination of molecular structure by means of rotational spectroscopy have been mentioned in the book *Microwave Molecular Spectra* [1], which offers a great deal of information on the traditional methods for the determination of the structure of molecules from their inertial moments. More recent and rather detailed reviews on the work done in this field have been published in *Advances in Molecular Structure Research* by Rudolph [2] and by Demaison, Wlodarczak, and Rudolph [3]. The most recent report appears to be a contribution by Groner [4], who reviews the traditional methods as well as the still-developing modern efforts, about which the review is particularly detailed. This chapter is intended not only to present and explain as far as possible the more recent methods but also to direct the reader's attention to the variety and diversity of ideas that have been brought to bear on this task even if their application has remained limited for different reasons.

5.2 RELATIONSHIP BETWEEN STRUCTURE AND MOMENTS OF INERTIA

5.2.1 Structural Data to Be Determined

The basic set of molecular structural data that the application of molecular spectroscopy strives to determine is the minimum set of p independent internal bond coordinates, $\beta_j, j = 1, \ldots, p$, required to build up the molecule (distances between two atoms, angles defined by connecting three atoms, and dihedral angles formed by four connected atoms for torsional rotation about the middle connection). This set is also known as the Z-matrix of the molecule. Since these "bonds" need not necessarily correspond to real chemical bonds, this set of independent internal bond coordinates is not unique. It may occasionally be convenient to also use internal coordinates that do not correspond to actual chemical bonds but are better adapted to the problem (annular molecules are an example case). Massless (often called "dummy") "atoms" are allowed if they help describe the interatomic geometrical structure. These internal bond coordinates are the parameters to be determined by the evaluation of the spectra. The set of independent internal bond parameters is assumed identical for all isotopologues of a molecule (a possible exception for X–H bonds is discussed in Section 5.5.4.2).

The positions of the n atoms can also be expressed in Cartesian coordinates as the vectors $\mathbf{r}_i = (x_i, y_i, z_i), i = 1, \ldots, n$, where n is the number of atoms in the molecule [5]. When the origin and the orientation of the axes have been fixed with respect to the molecular structure, the number of independent coordinates required to describe the rigid molecule is generally $3n - 6$ (e.g., $3 \times n$ Cartesian coordinates -6 constraints; constraints include 3 first-moment equations [Equation 5.4] and 3 second-moment equations [Equation 5.8]). The number is less for specific configurations (planar, linear) and when symmetries are present. The complete result of a structural investigation (by least-squares methods, see Chapter 2) should always be understood as a $p \times 1$ vector $\hat{\boldsymbol{\beta}}$ of the p required internal bond coordinates $\beta_j, j = 1, \ldots, p$ plus their covariance matrix $\Theta_{\hat{\beta}}$ because coordinates $\hat{\beta}$ when obtained by experimental methods are necessarily afflicted by errors and correlations.

5.2.2   BASIC MODEL

The simplest model for a rotating molecule and its isotopologues is that of a rotating rigid body (rigid top) consisting of points with known atomic mass fixed in a geometry that does not change when the mass of one or more of the atomic mass points is changed. In mechanics, the inertial properties of a rotating rigid body are fully described by its inertial moment tensor $\mathbf{I}$. The inertial moment (also known as the moment of inertia) $\mathbf{I}$ of a rotating rigid body is a measure of its resistance to changes in its rotation rate. The subsequent equations can be somewhat simplified if, instead of $\mathbf{I}$, the planar inertial moment tensor $\mathbf{P}$ is employed, which is equivalent to $\mathbf{I}$. The total mass of the molecule is $M = \sum\limits_{i=1}^{n} m_i$. Let $\mathbf{r}''_i = (x''_i, y''_i, z''_i)$ be the vector to the mass point i with mass m_i in an arbitrary Cartesian coordinate system (we start with a doubly primed system to successively arrive at a "neat" unprimed system for the basic Equation 5.6). The position of the center of mass is given by

$$\mathbf{r}''_{cm} = \frac{1}{M} \sum_{i=1}^{n} m_i \mathbf{r}''_i \tag{5.1}$$

The planar inertial moment of this isotopologue can be compactly written as the 3×3 dyadic (i.e., a square matrix obtained by multiplying the column vector $\mathbf{r}$ by the row vector $\mathbf{r}^T$, where ^{T} means transpose) as follows:

$$\mathbf{P}'' = \sum_{i=1}^{n} m_i \mathbf{r}''_i \mathbf{r}''^T_i - M \mathbf{r}''_{cm} \mathbf{r}''^T_{cm} \tag{5.2}$$

Using Equation 5.1, the planar moment is explicitly

$$\mathbf{P}'' = \sum_{i=1}^{n} m_i \mathbf{r}''_i \mathbf{r}''^T_i - \frac{1}{M} \sum_{i=1}^{n} \sum_{j=1}^{n} m_i m_j \mathbf{r}''_i \mathbf{r}''^T_j \tag{5.3}$$

Let the coordinate system $\mathbf{r}'$ refer to the center of mass as origin, $\mathbf{r}'_i = \mathbf{r}''_i - \mathbf{r}''_{cm}$, all i,

$$\mathbf{r}'_{cm} = \frac{1}{M} \sum_{i=1}^{n} m_i \mathbf{r}'_i = 0 \tag{5.4}$$

Then the coordinates $\mathbf{r}'$ satisfy the center-of-mass conditions or first-moment equations (Equation 5.4), the second term of Equations 5.2 and 5.3 vanishes, and the planar moment is represented by a symmetric, in general nondiagonal, matrix, as follows:

$$\mathbf{P'} = \sum_{i=1}^{n} m_i \mathbf{r}'_i \mathbf{r}'^{\mathrm{T}}_i = \begin{pmatrix} \sum_i m_i x_i'^2 & \sum_i m_i x_i' y_i' & \sum_i m_i x_i' z_i' \\ & \sum_i m_i y_i'^2 & \sum_i m_i y_i' z_i' \\ \text{symm.} & & \sum_i m_i z_i'^2 \end{pmatrix} \tag{5.5}$$

The elements depend on the orientation of the reference axes of the coordinate system. These reference axes can be rigidly rotated about the origin by an orthogonal transformation $\mathbf{T}$ (with $\mathbf{T}^{\mathrm{T}} = \mathbf{T}^{-1}$); let $\mathbf{r}_i = \mathbf{T}\mathbf{r}'_i$ (all i), and then transposed as $\mathbf{r}^{\mathrm{T}}_i = \mathbf{r}'^{\mathrm{T}}_i \mathbf{T}^{\mathrm{T}}$. Accordingly, the transformed planar moment tensor is

$$\mathbf{P} = \mathbf{T}\mathbf{P'}\mathbf{T}^{\mathrm{T}} = \sum_{i=1}^{n} m_i \mathbf{r}_i \mathbf{r}^{\mathrm{T}}_i \tag{5.6}$$

The most important orthogonal transformation is the principal axis transformation of the moment tensor, which rotates the general center-of-mass system into the eigensystem of $\mathbf{P}$, called the *principal axis system* (PAS), and hence diagonalizes the tensor $\mathbf{P}$.

The diagonalized tensor showing the eigenvalues, the principal planar moments, is explicitly given in Equation 5.7, where the position vectors are now called $\mathbf{r}_i = (a_i, b_i, c_i)$, $i = 1, \ldots, n$,

$$\mathbf{P}^{\mathrm{PAS}} = \begin{pmatrix} \sum_i m_i a_i^2 & 0 & 0 \\ 0 & \sum_i m_i b_i^2 & 0 \\ 0 & 0 & \sum_i m_i c_i^2 \end{pmatrix} \tag{5.7}$$

The vanishing off-diagonal elements in the PAS are known as the second-moment equations:

$$\sum_i m_i a_i b_i = 0, \quad \sum_i m_i b_i c_i = 0, \quad \sum_i m_i c_i a_i = 0 \tag{5.8}$$

The diagonal elements of Equation 5.7, the eigenvalues, are denoted by $\sum_i m_i a_i^2 = P_a$, a, b, c cyclical.

The coordinates x_i, y_i, z_i are replaced by a_i, b_i, c_i, which are ordered according to the relative magnitudes of the eigenvalues. By convention, we have $P_a \geq P_b \geq P_c$.

The planar moment tensor **P** and the inertial moment tensor **I** are related by

$$\mathbf{P} = \frac{1}{2}\,\mathrm{tr}(\mathbf{I})\cdot\mathbf{1} - \mathbf{I} \quad \mathbf{I} = \mathrm{tr}(\mathbf{P})\cdot\mathbf{1} - \mathbf{P} \tag{5.9}$$

where **1** is the unit matrix and the scalar $\mathrm{tr}(\mathbf{I})$ is the trace of the tensor **I**. This relation holds no matter which (common) reference system has been used for **I** and **P**. Equation 5.9 allows us to immediately transform **P′** of Equation 5.5 into the inertial moment tensor **I′**, both in the center-of-mass (but non-eigen-) system **r′**, as

$$\mathbf{I}' = \begin{pmatrix} \sum_i m_i\left(y_i'^2 + z_i'^2\right) & -\sum_i m_i x_i' y_i' & -\sum_i m_i x_i' z_i' \\[2mm] & \sum_i m_i\left(z_i'^2 + x_i'^2\right) & -\sum_i m_i y_i' z_i' \\[2mm] \text{symm.} & & \sum_i m_i\left(x_i'^2 + y_i'^2\right) z_i'^2 \end{pmatrix} \tag{5.10}$$

From Equation 5.9, it is also clear that if one tensor is represented in its eigensystem (**P**^{PAS}; Equation 5.7), then so is the other tensor,

$$\mathbf{I}^{\mathrm{PAS}} = \begin{pmatrix} \sum_i m_i\left(b_i^2 + c_i^2\right) & 0 & 0 \\[2mm] & \sum_i m_i\left(c_i^2 + a_i^2\right) & 0 \\[2mm] \text{symm.} & & \sum_i m_i\left(a_i^2 + b_i^2\right) \end{pmatrix} \tag{5.11}$$

The relations between the eigenvalues of the moments, Equations 5.7 and 5.11, shown here as transformations between vectors are

$$\begin{pmatrix} P_a \\ P_b \\ P_c \end{pmatrix} = \frac{1}{2}\begin{pmatrix} -1 & 1 & 1 \\ 1 & -1 & 1 \\ 1 & 1 & -1 \end{pmatrix}\cdot\begin{pmatrix} I_a \\ I_b \\ I_c \end{pmatrix} \quad \text{and} \quad \begin{pmatrix} I_a \\ I_b \\ I_c \end{pmatrix} = \begin{pmatrix} 0 & 1 & 1 \\ 1 & 0 & 1 \\ 1 & 1 & 0 \end{pmatrix}\cdot\begin{pmatrix} P_a \\ P_b \\ P_c \end{pmatrix} \tag{5.12}$$

It follows from this equation that $P_a \geq P_b \geq P_c$ corresponds to $I_a \leq I_b \leq I_c$. Obviously, the constant matrices of Equation 5.12 also transform the corresponding increments of the respective eigenvalues

$$\mathbf{P} = \mathbf{C}_{\mathrm{P\leftarrow I}}\cdot\mathbf{I} \quad \text{and} \quad \delta\mathbf{P} = \mathbf{C}_{\mathrm{P\leftarrow I}}\cdot\delta\mathbf{I}; \quad \mathbf{I} = \mathbf{C}_{\mathrm{I\leftarrow P}}\cdot\mathbf{P} \quad \text{and} \quad \delta\mathbf{I} = \mathbf{C}_{\mathrm{I\leftarrow P}}\cdot\delta\mathbf{P} \tag{5.13}$$

which means that the transformations $\mathbf{C}_{\mathrm{P\leftarrow I}}$ and $\mathbf{C}_{\delta\mathrm{P}\leftarrow\delta\mathrm{I}}$ are identical (and so are $\mathbf{C}_{\mathrm{I\leftarrow P}}$ and $\mathbf{C}_{\delta\mathrm{I}\leftarrow\delta\mathrm{P}}$). Note that in Equation 5.13 we have used the notations **I** and **P** for the 3×1 vectors of the eigenvalues of the inertial moment tensors and not for the tensors

I and **P** themselves to avoid the introduction of new notations. We shall do so also in the remainder of the chapter, when confusion is not to be expected. The covariance matrix of the measured inertial moments Θ_I is composed of elements of incremental type (see Equation 2.11). Due to Equation 5.13, we can, nonetheless, transform

$$\Theta_P = C_{P \leftarrow I} \cdot \Theta_I \cdot C_{P \leftarrow I}^T \quad \text{and} \quad \Theta_I = C_{I \leftarrow P} \cdot \Theta_P \cdot C_{I \leftarrow P}^T \tag{5.14}$$

For later use, we define the inertial defect Δ, which is a useful quantity for planar molecules,

$$\Delta = I_c - I_a - I_b \tag{5.15a}$$

When the equilibrium moments of inertia of a planar molecule are used in Equation 5.15a, Δ should be 0 because for a planar molecule a, b is always the molecule plane and all $c_i = 0$. The applications of the inertial defect are discussed in Section 7.4. It is also useful to define the "pseudoinertial defects" as follows:

$$2P_a = I_b + I_c - I_a = -\Delta_a \quad a,b,c \text{ cyclical} \tag{5.15b}$$

For instance, for a substitution on the a,b principal plane of isotopologue k giving isotopologue k', the a,b principal plane is the same for both isotopologues and all c_i coordinates remain invariant, $c_i(k) = c_i(k')$, which means for the principal moments

$$2[P_c(k') - P_c(k)] = 2\Delta P_c = \Delta I_a + \Delta I_b - \Delta I_c = 0 \tag{5.16a}$$

For a substitution on the principal axis c, the following two similar equations are found:

$$
\begin{aligned}
2\Delta P_a &= \Delta I_b + \Delta I_c - \Delta I_a = 0 \\
2\Delta P_b &= \Delta I_a + \Delta I_c - \Delta I_b = 0
\end{aligned}
\tag{5.16b}
$$

Obviously, these relations are strictly true only for the equilibrium moments. The pseudoinertial defect is further discussed in Section 7.4.4.

5.2.3 Observables

For most practical work, the principal inertial moments I_g, $g = a, b, c$, are preferred over the planar moments P_g because the original quantities evaluated from the observed spectra are the inertial moments I_g. Also, all three components of I_g may not be able to be determined from the spectra for an isotopologue k, which would then preclude the construction of the planar moments P_g for this isotopologue (see Equation 5.12).

Most often, the rotational constants $B_g = K/I_g$ (where conversion factor $K = I[\text{uÅ}^2] \cdot B[\text{MHz}] = h/8\pi^2 = 505\ 379.005\ \text{uÅ}^2 \cdot \text{MHz}$) are the data reported when experimental rotational (or rovibrational) spectra are evaluated. We may therefore also combine the three rotational constants B_g to the 3×1 vector $\mathbf{B}$ ($A \geq B \geq C$). The rotational constants are usually expressed in MHz (or GHz) and sometimes in cm⁻¹ ($1\ \text{cm}^{-1} = 29979.2458\ \text{MHz}$).

While the same constant matrices (Equation 5.13) transform between vectors $\mathbf{I}$ and $\mathbf{P}$ and between their increments $\delta\mathbf{I}$ and $\delta\mathbf{P}$, significantly nonconstant and different matrices are required for the transformation between vectors $\mathbf{B}$ and $\mathbf{I}$ (or $\mathbf{P}$) and between vectors $\delta\mathbf{B}$ and $\delta\mathbf{I}$ (or $\delta\mathbf{P}$):

$$\mathbf{I} = \begin{pmatrix} K/B_a^2 & 0 & 0 \\ 0 & K/B_b^2 & 0 \\ 0 & 0 & K/B_c^2 \end{pmatrix} \mathbf{B} = \mathbf{C}_{\mathrm{I \leftarrow B}} \cdot \mathbf{B}$$

$$\delta\mathbf{I} = \begin{pmatrix} -K/B_a^2 & 0 & 0 \\ 0 & -K/B_b^2 & 0 \\ 0 & 0 & -K/B_c^2 \end{pmatrix} \delta\mathbf{B} = \mathbf{C}_{\delta\mathrm{I \leftarrow \delta B}} \cdot \delta\mathbf{B} \tag{5.17}$$

In the iterated linear least-squares method (Section 2.3), the observables are, after each iteration, the increments $\delta\mathbf{I}$ with elements $\delta I_g = I_g^{\mathrm{exp}} - I_g(\beta^{\mathrm{calc}})$ and for the transformation of the covariance matrix $\boldsymbol{\Theta}_\mathbf{B}$ of the measured rotational constants $\mathbf{B}$, the matrix that must be used is

$$\boldsymbol{\Theta}_\mathbf{I} = \mathbf{C}_{\delta\mathrm{I \leftarrow \delta B}} \cdot \boldsymbol{\Theta}_\mathbf{B} \cdot \mathbf{C}_{\delta\mathrm{I \leftarrow \delta B}}^{\mathrm{T}} \tag{5.18}$$

The set of the q mass point ensembles (i.e., isotopologues), $k = 1, \ldots, q$, of a molecule is usually referred to as the substitution data set (SDS), where $k = 1$ is designated as the parent molecule. When calculations including all isotopologues k of the SDS are intended, it is practical to rearrange the q individual 3×1 inertial moment vectors $\mathbf{I}(k)$ with components $I_g(k)$, $g = a, b, c$, as one "long" $3q \times 1$ vector $\mathbf{I}$ with the components denoted by I_t, $t = 3 \times k - 3 + g$, and $t = 1, \ldots, 3q$. The alternative 3×1 vectors $\mathbf{B}(k)$ and $\mathbf{P}(k)$ can likewise be suitably rearranged as long $3q \times 1$ vectors $\mathbf{B}$ and $\mathbf{P}$. For the present discussion, this amounts to a new definition of the vectors $\mathbf{I}$, $\mathbf{B}$, and $\mathbf{P}$; the actual meaning, however, should be clear from the context.

The $3q \times 3q$ matrices $\mathbf{C}$, which transform any of these long vectors into another vector, are block diagonal with the appropriate 3×3 matrices $\mathbf{C}$ of Equations 5.12 or 5.17 along the diagonal. The errors and correlations of the long vectors $\mathbf{I}$, $\mathbf{B}$, and $\mathbf{P}$ are then likewise summarized in the $3q \times 3q$ covariance matrices $\boldsymbol{\Theta}_\mathbf{I}$, $\boldsymbol{\Theta}_\mathbf{B}$, and $\boldsymbol{\Theta}_\mathbf{P}$, respectively, where all three of them are block-diagonal matrices with 3×3 dimensioned blocks $\boldsymbol{\Theta}_{\mathbf{I}(k)}$, $\boldsymbol{\Theta}_{\mathbf{B}(k)}$, $\boldsymbol{\Theta}_{\mathbf{P}(k)}$, $k = 1, \ldots, q$, along the diagonal because correlations between the experimentally measured $\mathbf{B}(k)$ or $\mathbf{I}(k)$ values for different isotopologues are generally (by convention) assumed to vanish (which is a very disputable approximation). If one or more inertial moment components are missing the dimension of $\boldsymbol{\Theta}_\mathbf{I}$

and $\mathbf{\Theta_B}$ is less than $3q \times 3q$ and $\mathbf{\Theta_P}$ remains undetermined. The assumption of vanishing correlations between different isotopologues may no longer hold true when the $\mathbf{B}(k)$ or $\mathbf{I}(k)$ values have been subjected to certain processes connecting different k prior to their use for structural determination. The r_m^ρ method (see Section 5.5.3) is a good example in which the $I_g(k)$ for *all* isotopologues k are scaled by a common and unavoidably error-afflicted factor $(2\rho_g - 1)$ before they are committed to least-squares fitting.

5.2.4 CLASSES OF MOLECULES

When the three principal moments of inertia are different, the molecule is called an *asymmetric top*. When two moments of inertia are equal, the molecule is called a *symmetric top*; $P_a = P_b$ (or $I_b = I_c$), is a *prolate symmetric top*, and $P_b = P_c$ ($I_a = I_b$) is an *oblate symmetric top*. A symmetric top is characterized by a threefold or higher symmetry axis. This symmetry axis forces two of the principal moments of inertia to be equal. A linear molecule is a special form of symmetric top with an axis of infinitefold symmetry. Molecules with more than one threefold axis of symmetry are called *spherical tops* because all three principal moments of inertia are equal (see also Sections 4.2.3 through 4.2.6).

An asymmetric top may possess at most twofold axes and at most two planes of symmetry. Whenever a molecule possesses a symmetry axis (twofold or higher), that axis is necessarily a principal axis of the molecule. If a plane of symmetry is present, one principal axis is perpendicular to the plane with the other two principal axes being on the plane. For some simple molecules, the moments of inertia may be readily expressed in terms of bond lengths and bond angles of the molecule (see Appendix V, Table V.1, for a few examples).

EXAMPLE 5.1: PRINCIPAL AXIS TRANSFORMATION OF A PLANAR MOLECULE

The transformation matrix **T**, which diagonalizes both the 3×3 tensors of Equations 5.5 and 5.10, **I**′ and **P**′, in a center-of-mass (but non-eigen-) system, has a very simple form for the special case in which all but one of the products of inertia (non-diagonal matrix elements) vanish. This is the case when at least one principal axis is known from symmetry considerations (for a planar molecule in the x, y plane, we have for all atoms $z = 0$, and hence, see Equation 5.10, only $I_{xy} \neq 0$). In this case, the transformation matrix corresponds to a single rotation about the z axis (origin at the center of mass) and the result is (correct ordering a, b, c assumed) as follows:

$$
\mathbf{TI'T}^\mathsf{T} = \begin{pmatrix} \cos\theta & \sin\theta & 0 \\ -\sin\theta & \cos\theta & 0 \\ 0 & 0 & 1 \end{pmatrix} \begin{pmatrix} I'_{x'x'} & I'_{x'y'} & 0 \\ I'_{x'y'} & I'_{y'y'} & 0 \\ 0 & 0 & I_{zz} \end{pmatrix} \begin{pmatrix} \cos\theta & -\sin\theta & 0 \\ \sin\theta & \cos\theta & 0 \\ 0 & 0 & 1 \end{pmatrix}
$$

$$
= \begin{pmatrix} I_a & 0 & 0 \\ 0 & I_b & 0 \\ 0 & 0 & I_c \end{pmatrix} = \mathbf{I}
$$

(5.19)

and the matrix **T** diagonalizes **I'** for

$$\tan 2\theta = \frac{2I'_{x'y'}}{I'_{x'x'} - I'_{y'y'}} \tag{5.20}$$

The angle θ is the angle between the axis systems x, y and a, b. The principal moment about the z axis is $I_c = I_{zz}$, and the other two principal moments may be expressed using Equation 5.19

$$I_a = \frac{I'_{x'x'} + I'_{y'y'}}{2} + \frac{I'_{x'x'} - I'_{y'y'}}{2}\sqrt{1 + \tan^2 2\theta}$$

$$I_b = \frac{I'_{x'x'} + I'_{y'y'}}{2} - \frac{I'_{x'x'} - I'_{y'y'}}{2}\sqrt{1 + \tan^2 2\theta} \tag{5.21}$$

5.2.5 METHODS

In order to determine the true molecular equilibrium structure by any method from the moments of inertia, these moments should ideally be those of the equilibrium structure, I_g^e (equivalently B_g^e or P_g^e). However, the best observables that can be provided by experimental molecular spectroscopy for this purpose are in most cases the vibrational ground-state constants, I_g^0, B_g^0, or P_g^0, perhaps after the more easily accessible corrections for the small remaining effects of centrifugal distortion (Sections 3.2.3.2 and 4.3.3) and possible electronic contributions (Sections 3.2.3.3 and 4.3.2) have been made. But, they will then still be afflicted with the relatively large but unknown rovibrational contributions. For this reason, it is useful to investigate the behavior of the rovibrational contribution $\varepsilon_g = I_g^0 - I_g^e$.

Table 5.1 shows that the rovibrational contribution is small compared to the moments of inertia, typically less than 1%. What is still more interesting is that the rovibrational contribution does not vary much upon isotopic substitution, the range being about only 7% when no hydrogen is involved. Further, Watson [6] showed that ε is a homogeneous function of degree 1/2 in the masses (whereas the degree is 1 for the moments of inertia; see Section 5.5.1). Empirically, it was later verified [7] that

---

TABLE 5.1
Variation of the Vibrational Correction $\varepsilon = I_0 - I_e$ as a Function of the Moment of Inertia I^0 (in uÅ²)

Molecule	Number of Isotopologues	I^0	ε Mean	ε Range	ε/I^0 (%)
HCN	11	11.707	0.049(1)	0.007	0.4
N$_2$O	12	40.232	0.206(4)	0.015	0.5
OCS	12	83.101	0.250(6)	0.018	0.3
OCSe	27	125.019	0.349(9)	0.026	0.3

Source: Demaison, J. 2007. *Mol Phys* 105:3109–38. With permission.

the relationship between $\log \varepsilon$ and $\log I_0$ is mainly linear with a slope close to 1/2 for 76 diatomic molecules and for a range of almost three orders of magnitude for I_0. Later, this correlation was extended to polyatomic molecules [8,9]. In conclusion, it may be stated that ε varies approximately as the square root of I_e.

The available methods can be distinguished as follows by the way these unknown contributions ε_g are treated:

- They are completely disregarded, which gives the empirical "effective" or r_0 structure (Section 5.3).
- Assuming them to be of similar magnitude for all isotopologues k, these contributions are partly compensated by using methods in which the moment differences between a reference isotopologue (the parent) and the remaining isotopologues numerically dominate the result; this gives the empirical "substitution" or r_s structure (Section 5.4).
- They are explicitly accounted for, though in a more or less simplified manner:
  - The rovibrational correction is determined experimentally from both rotational as well as rovibrational spectroscopy (Section 4.3.1); this gives the experimental equilibrium structure r_e^{EXP}, generally denoted as r_e.
  - The rovibrational correction is calculated from an ab initio anharmonic force field (Chapter 3); this gives the semiexperimental equilibrium structure r_e^{SE}.
  - The rovibrational correction is approximated by a model function; this gives the mass-dependent structures, r_m, r_c, r_m^{p}, $r_m^{(1)}$, and $r_m^{(2)}$ (Section 5.5).

5.3   LEAST-SQUARES-FIT STRUCTURES AND r_0 OR EFFECTIVE STRUCTURE

5.3.1   PRINCIPLE OF THE METHOD

If the rovibrational contributions are completely disregarded, the application of the least-squares method using as observables the experimental $I_g^0(k)$, provided all required isotopologues k could be measured, gives the effective or r_0 structure usually expressed in internal coordinates. Except for very simple molecules in which the number of independent structural parameters is small enough to be determinable from the inertial moments of the parent alone, the required set of isotopologues encompasses the parent species of the molecule and, if possible, all singly substituted species. For atoms that have not been substituted, fixed values of the internal coordinates must be introduced to the best of one's knowledge (if the computer program allows it, with assumed errors). On the other hand, surplus, that is, multiply substituted isotopologues are (within limits) welcome.

The main argument in favor of this method is that the rovibrational contributions are quite small, in most cases of the order of 1% or less, compared to their moments of inertia (see Table 5.1). As for all least-squares-based methods discussed in this chapter, it is essential for the accuracy of the result that the condition number of the least-squares system is not critical, that is, sufficiently low (see Section 2.2.2.2).

The $3q \times 1$ rotational quantities $\mathbf{I}$ and $\mathbf{P}$ (and also $\mathbf{B}$ only within narrow limits) can be transformed among one another by nonsingular, constant, block-diagonal transformation matrices composed of blocks corresponding to Equation 5.13, and the same holds true for the increments $\delta\mathbf{I}$ and $\delta\mathbf{P}$ used as observables of the least-squares procedure (see Equations 5.12 and 5.17). Therefore, least-squares r_0-fits yield identical results no matter which of the equivalent types of experimental rotational quantities, I_g^0, P_g^0, or (for most practical cases also) B_g^0, are fitted, *provided* the covariance matrices, an integral part of the required information, have also been appropriately transformed (see Appendix V.2.1). The requirement of a constant transformation matrix is not strictly met for the transformation of $\mathbf{B}$ because the transformation $\mathbf{C}$ (see second line of Equation 5.17) depends on $\mathbf{B}^0$; however, the nearer $\mathbf{B}^0$ comes to the final $\mathbf{B}^{\text{calc}}$ with advancing iterations in a least-squares calculation, so much the better is $\mathbf{C}$ a constant matrix. For a résumé, let us denote the methods fitting $\mathbf{I}$, $\mathbf{P}$, or $\mathbf{B}$ symbolically and in a uniform way as $r_0(I)$, $r_0(P)$, and $r_0(B)$, respectively, because all three of them yield essentially identical "effective" or r_0 structures provided their covariance matrices $\mathbf{\Theta_I}$, $\mathbf{\Theta_P}$, and $\mathbf{\Theta_B}$ have been correctly transformed.

5.3.2 VARIANTS USING DIFFERENCES OF MOMENTS OF INERTIA

The nonsingular transformation matrices $\mathbf{C}$ need not be block-diagonal like those discussed in the previous section for the transformations among the "long" vectors $\mathbf{I}$, $\mathbf{P}$, and $\mathbf{B}$. For the moment, let $\mathbf{1}$ and $\mathbf{0}$ be unit and null matrices, respectively, of dimension 3×3 each; let $\mathbf{I}(k)$ be the vector of the 3×1 of principal inertial moments of the isotopologue k; and let the nonsingular constant transformation matrix $\mathbf{C}$, which is not block-diagonal, be composed as follows:

$$
\begin{pmatrix} \mathbf{I}(1) \\ \mathbf{I}(2)-\mathbf{I}(1) \\ \mathbf{I}(3)-\mathbf{I}(1) \\ \vdots \\ \mathbf{I}(q)-\mathbf{I}(1) \end{pmatrix} = \begin{pmatrix} \mathbf{1} & \mathbf{0} & \mathbf{0} & \cdots & \mathbf{0} \\ -\mathbf{1} & \mathbf{1} & \mathbf{0} & \cdots & \mathbf{0} \\ -\mathbf{1} & \mathbf{0} & \mathbf{1} & \cdots & \mathbf{0} \\ \vdots & \vdots & \vdots & \ddots & \vdots \\ -\mathbf{1} & \mathbf{0} & \mathbf{0} & \cdots & \mathbf{1} \end{pmatrix} \cdot \begin{pmatrix} \mathbf{I}(1) \\ \mathbf{I}(2) \\ \mathbf{I}(3) \\ \vdots \\ \mathbf{I}(q) \end{pmatrix} \qquad (5.22)
$$

Then the observables are the principal moments of only the parent $I_g^0(1)$, while the remaining observables are the differences $I_g^0(k) - I_g^0(1)$, $g = a, b, c$; $k = 2, \ldots, q$. When observables of this type are least-squares fitted, one might erroneously believe that by fitting predominantly differences, at least part of the rovibrational contributions to the inertial moments could be compensated. However, as shown in Appendices II.1.5 and V.2.1, the result would be identical to the r_0 structure obtained by directly fitting $I_g^0(1), \ldots, I_g^0(q)$, $g = a, b, c$. The observables prepared by Equation 5.22 would hence result in an r_0 structure symbolically designated as $r_0(I, \Delta I)$ with subscript 0 retained; the same is true for the equivalent structure $r_0(P, \Delta P)$.

In fact, *any* nonsingular, constant transformation $\mathbf{C}$ applied to the vector of principal inertial moments $\mathbf{I}$ and to the vector $\delta\mathbf{I}$ of increments (or equivalently to $\mathbf{P}$ and $\delta\mathbf{P}$) yields observables that lead to the same r_0 structure, *provided* the respective covariance matrix has been correctly transformed (Appendix V.2.1). If this transformation is disregarded, for example, by choosing unweighted (unity-weighted) fits for $\mathbf{I}$ as well as for the equivalent $\mathbf{P}$, that is, both with covariance matrices $\mathbf{\Theta}_{\mathbf{I}} = s^2\mathbf{1}$ and $\mathbf{\Theta}_{\mathbf{P}} = s^2\mathbf{1}$, the resulting structures will be different because these two covariance matrices are not connected by the transformation given by Equation 2.11, or Equation 5.14. Therefore, it is good practice to not only report which types of observables have been employed for a least-squares fit but also state that the proper type of covariance matrix has been used.

If in Equation 5.22 the first component of the vector on the left-hand side and the first row of the transformation matrix $\mathbf{C}$ is removed, the remaining part of $\mathbf{C}$ is no longer a square matrix and the transformation is hence singular. The observables to be fitted are now invariably differences of the type $I_g^0(k) - I_g^0(1)$, $g = a, b, c$; $k = 2, ..., q$ and the least-squares result is certainly different from the r_0 structure and not simply a variant of the r_0 method in Section 5.3.1 and the beginning of this section. If the components $g = a, b, c$ of the rovibrational contributions could be assumed identical for all isotopologues k and different only for $g = a, b, c$, these contributions would indeed be fully compensated when differences of the type $I_g^0(k) - I_g^0(1)$ are least-squares fitted.

Arguments similar to these show that a least-squares fit of planar moment differences $P_g^0(k) - P_g^0(1)$, $g = a, b, c$; $k = 2, ..., q$ yields an identical result because the transformation from $P_g^0(k) - P_g^0(1)$ to $I_g^0(k) - I_g^0(1)$ and vice versa can be performed by the same nonsingular constant matrices that are shown in Equation 5.12. We shall symbolically call the structures obtained by fitting only the differences of moments $r(\Delta I)$ and $r(\Delta P)$ structures, respectively. Because for Kraitchman's r_s structure, the moment differences numerically dominate the results (see discussion of Equation 5.35), the $r(\Delta I)$ and $r(\Delta P)$ structures have also been called *pseudo-Kraitchman* or ps-Kr structures. A nonsingular constant matrix that transforms the differences $B_g^0(k) - B_g^0(1)$ into differences $I_g^0(k) - I_g^0(1)$ does not exist. The $r(\Delta B)$ structure will, therefore, be different from the $r(\Delta I)$ structure (see Appendix V.3.3, discussion of Equation V.29).

The least-squares fitting of differences of moments is generally believed to produce solutions that are nearer to the true internal coordinate values than the aforementioned r_0 structure. Nösberger, Bauder, and Günthard [10] have suggested a refined least-squares fitting of moment differences $I_g^0(k) - I_g^0(1)$ using the singular-value decomposition algorithm (Section 2.2.2) with considerable benefits for the stability of the solution. Unfortunately, near-instability by high condition numbers is a situation frequently met with when fitting inertial moments to determine molecular structures, a main reason being the fact that often some or even many of the $I_g^0(k)$ (or equivalents) to be fitted are, for the different members k of the SDS and separately for $g = a, b, c$, not "sufficiently different" functions of the bond parameters β_j. As a consequence, several column vectors of the Jacobian matrix $\mathbf{X}$ of derivatives in Equation 2.39 tend to be collinear (see Section 2.2.2.2). This is an important point that should be systematically checked.

5.3.3 Variants Using Constant Rovibrational Contributions

If the true ground-state inertial moments and their isotopic differences

$$I_g^0(k) = I_g^e(k) + \varepsilon_g(k), \quad g = a,b,c; \quad \text{all } k$$
$$I_g^0(k') - I_g^0(k) = \Delta I_g^0 = \Delta I_g^e + \Delta\varepsilon_g$$

(5.23)

are approximated by no longer neglecting the rovibrational contributions but replacing them by isotopologue-independent constants c_g, we obtain the model (ps-rigid = pseudorigid),

$$I_g^0(k) = I_{g,\text{ps-rigid}}(k) + c_g, \quad g = a,b,c; \quad \text{all } k$$

(5.24)

Note that if the $I_{g,\text{ps-rigid}}(k)$ could be assumed to equal the equilibrium moments of inertia $I_g^e(k)$, the constants c_g should be denoted ε_g as in the original paper [11] in which the ε_g are to be understood as an approximation of the true $\varepsilon_g^0(k)$. If, however, the $I_{g,\text{ps-rigid}}(k)$ are taken to resemble the "substitution moments of inertia" $I_g^s(k)$ (for the construction of $I_g^s(k)$ from r_s-type parameters, see Section 5.4), the constants c_g would approximate $\varepsilon_g^0(k)/2$ (see equations 5, 20, and 21 of Watson, Roytburg, and Ulrich [12] and Equation 5.48 in this text). To avoid confusion, we have chosen the present form of Equation 5.24 and we call the structure obtained by the application of Equation 5.24 the $r(I, c)$ structure (former notations were $r(I, \varepsilon)$ or $r(\varepsilon, I)$ and should not be used).

The $I_{g,\text{ps-rigid}}(k)$ of a hereby created pseudorigid concept, the $r(I, c)$ method, should approximate $I_g^e(k)$ better than $I_g^0(k)$ does. Although the $r(\Delta I)$ method is based on the pure differences $\Delta I_g^0 = \Delta I_{g,\text{ps-rigid}} \approx \Delta I_g^e$ (no c_g) and is generally believed to yield better results than the r_0 method (exceptions are known in particularly ill-conditioned cases), this $r(\Delta I)$ method does not fully exploit the model Equation 5.24 because the additional constants c_g are not used. Therefore, we expect the $r(I, c)$ method to be advantageous over the $r(\Delta I)$ method. The $r(I, c)$-fit determines the p internal bond coordinates $\beta_j, j = 1, \dots, p$ plus the $c_g, g = a, b, c$. The number of input data required by the $r(I, c)$ method is larger by 3 [plus 3 $I_g^0(1)$] than that required by the $r(\Delta I)$ method, and so is the number of parameters to be determined (plus 3 c_g). The $r(I, c)$ method has been proven to predict much better inertial moments for isotopologues not yet measured than the $r(\Delta I)$ method [13]. The c_g values give at least a hint of the magnitude of the rovibrational contributions. However, it has been shown that $r(I, c)$ and $r(\Delta I)$ methods yield identical structural parameters [11], and they are expected to be of generally better quality than the r_0 structure, though exceptions exist.

5.3.4 Accuracy

As the order of magnitude of the rovibrational contribution is known, it is, in principle, possible to estimate the accuracy of the r_0 structure. Due to the magnitude of these neglected rovibrational contributions, one cannot usually expect bond lengths to be nearer to the equilibrium value than approximately 0.02 Å or angles to be better than approximately $0.2°$. Compared to available r_e values, the angles occasionally appear to be of somewhat higher accuracy than the distances, although exceptions exist. Since the rovibrational contributions to the inertial moments, at least for the larger molecules, are predominantly positive, a molecule in its r_0 structure (excepting perhaps very small molecules) might appear a bit expanded compared to the equilibrium structure if the rovibrational contributions are of roughly comparable magnitudes for all isotopologues. The accuracy is quite sensitive to the conditioning of the system of normal equations. If the system is well conditioned, the accuracy is often rather good—see for instance the case of SO_2 in Table 5.2, where the r_0 structure is very close to the r_e structure (0.0014 Å for the length and $0.11°$ for the angle).

On the other hand, if the system is ill-conditioned, the r_0 structure may be widely different from the equilibrium structure—see for instance Table 5.3, which compares different structures of vinyl fluoride ($H_2C=CHF$). There are three different r_0 values for the angle $\angle(CCH_g)$, which varies from $120.9°$ to $129.2°$. Further, an analysis of the residuals of the fit shows that they are, in all cases, highly correlated instead of being random—see for instance Figure 2.2, for the case of OCSe. It indicates that the model is not correct and that the standard deviations of the parameters are not reliable.

Inspection of Table 5.2 shows that there is another point worth discussing. For a planar molecule, it is possible to use as observables any of the following: I_a and I_b, I_b and I_c, or I_a and I_c. If the equilibrium moments of inertia are used, identical results should be obtained. But, when ground-state moments of inertia are used, significantly different results are obtained. This is because the inertial defect, Equation 5.14, is different from zero. As the rovibrational contributions increase with the values of the moments of inertia, the best results are obtained when the smallest two moments of inertia are used, for example, I_a^0 and I_b^0. This behavior is general. Table 5.4 gives the example of phosgene; $OCCl_2$.

TABLE 5.2
Structures of SO_2

Method	$r(SO)$	$\angle(OSO)$	Isotopologues	Comment
r_e	1.430782(2)	119.330(3)	16.32.16	
r_0 (from I_a, I_b)	1.4322	119.535	16.32.16	
r_0 (from I_b, I_c)	1.4351	119.129	16.32.16	
r_0 (from I_a, I_c)	1.4337	119.604	16.32.16	
r_s from Equation 5.35	1.4328	119.356	16.32.16	General Kraitchman
			16.34.16, 18.32.16	
r_s from Equation 5.37	1.4318	119.452	16.32.16	Planar Kraitchman
			16.34.16, 18.32.16	
r_s from Table 5.5	1.4308	119.330	16.32.16	Chutjian
			16.34.16, 18.32.18	.

Source: The r_e data are based on Flaud, J. M., and W. Lafferty. 1993. *J Mol Spectrosc* 161:396–402.
Note: Distances are in Å angles in °.

TABLE 5.3

Value of the Angle $\angle(CCH_g)$ in $H_2C=CH_gF^a$ (in °)

Year	Method	Value
1958	r_0	123.7
1961	r_0	129.2
1961	r_0	120.9
1974	r_g^b	127.7(7)
1979	r_z^b	130.8(25)
1989	r_s	124.35(63)
1992	r_0	127.6(42)
2006	r_e	125.95(20)
Range		9.9

Source: Demaison, J. 2007. *Mol Phys* 105:3109–38. With permission.

[a] H_g is the *geminal* hydrogen.

[b] From electron diffraction.

TABLE 5.4

Comparison of Different Structures of Phosgene

	$r(C=O)$	$r(C–Cl)$	$\angle(ClCCl)$
$r_0(a, b, c)^a$	1.166(10)	1.746(10)	111.3(10)
$r_0(a, b)^b$	1.1794(17)	1.7401(8)	111.93(8)
$r(I, c)\,(a, b, c)^a$	1.1871(22)	1.7347(13)	112.46(14)
$r(I, c)\,(a, b)^b$	1.1825(8)	1.7371(6)	112.17(5)
r_s	1.185(2)	1.736(1)	112.2(1)
r_e	1.1756(23)	1.7381(19)	111.79(24)

Source: Demaison, J., G. Wlodarczak, and H. D. Rudolph. 1997. Determination of reliable structures from rotational constants. In *Advances in Molecular Structure Research*, ed. M. Hargittai and I. Hargittai, vol. 3, 1–51. Greenwich, CT: JAI Press.

Note: Distances are in Å angles in °.

[a] Fit of the moments of inertia I_a, I_b, and I_c.

[b] Fit of the moments of inertia I_a and I_b only.

The r_0 bond length is compared to the r_s and r_e bond lengths in Section 5.4.4. The $\angle_0$ and $\angle_e$ bond angles are compared in Table V.2 of Appendix V. The sign of $\angle_e - \angle_0$ may be positive or negative although it is negative for almost all triatomic molecules. The median absolute deviation is 0.22°, which is comparable to the uncertainty of $\angle_0$ values. There are, however, a few large deviations that might be due in most cases to an inaccurate r_0 structure (see discussion in Appendix V.2.2).

5.4 SUBSTITUTION, OR r_s METHOD

5.4.1 KRAITCHMAN'S GENERAL EQUATIONS

Following an early suggestion by Costain [14], the substitution or r_s method introduced by Kraitchman [15] has for many decades probably been the method most often used to determine molecular structures from inertial moments, and it is still important. In its basic form, the method requires the principal planar moments (in practice, the ground-state planar moments) of a reference or parent molecule (1), $P_a(1), P_b(1), P_c(1)$, and those of only one isotopologue (2), $P_a(2), P_b(2), P_c(2)$, in which exactly one atom i has been substituted, the mass $M(1)$ of the parent molecule, and the mass increment Δm_i upon substitution. The method gives the squares of the Cartesian coordinates of this atom in the PAS of the parent $(a_i^{[1]})^2, (b_i^{[1]})^2, (c_i^{[1]})^2$. The structure of the molecule need not be known, neither the positions nor even the number of the remaining atoms.

To determine a complete molecular r_s structure, the r_s method must be separately applied to each atom although exceptions exist for atoms in symmetry-equivalent positions. Also, the user must have a clear enough conception of the molecular structure in order to associate the correct sign with the root of the squared coordinates given by the method (see Section 5.4.3). The derivation of Kraitchman's equations assumes rigid molecules of fixed geometry and would hence be exactly applicable only in cases where the equilibrium planar moments P_g^e of the parent and the isotopologue are known. For lack of this knowledge, ground-state planar moments must be used in practice, which necessarily leads to error-afflicted results and only approximate equalities where exact ones are expected.

Consider an example: If all positions of a molecule could be determined by the repeated application of the r_s method to locate all atoms using ground-state planar moments, the Cartesian coordinates will only approximately satisfy the first- and second-moment equations, Equations 5.4 and 5.8, and not exactly as would be true if equilibrium planar moments had been available. In apparent contrast, the two types of moment equations hold exactly for all r_0 methods and their variants discussed in Section 5.3, however, only because these moment equations are an integral part of the r_0 procedure no matter how much the input observables, ground-state inertial moments (or their equivalents), deviate from the respective equilibrium quantities. In r_0-type methods, this property can be used to find the position of an atom that perhaps could not be substituted (e.g., F, P, As, I, …). On the other hand, the r_s method allows the determination of only a partial structure without it being necessary to assign positions as accurately as possible to all the rest of the molecule, which would be a requirement of the r_0 method in this case.

The r_s method can be described as follows:

Let $k = 1$ be the parent isotopologue in its PAS and $\mathbf{r}_i$ the position vectors of the atoms. The diagonal planar moment tensor of the parent in its own PAS is, from Equation 5.6,

$$\mathbf{P}^{[1]}(1) = \sum_{i=1}^{n} m_i(1)\mathbf{r}_i^{[1]}\mathbf{r}_i^{[1]\mathrm{T}} \tag{5.25}$$

where the superscript $^{[1]}$ means that the axis system is the PAS of the isotopologue $k = 1$. Let $k = 2$ be an isotopologue with masses $m_i(2) = m_i(1) + \Delta m_i(1)$, $i = 1, …, n$

(most of the $\Delta m_i(1)$ or even all but one may be 0), and position this isotopologue in such a way in the PAS of the parent that its atoms coincide with those of the parent; the position vectors will then be exactly those of the parent, $\mathbf{r}_i^{[1]}$. However, its center of mass will not coincide with that of the parent ($r_{cm(2)}^{[1]} \neq 0$) nor will, in general, its principal axes. Hence, an equation of the type of Equation 5.2 has to be used to describe the planar moment of isotopologue 2 in the PAS of the parent:

$$\mathbf{P}^{[1]}(2) = \sum_{i=1}^{n} m_i(2)\mathbf{r}_i^{[1]}\mathbf{r}_i^{[1]\mathrm{T}} - M(2)\mathbf{r}_{cm(2)}^{[1]}\mathbf{r}_{cm(2)}^{[1]\mathrm{T}} \tag{5.26}$$

Since $\sum_{i=1}^{n} m_i(1)\mathbf{r}_i^{[1]} = 0$ for the parent, see Equation 5.4, the center of mass of isotopologue 2 is given by

$$
\begin{aligned}
\mathbf{r}_{cm(2)}^{[1]} &= \frac{1}{M(2)} \sum_{i=1}^{n} m_i(2)\mathbf{r}_i^{[1]} = \frac{1}{M(2)} \sum_{i=1}^{n} [m_i(1) + \Delta m_i(1)]\mathbf{r}_i^{[1]} \\
&= \frac{1}{M(2)} \sum_{i=1}^{n} \Delta m_i(1)\mathbf{r}_i^{[1]}
\end{aligned}
\tag{5.27}
$$

The first member on the right-hand side of Equation 5.26 can be separated as follows:

$$
\begin{aligned}
\sum_{i=1}^{n} m_i(2)\mathbf{r}_i^{[1]}\mathbf{r}_i^{[1]\mathrm{T}} &= \sum_{i=1}^{n} m_i(1)\mathbf{r}_i^{[1]}\mathbf{r}_i^{[1]\mathrm{T}} + \sum_{i=1}^{n} \Delta m_i(1)\mathbf{r}_i^{[1]}\mathbf{r}_i^{[1]\mathrm{T}} \\
&= \mathbf{P}^{[1]}(1) + \sum_{i=1}^{n} \Delta m_i(1)\mathbf{r}_i^{[1]}\mathbf{r}_i^{[1]\mathrm{T}}
\end{aligned}
\tag{5.28}
$$

Substituting Equations 5.28 and 5.27 in Equation 5.26, the planar moment of isotopologue 2 is

$$\mathbf{P}^{[1]}(2) = \mathbf{P}^{[1]}(1) + \sum_{i=1}^{n} \Delta m_i(1)\mathbf{r}_i^{[1]}\mathbf{r}_i^{[1]\mathrm{T}} - \frac{1}{M(2)}\left(\sum_{i=1}^{n} \Delta m_i(1)\mathbf{r}_i^{[1]}\right)\left(\sum_{i=1}^{n} \Delta m_i(1)\mathbf{r}_i^{[1]}\right)^{\mathrm{T}} \tag{5.29}$$

It is important to realize that $\mathbf{P}^{[1]}(2)$ in Equation 5.29 depends explicitly only on those atoms that have been substituted in isotopologue 2.

Suppose that in isotopologue 2 only one atom has been substituted (Kraitchman's original proposition), say atom i, then the last two terms of Equation 5.29, which depend explicitly on this atom, reduce to

$$\left(\Delta m_i(1) - \frac{1}{M(1) + \Delta m_i(1)}\Delta m_i(1)^2\right)\mathbf{r}_i^{[1]}\mathbf{r}_i^{[1]\mathrm{T}} = \frac{M(1)\cdot\Delta m_i(1)}{M(1) + \Delta m_i(1)}\mathbf{r}_i^{[1]}\mathbf{r}_i^{[1]\mathrm{T}} \equiv \mu\mathbf{r}_i^{[1]}\mathbf{r}_i^{[1]\mathrm{T}} \tag{5.30}$$

where μ is the "reduced mass of substitution." The nondiagonal planar moment tensor of isotopologue 2, $\mathbf{P}^{[1]}(2)$ of Equation 5.29, has a diagonal and a nondiagonal term,

$$\mathbf{P}^{[1]}(2) = \mathbf{P}^{[1]}(1) + \mu\mathbf{r}_i^{[1]}\mathbf{r}_i^{[1]\mathrm{T}} \tag{5.31}$$

When both terms are combined, this is explicitly

$$
\mathbf{P}^{[1]}(2) = \begin{pmatrix} P_a(1) + \mu \left[a_i^{[1]} \right]^2 & \mu a_i^{[1]} b_i^{[1]} & \mu a_i^{[1]} c_i^{[1]} \\ \mu a_i^{[1]} b_i^{[1]} & P_b(1) + \mu \left[b_i^{[1]} \right]^2 & \mu b_i^{[1]} c_i^{[1]} \\ \mu a_i^{[1]} c_i^{[1]} & \mu b_i^{[1]} c_i^{[1]} & P_c(1) + \mu \left[c_i^{[1]} \right]^2 \end{pmatrix} \tag{5.32}
$$

Since the primary input data to Kraitchman's equations are the principal planar inertial moments of the parent $k = 1$, $P_a(1)$, $P_b(1)$, $P_c(1)$, and the isotopologue $k = 2$, $P_a(2), P_b(2), P_c(2)$, with different PASs a, b, c (in general, different origins and axis directions), we have retained in Equation 5.32 the superscript [1] to clearly indicate that these coordinates refer to the PAS of the parent, exactly those we wish to determine.

Kraitchman solved the problem in the following way: to reduce $\mathbf{P}^{[1]}(2)$ to diagonal form, the system of the three eigenvalue–eigenvector equations, Equation 5.33, must be solved to obtain the principal planar moments $P_g(2)$ (1 is the 3×3 unit matrix)

$$
[P(2) \cdot \mathbf{1} - \mathbf{P}^{[1]}(2)]\mathbf{t} = 0 \tag{5.33}
$$

where $P(2)$ represents any of the three eigenvalues of $\mathbf{P}^{[1]}(2)$, and $\mathbf{t}$ the corresponding eigenvector. Note that the eigenvalues of $\mathbf{P}^{[1]}(2)$ are necessarily those of $\mathbf{P}(2)$, that is, $P_a(2), P_b(2), P_c(2)$, because an eigenvalue of a tensor does not depend on the coordinate system in which the tensor is represented. When the three column eigenvectors $\mathbf{t}$ are combined to the 3×3 matrix $\mathbf{T}(2)$, this matrix diagonalizes $\mathbf{P}^{[1]}(2)$: $\mathbf{T}^{\mathrm{T}}(2)\mathbf{P}^{[1]}(2)\mathbf{T}(2) = \mathbf{P}^{[2]}(2) = \mathrm{diag}\{P_a(2), P_b(2), P_c(2)\}$; however, no use is directly made of this latter property when deriving Kraitchman's equations. Expanding the characteristic (or secular) equation of Equation 5.33 as follows, it is immediately seen that only squares of the components of $\mathbf{r}_i^{[1]} = (a_i^{[1]}, b_i^{[1]}, c_i^{[1]})$ occur:

$$
\begin{aligned}
&| P(2) \cdot \mathbf{1} - \mathbf{P}^{[1]}(2) | \\
&= \left(P_a(1) + \mu \left(a_i^{[1]} \right)^2 - P(2) \right)\left(P_b(1) + \mu \left(b_i^{[1]} \right)^2 - P(2) \right)\left(P_c(1) + \mu \left(c_i^{[1]} \right)^2 - P(2) \right) \\
&\quad + 2\mu^3 \left(a_i^{[1]} \right)^2 \left(b_i^{[1]} \right)^2 \left(c_i^{[1]} \right)^2 - \left(P_a(1) + \mu \left(a_i^{[1]} \right)^2 - P(2) \right)\mu^3 \left(b_i^{[1]} \right)^2 \left(c_i^{[1]} \right)^2 \\
&\quad - \left(P_b(1) + \mu \left(b_i^{[1]} \right)^2 - P(2) \right)\mu^3 \left(c_i^{[1]} \right)^2 \left(a_i^{[1]} \right)^2 \\
&\quad - \left(P_c(1) + \mu \left(c_i^{[1]} \right)^2 - P(2) \right)\mu^3 \left(a_i^{[1]} \right)^2 \left(b_i^{[1]} \right)^2 \\
&= 0
\end{aligned} \tag{5.34}
$$

In the general case of a nonplanar asymmetric-top molecule, this is a cubic equation in the variable $P(2)$ whose three solutions are already known: $P_a(2), P_b(2), P_c(2)$; what are not known, however, are all quantities contained in the coefficients of the equation. When arranged according to powers of $P(2)$, the comparison of coefficients (theory of equations) leads to the following conclusions:

- Coefficient of $P^2(2)$ equals $P_a(2) + P_b(2) + P_c(2)$.
- Coefficient of $P(2)$ equals $P_a(2)P_b(2) + P_b(2)P_c(2) + P_c(2)P_a(2)$.
- Constant term equals $P_a(2)P_b(2)P_c(2)$.

The right-hand sides of the three equalities are combinations of the known roots of the secular equation, the principal planar moments of the isotopologue $P_a(2)$, $P_b(2)$, $P_c(2)$, and the left-hand coefficients (not shown) are known, but rather lengthy, functions of the known principal planar moments of the parent $P_a(1)$, $P_b(1)$, $P_c(1)$ and the unknown squared position coordinates of the substituted atom $(a_i^{[1]})^2$, $(b_i^{[1]})^2$, $(c_i^{[1]})^2$. These three equalities can be solved to give the three solutions for $(a_i^{[1]})^2$, $(b_i^{[1]})^2$, $(b_i^{[1]})^2$, Kraitchman's equations [15],

$$
\begin{aligned}
\left(a_i^{[1]}\right)^2 &= \frac{1}{\mu}(P_a(2) - P_a(1)) \frac{P_b(2) - P_a(1)}{P_b(1) - P_a(1)} \times \frac{P_c(2) - P_a(1)}{P_c(1) - P_a(1)} \\
&= \frac{1}{\mu} \Delta P_a \left(1 + \frac{\Delta P_b}{P_b(1) - P_a(1)}\right)\left(1 + \frac{\Delta P_c}{P_c(1) - P_a(1)}\right) \\
&= \frac{1}{2\mu}(-\Delta I_a + \Delta I_b + \Delta I_c)\left(1 + \frac{\Delta I_a - \Delta I_b + \Delta I_c}{2(\Delta I_a - \Delta I_b)}\right)\left(1 + \frac{\Delta I_a + \Delta I_b - \Delta I_c}{2(\Delta I_a - \Delta I_c)}\right)
\end{aligned}
\tag{5.35}
$$

where $\Delta P_a = P_a(2) - P_a(1)$. The squared coordinates $(b_i^{[1]})^2$ and $(c_i^{[1]})^2$ follow a common cyclical permutation of a, b, c. Since the differences upon substitution in the numerator $P_g(2) - P_g(1) = \Delta P_g$ for the same g are expected to be usually much smaller than the differences $P_g(k) - P_{g'}(k)$ for different g in the denominator, the last two factors in Equation 5.35, second row, will be near unity and the resulting Cartesian coordinate (g) squared depends mainly on an isotopic difference ΔP_g whereby at least the isotope-independent part of the rovibrational contribution is approximately compensated. This is the reason why Kraitchman's r_s method has for a long time been the preferred and traditional method of molecular structure determination by rotational spectroscopy in the earlier decades of research and even up to the present day when only a partial structure can be determined.

   We have so far considered the general case. The solutions for planar asymmetric and symmetric tops and for linear molecules are simpler and easily expressible in inertial moments I_g. For instance, for a linear molecule or for the location of an atom on the symmetry axis of a prolate symmetric-top molecule (where formally $b_i^{[1]} = c_i^{[1]} = 0$; $I_b = I_c$), Equation 5.35 reduces to Equation 5.36 after making adequate use of Equation 5.12:

$$
\begin{aligned}
a^2 &= \frac{1}{\mu}[I_b(2) - I_b(1)] = \frac{1}{\mu}[I_c(2) - I_c(1)] \\
&= \frac{\Delta I_b}{\mu} \qquad\qquad = \frac{\Delta I_c}{\mu}
\end{aligned}
\tag{5.36}
$$

where the equality of the b- and c-terms is purely formal (for an oblate symmetric top, change a, b, $c \to c$, a, b). A more direct (and simpler) demonstration may be found in Appendix V.3.1.1.

For a planar asymmetric molecule where the c axis is perpendicular to the molecular plane,

$$a^2 = \frac{[I_b(2) - I_b(1)][I_a(2) - I_b(1)]}{\mu[I_a(1) - I_b(1)]} = \frac{\Delta I_b}{\mu} \frac{[I_a(2) - I_b(1)]}{[I_a(1) - I_b(1)]}$$

$$b^2 = \frac{[I_a(2) - I_a(1)][I_b(2) - I_a(1)]}{\mu[I_b(1) - I_a(1)]} = \frac{\Delta I_a}{\mu} \frac{[I_b(2) - I_a(1)]}{[I_b(1) - I_a(1)]}$$

(5.37)

The particular case of the substitution structure of a symmetric top is discussed in Appendix V.3.1.

5.4.2 FURTHER ASPECTS

Chutjian [16] has given the special forms that Kraitchman's equations assume for the common substitution of all partners of a symmetry-equivalent group of atoms (two or three according to the symmetry group). This is useful because the chemical preparation of these species may sometimes be easier than that of the corresponding singly substituted species. Furthermore, it tends to avoid the effect of large axes rotation (see Appendix V.3.2).

The results have been partly simplified and supplemented by Nygaard [17]. The equations for the disubstitution are reported in Table 5.5. The substitution of three equivalent atoms is discussed in Appendix V.3.1.2. As for all methods presented here, the relations given are strictly true only for equilibrium moments. If ground-state moments are used, inaccuracies must be expected. When Kraitchman's equations, shown in Equation 5.35, are fully expressed in inertial moments instead of planar moments together with the conditions, shown in Equation 5.16, for substitution on a principal plane or axis, several possibilities exist for the elimination of either one or two of the $\Delta I_g^0(k)$ values.

Ambiguous results are obtained in practice as a consequence of the zero-point effects, depending on which $\Delta I_g^0(k)$ had been eliminated. To avoid ambiguities, one may be tempted also in the case of planar molecules to adhere to Kraitchman's equations in the form given in Equation 5.35 and follow the procedure [18]: for a substitution on a principal plane, remove either of the last two factors of Equation 5.35 (depending on the principal plane on which the substitution occurs) to obtain the two nonzero coordinates and remove both factors to obtain the only nonzero coordinate for a substitution on a principal axis. Then, the ambiguities can be demonstrated as having been "averaged out." However, in the particular case of a planar molecule, we must note that Equations 5.37 may probably give a result closer to the equilibrium structure, because the largest inertial moment I_c^0, which is possibly contaminated by the largest rovibrational contribution (see the discussion of the contents of Table 5.2 in section 5.3.2), is then left out. In this case, the inclusion of terms containing I_c^0 may then not lead to the desired result of the averaging effect. For instance, when the molecular plane is the ab plane, only the I_a and I_b moments of inertia should be used—a recommendation also true for the r_0 structure (see Section 5.3). The example of SO_2 is given in Table 5.2.

Although the r_s method is based on the moment differences of only one pair of species, the parent and an isotopologue with only one substituted position, Typke

TABLE 5.5

Chutjian's Equations for Substitution of Two Equivalent Atoms[a]

Symmetry	Substitution in Position	x^2	y^2	z^2	Examples
C_{2v}^z	$\pm x \quad 0 \quad z$	$\dfrac{\Delta P_x}{2\Delta m}$	0	$\dfrac{\Delta P_z}{\mu_2}$	H_2O, SO_2
C_{2h}^z	$\begin{bmatrix} x & y & 0 \\ -x & -y & 0 \end{bmatrix}$	$\dfrac{\Delta P_x}{2\Delta m}\left(1 + \dfrac{\Delta P_y}{P_y - P_x}\right)$	$\dfrac{\Delta P_y}{2\Delta m}\left(1 + \dfrac{\Delta P_x}{P_x - P_y}\right)$	0	*trans*-1,2-difluoroethene
C_2^z	$\begin{bmatrix} x & y & z \\ -x & -y & z \end{bmatrix}$	$\dfrac{\Delta P_x}{2\Delta m}\left(1 + \dfrac{\Delta P_y}{P_y - P_x}\right)$	$\dfrac{\Delta P_y}{2\Delta m}\left(1 + \dfrac{\Delta P_x}{P_x - P_y}\right)$	$\dfrac{\Delta P_z}{\mu_2}$	H_2O_2
C_s^{xy}	$x \quad y \quad \pm z$	$\dfrac{\Delta P_x}{\mu_2}\left(1 + \dfrac{\Delta P_y}{P_y - P_x}\right)$	$\dfrac{\Delta P_y}{\mu_2}\left(1 + \dfrac{\Delta P_x}{P_x - P_y}\right)$	$\dfrac{\Delta P_z}{2\Delta m}$	acetaldehyde

Source: Nygaard, L. 1976. *J Mol Spectrosc* 62:292–3.

[a] Notations: $\Delta P_g = P_g(2) - P_g(1)$, $2\Delta m = M(2) - M(1)$, $\mu_2 = \dfrac{2\Delta m\, M(1)}{M(2)}$.

[19] has described and coded a least-squares computer program that, while fully retaining the basic r_s principles, allows the inclusion of any number of isotopologues, also multiply substituted ones. Even the first- and second-moment equations can be included as constraints with a variable weight factor when all atomic positions have been substituted. This r_s-fit structure is expected to be more balanced than a structure composed of atomic positions obtained by the application of the r_s method suitable to each individual atom.

5.4.3 SIGN OF THE COORDINATES

The Kraitchman equations in Equation 5.35 only provide the square of the coordinates. However, the signs can often be guessed when the shape of the molecule is known. Furthermore, the signs of the coordinates have to be chosen so that the first-moment (Equation 5.4) and second-moment (Equation 5.8) equations are roughly obeyed. It is also possible to use the *isotopic pulling method* [20]. The coordinates of one atom are determined from several different reference isotopologues $k = 1, \ldots, q$. By observing either an increase or a decrease in the magnitude of a particular coordinate upon changing the reference isotopologue, the signs of a coordinate relative to those of another atom can be determined. The isotopic pulling method can also be used to make small coordinates larger, that is, to improve their precision (see the next section).

5.4.4 ACCURACY

When experimental ground-state inertial moments $I_g^0(k)$ are used (this is the common application of the substitution method), it is difficult to make concise statements regarding the errors of r_s-coordinates, except that the experimental measurement errors can well be neglected compared to the unknown inaccuracies introduced by the zero-point effects. We have argued that due to the differences $P_g(2) - P_g(1)$ being much smaller than the differences $P_g(k) - P_g(k)$ in Equation 5.35, a rough approximation of the square of the r_s-coordinate (writing P_z for P_a) is $z_i^2 \approx \Delta P_z/\mu$. Differentiating, we obtain *Costain's rule* [21],

$$| \delta z_i | \approx \frac{1}{2\mu} \left| \frac{\delta P_z}{z_i} \right| = \frac{C}{|z_i|} \tag{5.38}$$

Several values for the constant C have been suggested, starting with $C = 0.0012$ Å^2 by Costain, expressly for larger distances, which, with $\mu \approx 1$ u (mass unit), assumes an uncertainty of approximately 0.0024 uÅ^2 for δP_z. Schwendeman [22] has advocated a somewhat larger value, $C = 0.0015$ Å^2. A reasonable choice for the value of C appears to depend on the problem at hand—for a hydrogen/deuterium (H/D) substitution, one should certainly choose a much larger value. For this latter case, Van Eijck [23] recommended $C = 0.0032$ uÅ^2. It is important to realize that the error increases with decreasing coordinate value. This is probably the most serious drawback of the r_s method—small coordinates can be determined only with a large uncertainty. Consider an example: When using $C = 0.0015$ Å^2 as suggested by Schwendeman (see discussion of Equation 5.38), which is probably too small for estimating the errors of very small distances, r_s-coordinates less than 0.12 Å could only be obtained with an error larger than 10%. For very small coordinates, the r_s method may even produce a negative square, that is, an imaginary coordinate.

The accuracy of the r_s method has recently been critically studied [24]. The basic hypothesis of the r_s method is that the rovibrational contribution ε remains constant upon isotopic substitution. From inspection of Table 5.1, this seems to be a good assumption because the variation of ε upon isotopic substitution (its range) is much smaller than its mean value. However, it is obviously not valid in three cases:

1. When a hydrogen atom is substituted by deuterium because the change of mass (and of ε) is large
2. When there are large axis rotations upon isotopic substitution (see Appendix V.3.2)
3. For most molecular complexes, because the variation of ε upon substitution is large

In other cases, the resulting r_s structure is believed to have a high degree of validity for heavy atoms. Unfortunately, even this is not always true. In the particular case of a linear molecule, the Cartesian coordinate z of the substituted atom may be written using the Kraitchman equations in Equation 5.36 as follows (writing z for a):

$$z_s^2 = \frac{I_0(2) - I_0(1)}{\mu} = \frac{\Delta I_e + \Delta \varepsilon}{\mu} = z_e^2 + \frac{\Delta \varepsilon}{\mu}; \quad z_s \approx z_e \left(1 + \frac{\Delta \varepsilon}{2\mu z_e^2}\right) \quad (5.39)$$

This equation permits the following estimation of the uncertainty δz on z:

$$\delta z \approx |z_s - z_e| \approx \frac{1}{2} \frac{\Delta \varepsilon}{\mu |z|} \quad (5.40)$$

Equation 5.40 is identical to Equation 5.38, but Costain assumed that $\Delta \varepsilon$ is a constant, whereas Equation 5.40 shows that the uncertainty is large when $\Delta \varepsilon$ is large, that is, when hydrogen is substituted by deuterium or when there is a large axis rotation. These problems, already mentioned, are well known. What is less known is that $\Delta \varepsilon$ roughly increases with the mass of the molecule (see Figure 5.1); that is to say, it is difficult to obtain an accurate substitution structure for a large (i.e., heavy) molecule. An illuminating example is the structure of C_5O: although the equilibrium Cartesian coordinate of the C_2 atom is as large as 0.4922 Å. the error is $z_e - z_s = 0.0256$ Å, which is an order of magnitude larger than the uncertainty predicted by Equation 5.38, that is, 0.0029 Å [24].

This unveils the necessity for an isotopologue-dependent, that is, mass-dependent ε_g.

A first advance in this direction was made by Pierce [25], who proposed locating atoms near a principal axis (i.e., with small coordinates) by using the double

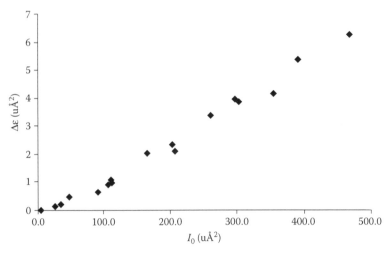

FIGURE 5.1 Plot of the variation of ε with isotopic substitution $X^{81}Br \leftarrow X^{79}Br$ for bromide derivatives (unit = $uÅ^2$). (Reprinted from Demaison, J., and H. D. Rudolph. 2002. *J Mol Spectrosc* 215:78–84. With permission from Elsevier.)

substitution method. However, this method is not easy to use and, furthermore, the results are sometimes disappointing; this is discussed in Appendix V.4.1.

For a diatomic molecule, it is possible to show that the following holds (see Section 5.5.1 and Appendix V.3.3):

$$r_s \approx \frac{(r_e + r_0)}{2} \tag{5.41}$$

No simple formula is available for polyatomic molecules. However, when reliable coordinates have been determined (i.e., for bonds involving heavy atoms with large coordinates), it has been found that the following inequality often holds:

$$r_0 \geq r_s \geq r_e \tag{5.42}$$

The r_0, r_s, and r_e heavy-atom bond lengths are compared for a few molecules in Table V.10 of Appendix V.3.4. It appears that $r_0 \geq r_e$ is well obeyed. In many cases, the full inequality in Equation 5.42 is also verified. However, there are exceptions regarding the placement of r_s in this inequality, mainly due to the smallness of the particular coordinate considered.

The case of a bond involving a hydrogen atom is more complicated, and it happens that the r_0 and r_s structures are often unreliable. The particular case of the C–H bond length has been analyzed using a few simple molecules [26]. It comes out that the respective property of a C–H bond may be divided into two classes: The first one involves an *sp* carbon atom (i.e., $\equiv$C–H) and the second one an sp^2 or sp^3 carbon. For the first class, $r_e - r_0 > 0$, and for the second one, $r_e - r_0 < 0$, and in both cases, $|r_e - r_0| \approx 0.005(2)$ Å.

5.5 MASS-DEPENDENT STRUCTURES

5.5.1 THE r_m STRUCTURE

Watson [6] has given an approximate relation, Equation 5.45, which has become the basis of the "mass-dependent" or r_m structure. Expanding the change of the rovibrational contribution $\Delta\varepsilon$ when (only) one particular atom λ is substituted with respect to the change of mass of this atom, he obtained

$$\Delta\varepsilon = \varepsilon(\dots,m_\lambda + \Delta m_\lambda,\dots) - \varepsilon(\dots,m_\lambda,\dots) = \frac{\partial\varepsilon(\dots,m_\lambda,\dots)}{\partial m_\lambda}\Delta m_\lambda + O\left(m_\lambda^2\right) \quad (5.43)$$

where $O\left(m_\lambda^2\right)$ means order of magnitude of m_λ^2. No assumption has been made as to the value of the rovibrational contribution ε itself. When the special case of Equation 5.36 for a linear molecule (along the a axis) is chosen ($I_a = 0$, $I_b = I_c$), the square of the substitution coordinate a_λ^2 of atom λ is given by (*note:* $\mu = \dfrac{M\Delta m_\lambda}{M + \Delta m_\lambda}$)

$$
\begin{aligned}
a_{\lambda,s}^2 &= \frac{1}{\mu}\left[I_b^0(m_\lambda + \Delta m_\lambda) - I_b^0(m_\lambda)\right] = \frac{1}{\mu}\left(\Delta I_b^e + \Delta\varepsilon\right) \\
&= a_{\lambda,e}^2 + \frac{\partial\varepsilon(m_\lambda)}{\partial m_\lambda} + O(\Delta m_\lambda)
\end{aligned}
\quad (5.44)
$$

Now, let atom λ be any of the atoms $i = 1, \dots, n$. Summing over i gives the "substitution inertial moment" with components I_g^s [*note:* summing does not change $O(\Delta m)$], as follows:

$$I_b^s = \sum_{i=1}^n m_i a_{i,s}^2 = I_b^e + \sum_{i=1}^n m_i \frac{\partial\varepsilon(m_i)}{\partial m_i} + O(\Delta m) \quad (5.45)$$

In his study, Watson shows that a large number of molecular parameters f are homogeneous functions of the atomic masses m_i of different degree α, as follows:

$$f(\rho m_1, \rho m_2, \rho m_3,\dots) = \rho^\alpha f(m_1, m_2, m_3,\dots) \quad (5.46)$$

for which Euler's theorem gives

$$\sum_{i=1}^n \frac{\partial f}{\partial m_i} m_i = \alpha f \quad (5.47)$$

The rovibrational contribution is a homogeneous function of the atomic masses of degree $\alpha = 1/2$. The sum in Equation 5.45 can hence be replaced by $\varepsilon/2$, which gives

$$2I_g^s = 2I_g^e + \varepsilon + O(\Delta m) = I_g^e + I_g^0 + O(\Delta m) \quad (5.48)$$

Note that the factor 2 does not change $O(\Delta m)$.

Defining a new type of inertial moment $I_g^m \equiv 2I_g^s - I_g^0$, we see that I_g^m differs from I_g^e only by terms of the order of Δm. Neglecting them, the introduction of I_g^m in Equation 5.48 should hence be an improvement compared to the use of I_g^0,

$$2I_g^s - I_g^0 + O(\Delta m) = I_g^m + O(\Delta m) = I_g^e; \quad I_g^m = 2I_g^s - I_g^0 \approx I_g^e \tag{5.49}$$

If the higher-order term, $O(\Delta m)$, is calculated explicitly, it gives [27]

$$I_g^m \equiv 2I_g^s - I_g^0 = I_g^e + \frac{1}{M}\sum_{i=1} m_i \Delta m_i \frac{\partial^2(M\varepsilon_g)}{\partial m_i^2} + \cdots \tag{5.50}$$

The superscript on I_g^m indicates that the molecular structure calculated by using only type I_g^m (k) inertial moments for all isotopologues k is called the r_m structure of the molecule. Note that the substitution moment I_g^s (k) of isotopologue k is no more than an expedient and must be calculated by strictly using the squares of all atomic position coordinates as individually obtained by the r_s method, even when the square is negative for a very small coordinate. To calculate the I_g^s (k) for any isotopologue k, the atomic positions must all be r_s-determined with this isotopologue k as the parent. Therefore, for all but the smallest molecules, the expense involved is great and the application of the r_m structure, has so far been limited exclusively to very small molecules. Equation 5.50 then permits an estimate of how far I_g^m may deviate from I_g^e by checking the order of magnitude of this higher-order term.

In a detailed discussion, Smith and Watson [27] came to the conclusion that for individual bond lengths the r_m structure is in general not closer than the r_s structure to the r_e structure. In particular, the r_m structure is not applicable to hydrogen-containing molecules because the large relative mass changes involved in the substitution of hydrogen by deuterium makes the term $O(\Delta m)$ in Equation 5.49 far from negligible. Finally, even if the molecule does not contain any hydrogen, the term $O(\Delta m)$ is not small and may therefore play an important role if the system is not well conditioned. Indeed, Watson [6] obtained a good agreement when the number of parameters is equal to the number of moments of inertia of a single isotopologue (i.e., when the system is well conditioned). Actually, for molecules as small as OCS or OCSe for which the experimental data are many and accurate, the difference between the r_m and r_e structures is serious. For instance, for OCS, $\Delta r(OC) = r_e - r_m = -0.0025$ Å and $\Delta r(CS) = 0.0021$ Å. Note that the distance OS = $r(OC) + r(CS)$ is well-determined, which is characteristic of a high negative correlation between $r(OC)$ and $r(CS)$. The origin of the discrepancy in the structure (r_e and r_m) has been discussed by Smith and Watson [27]. It is due to the use of finite changes in mass Δm_i, that is, the term $O(\Delta m)$. The extrapolation of r_s-coordinates to $\Delta m_i = 0$ (Equation 5.44) provides I_m moments free of the dependence on finite mass changes. However, this method necessitates that each atom be substituted by two different isotopes, which is not practical in most cases.

5.5.2 THE r_c STRUCTURE

Kuchitsu and coworkers [28,29] suggested an improvement of the r_m structure by composing "complementary sets" of isotopologues in such a way that at least the first term, proportional to Δm, of $O(\Delta m) = I_g^e - I_g^m$, shown explicitly in Equation 5.50, is (partially) compensated. For this purpose, the individual r_m-parameter values obtained by substitutions with positive Δm and those obtained with negative Δm must be averaged. The sign of Δm when the r_s-coordinates were determined for use in the I_g^s and thus for the I_g^m and eventually for the r_m-coordinates, decides on which complementary set a particular r_m-coordinate belongs to. The coordinates obtained by averaging both types of the respective r_m are called r_c-coordinates.

While Watson's r_m method for a diatomic molecule requires three isotopologues— XY(parent), X*Y, and XY* (* denotes a substituted atom)—to evaluate $I_m(XY)$ for the parent isotopologue by Equation 5.49, the r_c method requires the use of an additional "complementary set" of isotopologues—X*Y*(parent), XY*, and X*Y—in order to also calculate $I_m(X^*Y^*)$. Assuming m_i and $\partial^2(M\varepsilon_g)/\partial m_i^2$ in Equation 5.50 to be sufficiently invariant with respect to XY and X*Y*, the second-order terms of Equation 5.50 are then approximately equal in magnitude and opposite in sign to $I_m(XY)$ and $I_m(X^*Y^*)$. The bond lengths calculated are, therefore, $r_m(XY) = r_e + \Delta r$ and $r_m(X^*Y^*) = r_e - \Delta r^{**}$ with a similar but not exactly equal difference from r_e, from which the "complementary" bond length $r_c(XY) = r_e + 1/2 (\Delta r - \Delta r^{**}) \approx r_e$ is obtained by averaging.

An excellent agreement was found for Cl_2O (29) by averaging the r_m obtained for $^{16}O^{35}Cl_2$, and $^{18}O^{37}Cl_2$. The isotopic mass changes are positive for all atoms in the parent $XY = {}^{16}O^{35}Cl_2$ and negative in the parent $X^*Y^* = {}^{18}O^{37}Cl_2$. The same method was also applied with success to OCSe [30] and phosgene, $OCCl_2$ [28]. The details of the calculations are given in Appendix V.4.2.

Even more than the r_m method, the large number of isotopologues required again precludes a wide application of the proposal for all but the smallest molecules. To simplify the method somewhat, Nakata and Kuchitsu [31] devised a clever additivity rule to determine the moments of inertia of multiply substituted species from the moments of inertia of the monosubstituted species. This is also described in Appendix V.4.2.

5.5.3 r_m^ρ STRUCTURE

Harmony and coworkers [32–37] presented a more easily applicable variant of the r_m method—the number of isotopologues available must be sufficient to calculate no more than just one complete substitution structure. The authors noticed while testing the rewritten Equation 5.49

$$I_g^e \approx I_g^m = \left(2\frac{I_g^s}{I_g^0} - 1\right)I_g^0 = (2\rho_g - 1)I_g^0 \tag{5.51}$$

that ρ_g hardly depended on the isotopologues. When I_g^s was determined for one complete set of isotopologues, $k = 1$ being the parent, this $\rho_g = I_g^s(1)/I_g^0(1)$ could then be used to correct the I_g^0 of all isotopologues present by scaling them by the fixed factor

$$I_g^\rho = (2\rho_g - 1)I_g^0 \tag{5.52}$$

The structure obtained by the least-squares fitting of the I_g^ρ of the isotopologues is called the r_m^ρ structure, and it should be better than the r_s structure that had been required to obtain the scaling factor. The authors contended that because the r_m^ρ structure needs no more input data than the respective r_s structure, it should be used whenever a complete r_s structure is possible. They also introduced an approximate correction for the nonnegligible bond-length change resulting from an H/D substitution (see also Section 5.5.4.2). The r_m^ρ method has indeed found a wide range of applications.

Regarding the errors of this method, Demaison et al. [38] made some critical remarks. The scaling factor $\rho_g = I_g^s(1)/I_g^0(1)$ is afflicted with errors, which should be considered, and since the factor $(2\rho_g - 1)$ multiplies the inertial moments I_g^0 of all isotopologues present, considerable correlations among the I_g^ρ of different isotopologues will exist. The r_m^ρ method has also been criticized by Cooksy et al. [39] because the correction $(2\rho_g - 1)$ does not allow for the proper mass dependence of the rovibrational contribution ε in Equation 5.53. Using Equation 5.51, Equation 5.53 is found to be linear in the moments of inertia and hence of the order of 1 in the masses instead of order 1/2 as it should be [6] (see Section 5.5.1).

$$I_g^0 - I_g^e = \varepsilon_g = 2(1 - \rho_g)I_g^0 \tag{5.53}$$

Actually, since ε is small and the range of I_g^0 not too large in most cases, this "approximate" relation often gives satisfactory results. The invariance of ρ_g could be checked experimentally in the case of OCSe because many isotopologues are available for this molecule [30]. The heavy-atom r_m^ρ structure of some polyatomic organic molecules is given in Appendix V.4.3.

5.5.4   Mass-Dependent $r_m^{(1)}$ and $r_m^{(2)}$ Methods

5.5.4.1   Principle of the Method

Starting from the r_m method, Watson, Roytburg, and Ulrich [12] after checking several possibilities arrived at an approximation that yields a two-step model for an $r_m^{(1)}$ or an $r_m^{(2)}$ structure. The calculated ground-state principal inertial moments to be fitted to the experimental moments are shown in Equation 5.54 where, for the $r_m^{(2)}$ structure, the isotopologue-independent parameters c_g and d_g take care of the rovibrational contribution (see Equation 31 from reference [12]):

$$I_g^0 = I_g^m + c_g\sqrt{I_g^m} + d_g\left(\frac{m_1 m_2 \ldots m_n}{M}\right)^{1/(2n-2)} \quad ; \quad I_g^m = I^{\text{rigid}}(r_m) \tag{5.54}$$

where $g = a, b, c$; m_i are atomic masses of the particular isotopologue; and M is its molecular mass.

For the $r_m^{(1)}$ structure, the last term of Equation 5.54 is omitted. Both correction terms have the correct dependence of order 1/2 in the masses. The first correction term (in c_g) scales the moments of inertia accordingly. The second term (in d_g) is an attempt to correct for the problems encountered with small coordinates. It is based on the observation that the contribution to ε from atoms with small coordinates tends to be negative whereas it is positive for atoms with larger coordinates. It is worth mentioning that Le Guennec et al. [40] had, earlier and among other models, also used a model with $\varepsilon_g = c_g \sqrt{I_g^0}$ when reporting the structure of methyl cyanide; this model is very similar to $r_m^{(1)}$.

The $r_m^{(1)}$ or $r_m^{(2)}$ method described in this section does not take into account the possible rotation of axes upon isotopic substitution, which may have drastic consequences in some cases, particularly for oblate molecules (e.g., see Appendix V.3.2 for the effect on the r_s structure). For this reason, it seems desirable to improve the $r_m^{(1)}$ and $r_m^{(2)}$ models in such a way that they can take care of this effect also. Watson, Roytburg, and Ulrich [12] described a procedure to take into account this axis rotation (see Appendix Va). The rovibrational contributions to the inertial moments can be expected to depend more on the fixed geometrical structure of the molecule, in short on its shape or contour than on the small localized atomic mass changes in the different isotopologues. This molecular shape, identical for all isotopologues, can best be envisaged as the molecule represented by its "ball-and-stick" or "van der Waals radii" model. When the PASs of all isotopologues are superimposed to one uniform coordinate system, the shapes of the different isotopologues will generally be oriented differently. An improvement can be achieved by rotating the shapes of all the isotopologues into a position parallel to that of the parent before the two correction terms of Equation 5.54 are added to the inertial moment tensor of a particular isotopologue. After this rotation, the tensor is still symmetric but, in general, nondiagonal (likewise its square root) because the isotopologue has been rotated out of its PAS. This requires the introduction of "mixed" coefficients c_{ab}, c_{bc}, c_{ca}, multiplying the nondiagonal elements of the square root of the inertial tensor in addition to the coefficients c_a, c_b, c_c of Equation 5.54, which then modify the diagonal elements. In contrast, there are no new mixed elements d_g because the coefficient of d_g, which is a function of only the masses, is a constant independent of g. Structures that have been obtained using this expanded treatment have been called "$r_m^{(1r)}$" and "$r_m^{(2r)}$" structures.

A practical problem that arises is that the mixed coefficients c_{gh} increase the number of parameters to be determined and that they are parameters that do not occur in the inertial moment tensor of the parent, which is not rotated nor in the tensor of isotopologues with the mass symmetry of the parent and which hence need not be rotated. Nonetheless, in planar molecules (where only c_{ab} is present), improvements have been obtained.

5.5.4.2   Laurie Correction

In contrast to the fundamental assumption inherent in all models discussed so far, the statement that the structures are identical for all isotopologues of a molecule cannot generally be upheld for an X–H bond length. Laurie [41] had already pointed out that the sizable contraction of an X–H bond upon deuteration because of the large fractional change of mass may no longer be neglected when a more accurate structural determination by empirical methods is attempted. Using different effective bond lengths for X–H and X–D bonds, he found for a number of molecules a range of 0.002–0.004 Å for the difference $r_0(X–H) - r_0(X–D)$. The shrinkage of the X–H bond length upon deuteration is expected: the heavier deuterium atom lies lower in the asymmetric potential well of the X–H vibrational stretching mode than the hydrogen atom and, hence, nearer the atom X. The ensuing bond-length change has been shown [13] to remain often restricted to the X–H bond length considered, hardly affecting the remaining bond parameters, at least when the least-squares system is not ill-conditioned.

Watson, Roytburg, and Ulrich [12] proposed the introduction of this mass dependence in the $r_m^{(1)}$ or $r_m^{(2)}$ method by using the following "effective" X–H bond parameter (H/D: H or D):

$$r_m^{eff}(X-H/D) = r_m(X-H) + \delta_H \cdot \left(\frac{M_{H/D}}{m_{H/D}(M_{H/D} - m_{H/D})} \right)^{1/2} \qquad (5.55)$$

The mass coefficient of δ_H in the correction term is the square root of the reduced mass for the vibration of either H or D against the remainder of the molecule and hence different for the isotopologues containing X–H or X–D bonds. The least-squares fit must now determine two parameters for the effective X–H bond length, $r_m(X-H)$ and δ_H. Unfortunately, the necessity of including δ_H also in the set of parameters to be determined generally produces a high correlation between $r_m(X-H)$ and δ_H, which deteriorates the quality of the fit (high condition number). Structures including the Laurie correction, Equation 5.55, have been called $r_m^{(1L)}$ and $r_m^{(2L)}$ structures ($r_m^{(1rL)}$ and $r_m^{(2rL)}$ if applicable).

5.5.4.3 Accuracy

The particularly detailed presentation of the mass-dependent $r_m^{(1)}$ and $r_m^{(2)}$ methods given in Sections 5.5.4.1 and 5.5.4.2 indicate that these methods, whenever applicable, should be preferred over other empirical methods. However, at each step of the improvement of the model ($r_0 \rightarrow r_m^{(1)} \rightarrow r_m^{(2)} \rightarrow r_m^{(2r)}$), the number of parameters increases. Thus, the conditioning of the system is expected to deteriorate. See, for instance, the example of SO_2 given in Table 5.6. The r_0- and r_e-fits are extremely well conditioned because the two structural parameters can be determined from only one isotopologue.

On the other hand, the $r_m^{(1)}$- or the $r_m^{(2)}$-fits require the determination of three or six additional parameters, respectively, which considerably worsens the fit (less for planar molecules with only one additional "mixed" parameter c_{gh}). This is important because the $r_m^{(1)}$ (or $r_m^{(2)}$) model is still an approximate one—see Figure 2.3, which shows that the residuals are still not random in the case of an $r_m^{(2)}$-fit. If the rovibrational contribution is estimated at 0.5% of the inertial moment and its model error at 10% of its value (as derived from the standard deviations of the parameters c_g and d_g), then the model error is 0.05% of the inertial moment, which is more, by orders of magnitude, than the experimental error in modern spectroscopy. This model error, with repercussions on the accuracy of the parameters, is often amplified by ill conditioning. Recent work has shown that often the variations of the rovibrational contributions among isotopologues are more irregular than can be represented by the rather uniform mass dependence of even the more advanced empirical methods.

In conclusion, the $r_m^{(1)}$ (or $r_m^{(2)}$) method is well-suited when the number of available rotational constants is large compared to the number of structural parameters,

TABLE 5.6
Different Structures of SO_2[a]

	r_0	$r_m^{(1)}$	$r_m^{(2)}$	r_e
p[b]	2	5	8	2
σ[c]	0.0475	0.000763	0.000364	e
κ[d]	2.98	343	1486	8
$r(S{=}O)$	1.43358(17)	1.43084(24)	1.43068(12)	1.430782(15)
$\angle(OSO)$	119.420(30)	119.442(32)	119.340(20)	119.3297(30)

Note: Distances are in Å angles in °.

[a] Unit weighted fit of 17 isotopologues, except for r_e where only one isotopologue was used.

[b] Number of fitted parameters.

[c] Standard deviation of the fit.

[d] Condition number.

[e] No standard deviation is given, because only two moments of inertia are used.

a condition that is usually fulfilled only for small molecules. Nevertheless, the structures of several molecules (not all of them small) were recently determined using the $r_m^{(1)}$ (or $r_m^{(2)}$) method, and these experimental structures were found to be in good agreement with the corresponding ab initio structures (i.e., within 0.002–0.003 Å; see Table V.14 in Appendix V). The structure of the large molecule, phenylacetylene, is also discussed in Appendix V.4.4.2.

5.6   COMPUTER PROGRAMS

One of the first publicly accessible computer programs for calculating the r_0 structure from the inertial moments of a sufficiently large set of isotopologues was coded and described by Schwendeman [22]. A more recent computer program with many options and several auxiliary programs is that of Z. Kisiel [42]. The STRFIT program by Schwendeman provides an option using only differences of inertial moments, as does Kisiel's program.

Appendix Va presents all the important steps in composing a linearized least-squares program with true derivatives in the Jacobian (or design) matrix for determining the molecular structure $(r_0, r_m^{(1)}, \ldots, r_m^{(2rL)}$ types) from a sufficiently large number of isotopologues. This program has already been successfully used for several investigations. A commented source code will soon be available from the Chemical Information Systems Department of the University of Ulm, Germany (c/o Dr. J. Vogt) [43].

5.7   CONCLUSION FOR EMPIRICAL STRUCTURES

Empirical methods permit one to obtain a molecular structure using only the ground-state rotational constants. The results are useful and are seemingly easy to obtain. However, some expertise is required to determine a reliable structure, and often even this is not sufficient because the information furnished by the rotational constants of the available isotopologues does not differ sufficiently in functional dependence on the structural parameters that are to be determined. In other words, the system is ill-conditioned and thus sensitive to small errors originating from the approximation inherent in any of the methods used. In addition, as the number of data available is usually small, the results are sensitive to the weighting scheme, which makes the structure still more unreliable (see the example of NH_2F in Appendix V.2.2 [Table V.3]).

Nonetheless, appreciable progress has been made from the early r_0 structure to the contemporary $r_m^{(1)}$ or $r_m^{(2)}$ structures and their expansions $r_m^{(2r)}$ and $r_m^{(2L)}$ over decades of research. When the number of isotopologues available is insufficient for a complete structure analysis, the r_s method can still be used. At the same time, much effort has been devoted to the error problem, which is particularly serious. In the empirical methods, the unknown rovibrational contributions contained in the experimental ground-state inertial moments must be expressed by a very limited number of fittable parameters, or they must be compensated (even less successfully) by forming differences of inertial moments.

Although structural parameters can nowadays, at least occasionally, be obtained with amazingly high precision (differences with respect to [calculated] equilibrium parameters of the order of 0.002 Å for atomic distances and 0.2° for angles) their errors due to the imperfect approximation of the rovibrational contributions by the empirical methods are still higher than those due to the experimental measurement errors. Further progress can be expected by the increased application of correctly calculated ab initio isotopologue-dependent rovibrational contributions (see Chapter 3).

REFERENCES

1. Gordy, W., and R. L. Cook. 1984. *Microwave Molecular Spectra*. New York: Wiley.
2. Rudolph, H. D. 1995. Accurate molecular structure from microwave rotational spectroscopy. In *Advances in Molecular Structure Research*, ed. M. Hargittai and I. Hargittai, vol. 1, 63–114. Greenwich, CT: JAI Press.
3. Demaison, J., G. Wlodarczak, and H. D. Rudolph. 1997. Determination of reliable structures from rotational constants. In *Advances in Molecular Structure Research*, ed. M. Hargittai and I. Hargittai, vol. 3, 1–51. Greenwich, CT: JAI Press.
4. Groner, P. 2000. The quest for the equilibrium structure of molecules. In *Vibrational Spectra and Structure: Equilibrium Structural Parameters*, ed. J. R. Durig, vol. 24, 165–252. Amsterdam: Elsevier.
5. Thompson, H. B. 1967. *J Chem Phys* 47:3407–10.
6. Watson, J. K. G. 1973. *J Mol Spectrosc* 48:479–502.
7. Demaison, J., and L. Nemes. 1979. *J Mol Struct* 55:295–9.
8. Burczyk, K., H. Bürger, M. Le Guennec, G. Wlodarczak, and J. Demaison. 1991. *J Mol Spectrosc* 148:65–79.
9. Le Guennec, M., W. Chen, G. Wlodarczak, J. Demaison, R. Eujen, and H. Bürger. 1991. *J Mol Spectrosc* 150:493–510.
10. Nösberger, P., A. Bauder, and Hs. H. Günthard. 1973. *Chem Phys* 1:418–25.
11. Rudolph, H. D. 1991. *Struct Chem* 2:581–8.
12. Watson, J. K. G., A. Roytburg, and W. Ulrich. 1999. *J Mol Spectrosc* 196:102–19.
13. Epple, K. J., and H. D. Rudolph. 1992. *J Mol Spectrosc* 152:355–76.

14. Costain, C. C. 1958. *J Chem Phys* 29:864–74.
15. Kraitchman, J. 1953. *Am J Phys* 21:17–24.
16. Chutjian, A. 1964. *J Mol Spectrosc* 14:361–70.
17. Nygaard, L. 1976. *J Mol Spectrosc* 62:292–3.
18. Rudolph, H. D. 1981. *J Mol Spectrosc* 89:460–4.
19. Typke, V. 1978. *J Mol Spectrosc* 69:173–8.
20. Pasinski, J. P., and R. A. Beaudet. 1974. *J Chem Phys* 61:683–91.
21. Costain, C. C. 1966. *Trans Am Cryst Assoc* 2:157–64.
22. Schwendeman, R. H. 1974. Structural parameters from rotational spectra. In *Critical Evaluation of Chemical and Physical Structural Information*, ed. D. R. Lide and M. A. Paul, 94–115. Washington, DC: National Academy of Sciences.
23. Van Eijck, B. P. 1982. *J Mol Spectrosc* 91:348–62.
24. Demaison, J., and H. D. Rudolph. 2002. *J Mol Spectrosc* 215:78–84.
25. Pierce, L. 1959. *J Mol Spectrosc* 3:575–80.
26. Demaison, J., and G. Wlodarczak. 1994. *Struct Chem* 5:57–66.
27. Smith, J. G., and J. K. G. Watson. 1978. *J Mol Spectrosc* 69:47–52.
28. Nakata, M., T. Fukuyama, K. Kuchitsu, H. Takeo, and C. Matsumura. 1980. *J Mol Spectrosc* 83:118–29.
29. Nakata, M., M. Sugie, H. Takeo, C. Matsumura, T. Fukuyama, and K. Kuchitsu. 1981. *J Mol Spectrosc* 86:241–9.
30. Le Guennec, M., G. Wlodarczak, J. Demaison, H. Bürger, M. Litz, and H. Willner. 1993. *J Mol Spectrosc* 157:419–46.
31. Nakata, M., and K. Kuchitsu. 1994. *J Mol Struct* 320:179–92.
32. Harmony, M. D., and W. H. Taylor. 1986. *J Mol Spectrosc* 118:163–73.
33. Harmony, M. D., R. J. Berry, and W. H. Taylor. 1988. *J Mol Spectrosc* 127:324–36.
34. Berry, R. J., and M. D. Harmony. 1988. *J Mol Spectrosc* 128:176–94.
35. Berry, R. J., and M. D. Harmony. 1989. *Struct Chem* 1:49–59.
36. Tam, H. S., J.-I. Choe, and M. D. Harmony. 1991. *J Phys Chem* 95:9267–72.
37. Harmony M. D. 2000. Molecular structure determination from spectroscopic data using scaled moments of inertia. In *Vibrational Spectra and Structure: Equilibrium Structural Parameters*, ed. J. R. Durig, vol. 24, 1–83. Amsterdam: Elsevier.
38. Demaison, J., G. Wlodarczak, H. Rück, K. H. Wiedemann, and H. D. Rudolph. 1996. *J Mol Struct* 376:399–411.
39. Cooksy, A. L., J. K. G. Watson, C. A. Gottlieb, and P. Thaddeus. 1994. *J Chem Phys* 101:178–86.
40. Le Guennec, M., G. Wlodarczak, J. Burie, and J. Demaison. 1992. *J Mol Spectrosc* 154:305–23.
41. Laurie, V. W. 1958. *J Chem Phys* 28:704–6.
42. Kisiel, Z. http://info.ifpan.edu.pl/~kisiel/prospe.htm
43. http://www.uni-ulm.de/cheminfo/

CONTENTS OF APPENDIX V (ON CD ROM): DETERMINATION OF THE STRUCTURAL PARAMETERS FROM THE INERTIAL MOMENTS

CONTENTS OF APPENDIX VA (ON CD ROM): DETERMINATION OF THE STRUCTURAL PARAMETERS FROM THE INERTIAL MOMENTS: PRACTICAL CALCULATION OF THE MOMENTS OF INERTIA AND THE JACOBIAN MATRIX

6 Determining Equilibrium Structures and Potential Energy Functions for Diatomic Molecules

Robert J. Le Roy

CONTENTS

For diatomic molecules, we are able to obtain a much broader range of information from experimental data than is possible for larger molecules. In addition to equilibrium structures, we can often determine an accurate bond dissociation energy, an accurate potential curve for the whole potential well, and sometimes also the Born–Oppenheimer breakdown (BOB) radial strength functions, which define the small differences between the electronic and centrifugal potential energy functions for different isotopologues of a given species. Such results allow us to make realistic

159

predictions of the energies and properties of unobserved levels and of a wide range of other types of data not considered in the original data analysis, including the collisional properties of the atoms forming the particular molecular state. This is possible for two reasons: (1) modern experimental methods often provide very high-quality data for vibrational levels spanning a large fraction of the potential well, and (2) the relevant Schrödinger equation is effectively one-dimensional, and can be solved efficiently using standard numerical methods. This chapter takes the first point for granted and focuses on how the experimental information thus obtained can be used to determine both precise and accurate equilibrium properties, and reliable overall potential energy functions for diatomic molecules.

There are two basic approaches for determining diatomic molecule potential energy functions from experimental data. The first begins with a description of the patterns of molecular level energies as analytic functions (usually polynomials) of vibrational and rotational quantum numbers, and uses an inversion procedure based on a semiclassical quantization condition to determine an extremely precise and smooth pointwise potential energy function. The second is fully quantum mechanical, and uses direct fits of simulated spectra to experiment to determine parameters defining an analytic potential energy function. While the latter approach is better in principle, the associated fits are nonlinear, and hence require realistic initial trial parameter values (see Chapter 2). The simplest way to obtain the latter is usually from a preliminary analysis using the conventional parameter-fit/semiclassical-inversion methodology. This traditional approach also offers more direct insight regarding how various features of the data reflect the properties of the potential energy function, as well as insight regarding the nature and magnitude of isotope effects.

This chapter begins with a description of the "forward" problem of calculating vibrational-rotational level energies and some spectroscopic properties of an assumed-known potential energy function. Section 6.2 then describes the "traditional" semiclassical-based methods for the inverse problem of determining a potential energy function from experimental data, while the fully quantum mechanical direct-potential-fit (DPF) procedure for determining potentials is described in Section 6.3. Finally, the determination of BOB radial strength functions that account for the differences between the potentials for different isotopologues and for some nonadiabatic coupling are presented in Section 6.4.

6.1   QUANTUM MECHANICS OF VIBRATION AND ROTATION

If we ignore BOB and effects due to nonzero electronic and spin angular momentum, the vibration-rotation level energies of a diatomic molecule are the eigenvalues of the one-dimensional effective radial Schrödinger equation

$$-\frac{\hbar^2}{2\mu}\frac{d^2\psi(r)}{dr^2} + V_J(r)\psi(r) = E\psi(r) \tag{6.1}$$

in which r is the internuclear distance, $\hbar$ is Planck's constant divided by 2π, $\mu = m_A m_B/(m_A + m_B)$ is the reduced mass of the two atoms forming the molecule, and the overall effective potential energy function

$$V_J(r) = V(r) + \frac{\hbar^2}{2\mu r^2}[J(J+1)] \tag{6.2}$$

is the sum of the effective electronic potential $V(r)$ plus the centrifugal potential due to molecular rotation. If we ignore the effect of rotation, exact analytic solutions of Equation 6.1 are known for a number of simple analytic potential energy functions, the most familiar of which are the particle-in-a-box square-well potential, the harmonic oscillator potential, $V_{HO}(r) = 1/2\,k(r-r_e)^2$, and the Morse potential

$$V_M(r) = \mathfrak{D}_e[1 - e^{-\beta(r-r_e)}]^2 \tag{6.3}$$

in which r_e is the equilibrium internuclear distance associated with the potential minimum.

The eigenvalue expression for the particle-in-a-box, $E(v) = \left[\dfrac{(\pi^2)(\hbar^2)}{2\mu L^2}\right](v+1)^2$ tells us that all else being equal, vibrational level spacings decrease when the width of the box L increases, or when the effective mass μ increases. The energy equation for a harmonic oscillator, $E(v) = \omega_e\left(v + \dfrac{1}{2}\right)$ with $\omega_e = \sqrt{2k\left(\dfrac{\hbar^2}{2\mu}\right)}$, then tells us that if the width of the potential is exactly proportional to the square root of the energy, the level spacings will be constant. Thus, if the well width increases more rapidly than the square root of the energy, the level spacings will decrease with increasing energy. More generally, this means that the pattern of vibrational level spacings reflects or determines the rate at which the width of the potential well increases with energy. Unfortunately, none of the simple analytic potentials for which closed-form solutions are known has sufficient flexibility and sophistication to describe accurately the vibrational levels spanning a large fraction of the potential energy well of a real molecule. As a result, accurate treatments of real molecules necessarily depend on numerical methods.

For a realistic single-minimum potential energy function, Figure 6.1 illustrates the pattern of vibrational levels and the nature of the wave functions, and shows how these properties are affected by molecular rotation. Because of potential function anharmonicity, the level spacings systematically decrease with increasing energy, and for the rotationless molecule (background image), those spacings approach zero at the asymptote. Figure 6.1 also illustrates the extreme asymmetry of the wave functions for levels approaching the dissociation limit. In particular, we see that for $v = 13$ the maximum in the probability amplitude lies near 9.6 Å, and for the highest level supported by this potential, $v = 15$, which is bound by only 0.0008 cm^{-1}, the outermost wave function maxima lies past 22 Å! This extreme wave-function asymmetry can present challenges when one is performing eigenvalue calculations for levels lying very near dissociation.

The dotted curves in Figure 6.1 show the effective centrifugally distorted potential $V_J(r)$ for the case in which the angular momentum corresponds to $J = 60$, and the inset figure shows the associated vibrational level energies and wave functions.

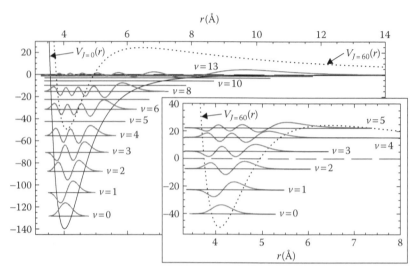

FIGURE 6.1 Level energies and wave functions of Kr_2 for $J = 0$ (background figure) and $J = 60$ (inset).

This illustrates the important point that the rotational level energies should not be thought of as a stack of sublevels associated with each pure vibrational ($J = 0$) level, but rather as vibrational levels of the centrifugally distorted potentials $V_J(r)$ for various J. Figure 6.1 also shows that metastable "quasibound" levels lying above the potential asymptote but below a potential barrier maximum have essentially the same qualitative properties as truly bound levels. In practice, most may be observed by sharp lines in experimental spectra, although the levels lying closest to a barrier maximum will be broadened by tunneling predissociation [1–3]. Figure 6.1 also shows that as J increases, the centrifugal potential will systematically spill vibrational levels out of the well until (at $J = 104$ for Kr_2) none remain.

In principle, a wide range of numerical methods may be used to solve Equation 6.1 to virtually any desired precision. In practice, however, many methods are unable to routinely treat quasibound levels and bound levels lying very near dissociation, as well as normal deeply bound levels. It is beyond the scope of this work to discuss details of this problem. However, the author's strong preference is for a Cooley-type implementation of the Numerov wave function propagator method [4], since it may readily be combined with a third-turning-point boundary condition which allows quasibound levels to be located as easily as truly bound states [1–3]. A "black box" computer code (accompanied by a manual) for determining any or all vibration-rotation eigenvalues and eigenfunctions of virtually *any* plausible radial potential $V(r)$ is freely available on the Internet [3].

From a given set of calculated eigenvalues and eigenfunctions, it is a straightforward matter to use the wave functions to compute expectation values of powers of r or of other functions, or to calculate matrix elements (overlap integrals) of properties between different levels of a given potential or between levels of two different potential energy functions [3]. One can also use such "forward" calculations to generate values of some conventional spectroscopic constants. In particular, it has been customary to express the rotational sublevels of a diatomic in terms of a power series expansion about the rigid-rotor limit using the expression [5], as follows:

$$E(v, J) = G_v + B_v[J(J+1)] - D_v[J(J+1)]^2 + H_v[J(J+1)]^3 + L_v[J(J+1)]^4 + \cdots \quad (6.4)$$

It was also known that if the centrifugal term in Equation 6.2 is treated as a perturbation, first-order perturbation theory allows the inertial rotation constant for any given vibrational level v to be defined as [5]

$$B_v = \frac{\hbar^2}{2\mu} \left\langle \psi_v(r) \left| \frac{1}{r^2} \right| \psi_v(r) \right\rangle \quad (6.5)$$

Then in 1981, Hutson showed that exact quantum mechanical values of the centrifugal distortion constants (CDCs) for all vibrational levels of a given potential well could be generated readily by solving inhomogeneous versions of Equation 6.1 in which the inhomogeneous term depends on lower-order solutions and on the centrifugal term $\hbar^2/(2\mu r^2)$. This quickly became a standard, widely adopted procedure, and it is now a routine matter to calculate the "band constants" $\{G_v, B_v, D_v, H_v, \ldots\}$ associated with all vibrational levels of any given diatomic molecule potential [3,6,7,8].

It has also long been customary to express the vibrational level energies as expansions about the harmonic oscillator limit,

$$G_v = \omega_e \left(v + \frac{1}{2} \right) - \omega_e x_e \left(v + \frac{1}{2} \right)^2 + \omega_e y_e \left(v + \frac{1}{2} \right)^2 + \omega_e z_e \left(v + \frac{1}{2} \right)^3 + \cdots \quad (6.6)$$

and analogous power series expansions in $(v + 1/2)$ are used to express the v-dependence of the various rotational constants $B_v, D_v, H_v, \ldots$ [5]. Consideration of the expression for the inertial rotational constant,

$$B_v = B_e - \alpha_e \left(v + \frac{1}{2} \right) + \gamma_e \left(v + \frac{1}{2} \right)^2 + \cdots \quad (6.7)$$

together with Equation 6.5, indicates that within the approximation that the potential minimum corresponds to $v = -1/2$, the expression

$$r_e \approx r_e^{(1)} = \sqrt{\frac{\hbar^2}{2\mu B_e}} \quad (6.8)$$

yields a good estimate of the equilibrium structure. Unfortunately, although $v = -1/2$ corresponds to the potential minimum for both the quantum mechanical harmonic and Morse oscillators, and for all potentials within the first-order semiclassical approximation (see Section 6.2.1), it is not precisely true for real molecules. Hence, use of this simple extrapolation of an empirical B_v function to obtain an estimate of the equilibrium bond length requires small corrections (see Section 6.2.1).

For the case of a potential function expressed as a harmonic oscillator with higher-order power-series terms treated as corrections, Kilpatrick and Kilpatrick used perturbation theory to show that power series in $(v + 1/2)$ are indeed a natural way of expressing the v-dependence of the various band constants [9,10]. However, the resulting expressions quickly grow to have an intimidating degree of complexity, and hence do not provide a practical way of defining unique values of parameters such as ω_e, $\omega_e x_e$, $\omega_e y_e$, ..., associated with a given potential energy function. This remains the situation today, and in practice, except for the very lowest-order terms, estimates of the $(v + 1/2)$ expansion coefficients associated with the band constants for a given potential energy function can only be obtained from empirical fits to calculated values of those band constants.

The perturbation theory–based expressions of Kilpatrick and Kilpatrick [9,10] proved difficult to work with and impractical to invert, and there exists no other fully quantum technique for directly inverting discrete spectroscopic data to obtain a potential energy function. This stimulated the development and very wide application of methods based on an approximate semiclassical way of solving the radial Schrödinger equation. While not as accurate as a full quantum treatment, their robustness and ease of use led these semiclassical methods to be the basis of most practical diatomic molecule spectroscopic data analyses for over half a century.

6.2 SEMICLASSICAL METHODS

6.2.1 THE SEMICLASSICAL QUANTIZATION CONDITION

Semiclassical or "phase integral" methods are approximate techniques for solving differential equations, which were well-known to mathematicians in the nineteenth century. They were well-developed in the early days of quantum mechanics, and while not exact, they often provide quite accurate results, particularly for species whose reduced mass is relatively large. The tractability of these methods also made them particularly useful before improvements in digital computers made the application of modern fitting/inversion methods (see Section 6.3) feasible, and the explicit expressions they yield relating patterns of level energies to the nature of the potential energy function are an important source of physical insight. Indeed, semiclassical theory is the basis for much of our understanding of isotope effects in molecular spectra, as well as for some of the most widely used data inversion methods in molecular physics [11,12].

The semiclassical approach to solving the one-dimensional radial Schrödinger equation of Equation 6.1 begins by writing the eigenfunction as

$$\psi(r) = e^{\frac{iS(r)}{\hbar}} \tag{6.9}$$

in which $i \equiv (-1)^{1/2}$. Substituting this expression into Equation 6.1 and removing the common factor $e^{\frac{iS(r)}{\hbar}}$ yields a differential equation for $S(r)$, which is precisely equivalent to the original Schrödinger equation,

$$i\hbar \frac{d^2 S}{dr^2} - \left(\frac{dS}{dr}\right)^2 + 2\mu[E - V_J(r)] = 0 \tag{6.10}$$

Since Planck's constant $\hbar$ is quite small, the first term in this differential equation will be much smaller than the second, so as a zeroth-order approximation (partly corrected for later) it is neglected, yielding a simple first-order equation whose solution is

$$S^{(0)}(r) = \pm\hbar\int^{r} Q(r')\,dr' \tag{6.11}$$

in which $Q(r') \equiv \sqrt{(2\mu/\hbar^2)[E - V_J(r')]}$. Using the second derivative of this result to replace the first term in Equation 6.10 yields an improved first-order differential equation for $S(r)$ whose solution yields the first-order semiclassical approximation for the wave function,

$$\psi^{(1)}(r) = \frac{A}{\sqrt{|Q(r)|}}\exp\left\{\pm i\int^{r} Q(r')\,dr'\right\} \tag{6.12}$$

Equation 6.12 is a fairly good approximation for the wave function except near classical turning points where $Q(r) = 0$ (e.g., near the points $r_1(E)$ and $r_2(E)$ in Figure 6.2a). However, in the immediate neighborhood of such turning points, it is a very good approximation to represent the potential as a linear function of r. The exact solutions of the Schrödinger equation for a linear potential are Airy functions, whose properties are well-known [13]. The combination of Airy functions near the turning points with the semiclassical wave functions of Equation 6.12 in other regions then provides a reasonably good representation for the wave function at all distances.

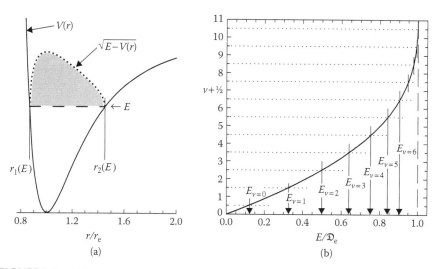

FIGURE 6.2 (a) Integrand of the quantization integral of Equation 6.13. (b) Application of the quantization condition of Equation 6.13 for defining first-order semiclassical eigenvalues.

Finally, we can show that the usual boundary conditions that the wave function must die off in the classically forbidden regions where $E < V_J(r)$ are only satisfied if the integral of $Q(r)$ over the classically allowed region between the inner and outer turning points $r_1(E)$ and $r_2(E)$ (see Figure 6.2a) is precisely equal to a half-integer multiple of π [12]. In other words, the eigenvalues of the given potential are the discrete energies for which

$$v + \frac{1}{2} = \frac{1}{\pi}\sqrt{\frac{2\mu}{\hbar^2}} \int_{r_1(E)}^{r_2(E)} [E - V_J(r)]^{\frac{1}{2}}\, dr \tag{6.13}$$

where v is the (integer!) vibrational quantum number of the level in question. This expression is known as the Bohr–Sommerfeld quantization condition.

Figure 6.2a illustrates the definition of the inner and outer turning points $r_1(E)$ and $r_2(E)$ of a potential energy curve $V(r)$ at a given energy E, and the shaded area shows the integrand of the integral of Equation 6.13. Figure 6.2b then shows how the right-hand side of Equation 6.13 varies as the energy increases from the potential minimum to the dissociation limit of a typical single-minimum potential. Within the first-order semiclassical approximation, the energies at which the dotted horizontal lines at half-integer values of $(v + 1/2)$ intersect this curve define the discrete vibrational level energies of this potential.

Because it is based only on the first-order semiclassical wave function of Equation 6.12, the quantization condition of Equation 6.13 lacks full quantum mechanical accuracy, and as a result it is rarely used for practical eigenvalue calculations. However, its simple form and explicit dependence on the potential energy function mean that for certain types of analytic potential it may be inverted to give explicit expressions for vibration-rotation level energies as functions of the potential function parameters. Moreover, we will see that it can also be inverted to yield a numerical procedure for calculating a pointwise potential function from a knowledge of experimental vibration-rotation level spacings.

The most famous and most widely used potential for which analytic level energy expressions may be obtained from Equation 6.13 is a *Dunham-type* potential, which is the following polynomial expansion about the equilibrium internuclear distance:

$$V_{\text{Dun}}(r) = a_0\xi^2(1 + a_1\xi + a_2\xi^2 + a_3\xi^3 + \cdots) \tag{6.14}$$

in which $\xi \equiv (r - r_e)/r_e$. Since $r = r_e(1 + \xi)$, the centrifugal potential may be expressed as the following power series in ξ:

$$\frac{[J(J+1)]\hbar^2}{2\mu r^2} = \frac{[J(J+1)]\hbar^2}{2\mu(r_e)^2}\left\{1 - 2\xi + 3\xi^2 - 4\xi^3 + \cdots\right\} \tag{6.15}$$

As a result, the overall potential $V_J(r)$ may also be written as a power series in ξ. In 1932, J. L. Dunham showed that on substituting such a polynomial potential into the quantization condition of Equation 6.13 and applying some clever manipulations, we obtain the following explicit power-series expression for the level energies:

$$E(v,J) = \sum_{m=0} \sum_{\ell=1} Y_{\ell,m} \left(v + \frac{1}{2} \right)^{\ell} [J(J+1)]^m \qquad (6.16)$$

in which the coefficients $Y_{\ell,m}$ are explicitly known functions of the potential parameters $\{a_0, a_1, a_2, a_3, \ldots\}$ [14]. This derivation showed that this double power series in $(v+1/2)$ and $[J(J+1)]$ was indeed a natural way to describe level energies [14]. However, until relatively recently (see Section 6.3), the complexity of the algebraic expressions for the higher-order $Y_{\ell,m}$ coefficients, and the even greater complexity arising if one uses and inverts a more accurate higher-order version of the quantization condition, discouraged practical use of these expressions for determining potential energy functions.

While the above discussion considered only the first-order semiclassical approximation, extended versions of Equation 6.13 have been derived, which are based on third-order, fifth-order, and even higher-order semiclassical approximations for the wave functions [15]. In particular, the third-order quantization condition has the form

$$v + \frac{1}{2} = \frac{1}{2\pi} \sqrt{\frac{2\mu}{\hbar^2}} \oint [E - V_J(r)]^{1/2} \, dr + \frac{1}{96\pi} \sqrt{\frac{\hbar^2}{2\mu}} \oint \frac{V''(r)}{[E - V_J(r)]^{3/2}} \, dr \qquad (6.17)$$

in which line integrals have been replaced by contour integrals. The additional power of $\hbar^2$ associated with each higher order of approximation tells us that the accuracy of such treatments quickly approaches that of the full quantum result [16,17].

Dunham's original derivation included some consideration of this leading higher-order correction term [14], and the resulting estimate of the value of this term at the potential minimum ($E = 0$) led to an improved estimate for the value of v associated with the potential minimum [5]

$$v_{min} = -\frac{1}{2} - \delta v_{min} = -\frac{1}{2} - \left\{ \frac{B_e - \omega_e x_e}{4\omega_e} + \frac{\alpha_e}{12B_e} + \frac{\omega_e}{B_e} \left(\frac{\alpha_e}{12B_e} \right)^2 \right\} \qquad (6.18)$$

This in turn allows us to obtain an improved "third-order semiclassical" estimate of the equilibrium bond length from the empirical knowledge of the v-dependence of the vibrational energies and inertial rotation constants represented by Equations 6.6 and 6.7:

$$r_e = r_e^{(3)} = \sqrt{\frac{\hbar^2}{2\mu \, B_{v=v_{min}}}} \qquad (6.19)$$

Equation 6.19 provides the best estimate of a diatomic molecule equilibrium bond length that can be extracted from the conventional empirical expressions for vibrational-rotational level energies (see Equations 6.4, 6.6, and 6.7). The methodologies described in Section 6.3 provide an alternate and arguably more direct way of determining such equilibrium bond lengths. However, it is important to remember that the existence of quantum-mechanical zero-point energy means that this equilibrium distance does not describe the actual effective bond length of any real molecule. Our best estimate of that distance would be the average bond length of the molecule in its ground vibration-rotation level,

$$\bar{r}_{0,0} = \left\langle \psi_{0,0}(r) | r | \psi_{0,0}(r) \right\rangle \tag{6.20}$$

where $\psi_{v,J}(r)$ is the radial wave function in vibration-rotation state (v, J). However, there is no direct empirical way of determining this quantity, and to calculate it requires a knowledge of the potential energy function, which would allow us to determine the radial wave function for the zero-point level. This indicates that determination of precise values for real diatomic molecule bond lengths cannot be achieved simply from manipulation of empirical molecular constants, but also requires a knowledge of the potential energy function.

In closing, note that the real ground-state bond length $\bar{r}_{0,0}$ usually differs significantly from both r_e and the average bond length implied by the inertial rotational constant for the ground vibrational level, $\left\langle r^{-2} \right\rangle^{-1/2} = \sqrt{\hbar^2/(2\mu B_{v=0})}$. For example, for the ground state of H_2 $\bar{r}_{0,0} = 1.034$ $r_e = 1.021 \left\langle r^{-2} \right\rangle^{-1/2}$ [18]. The very large differences between these values provide a sober warning about the danger of thinking of the type of average bond length obtained from empirical inertial rotation constants B_v as being a proper measure of the actual average bond length in a given vibrational level.

6.2.2   THE RYDBERG–KLEIN–REES INVERSION PROCEDURE

By 1929, even before Dunham's landmark papers, the drawbacks of trying to work with model potentials for which exact analytic quantum mechanical eigenvalue solutions existed were becoming evident, since the few such functions available were not flexible enough to account fully for the experimental data spanning a significant fraction of a potential well. As a result, researchers began to investigate the use of semiclassical methods.

A key pioneer in this area was O. Oldenberg, who noticed that, according to Equation 6.13, the rate at which v changed with vibrational energy E_v depended on the rate at which the width of the potential well increased with energy [19]. R. Rydberg was another pioneer, who noticed that the concomitant change in the inertial constant B_v constrained the asymmetry of that rate of growth [20].

A formal derivation based on these observations was reported by O. Klein in 1932 [21]. He began by taking the derivative of Equation 6.13 with respect to the vibrational energy and then broke the integral into two segments as follows:

$$\frac{dv}{dE} = \frac{1}{2\pi}\sqrt{\frac{2\mu}{\hbar^2}} \left\{ \int_{r_1(E)}^{r_e} \frac{dr}{[E - V(r)]^{1/2}} + \int_{r_e}^{r_2(E)} \frac{dr}{[E - V(r)]^{1/2}} \right\} \tag{6.21}$$

Replacing the integration over r by integration over values of the potential energy with increment du then yielded the expression

$$\frac{dv}{dE} = \frac{1}{2\pi}\sqrt{\frac{2\mu}{\hbar^2}}\int_0^E\left\{\frac{dr_2(u)}{du} - \frac{dr_1(u)}{du}\right\}\frac{du}{[E-u]^{1/2}} \tag{6.22}$$

The next steps consisted of replacing E in the above expression by E', multiplying both sides by the factor $dE'/[E - E']^{1/2}$, interchanging the order of the resulting double integration on the right-hand side, integrating over the variable E' from its minimum (the current value of u) to E, and making use of the mathematical identity

$$\int_u^E \frac{dE'}{\sqrt{(E-E')(E'-u)}} = \pi \tag{6.23}$$

When this is done, the remaining integral on the right-hand side of the equation collapses to the difference $r_2(E) - r_1(E)$. Finally, replacing the integration over u by integration over the vibrational quantum number v' yielded the first of the Rydberg–Klein–Rees (RKR) equations:

$$r_2(v) - r_1(v) = 2\sqrt{\frac{\hbar^2}{2\mu}}\int_{-1/2}^v \frac{dv'}{[G_v - G_{v'}]^{1/2}} = 2f \tag{6.24}$$

In this final result, we have taken the liberty of replacing the symbol for the energy E by the notation G_v normally used for vibrational energy, and made use of the fact that within the first-order quantization condition of Equation 6.13, the potential minimum corresponds to $v' = -1/2$.

The derivation of the second equation starts by taking the partial derivative of Equation 6.13 with respect to the factor $[J(J + 1)]$, which defines the strength of the centrifugal contribution to the potential, and then setting $J = 0$.

$$\left(\frac{\partial v}{\partial[J(J+1)]}\right)_E = \left(\frac{\partial v}{\partial E}\right)_J \left(\frac{\partial E}{\partial[J(J+1)]}\right)_v = B_v\left(\frac{\partial v}{\partial E}\right)_{J=0}$$

$$= -\frac{1}{2\pi}\sqrt{\frac{2\mu}{\hbar^2}}\int_{r_1(E)}^{r_2(E)} \frac{dr}{r^2[E-V(r)]^{1/2}} \tag{6.25}$$

Breaking the range of integration in two at r_e and applying the same manipulations described above then yields the second RKR equation,

$$\frac{1}{r_1(v)} - \frac{1}{r_2(v)} = 2\sqrt{\frac{2\mu}{\hbar^2}}\int_{-1/2}^v \frac{B_{v'}\,dv'}{[G_v - G_{v'}]^{1/2}} = 2g \tag{6.26}$$

Combining Equations 6.24 and 6.26 then yields the following final expressions for the turning points:

$$r_2(v) = \left(f^2 + \frac{f}{g} \right)^{1/2} + f \tag{6.27}$$

$$r_1(v) = \left(f^2 + \frac{f}{g} \right)^{1/2} - f \tag{6.28}$$

In spite of their elegance and obvious potential utility, Klein's equations saw little practical use for over three decades. One reason for this would have been the practical difficulty of evaluating the Klein integrals accurately before the availability of digital computers. The nature of this problem is illustrated by the plots for the ground electronic state of Ca_2 shown in Figure 6.3. Figures 6.3a and b shows the nature of the $G_{v'}$ and $B_{v'}$ functions, while Figure 6.3c shows the integrands of Equations 6.24 and 6.26 defined by those functions for a representative vibrational level, $v = 26$. Although the areas under these curves are finite, the fact that the integrands go to infinity at the upper bound makes the accurate evaluation of these integrals somewhat challenging.

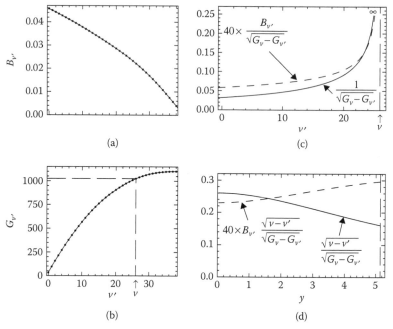

FIGURE 6.3 (a–b) Spectroscopic properties of Ca_2. (c) Integrands of the Klein integrals of Equations 6.24 and 6.26 for level $v = 26$ of Ca_2; a numerical factor of 40 has been introduced in order to place the two integrands on the same vertical scale. (d) Integrands of the transformed Klein integrals of Equations 6.29 and 6.30 for the case considered in (c). Units for energy are cm^{-1} in all figures.

In 1947, A.L.G. Rees pointed out that the two Klein integrals could be evaluated in closed form if $G(v)$ and B_v were represented by sets of quadratic polynomials in v for different segments of the range of integration [22]. This contribution led to his name being attached to the method, but the inconvenience of having to fit the data piecewise to sets of quadratics, and the associated lack of overall smoothness, meant that it still saw little use. Finally, by the early 1960s a number of groups had developed computer programs for evaluating these integrals for any user-selected expressions for $G(v)$ and B_v, and the RKR method quickly grew to become ubiquitously associated with diatomic molecule data analyses. However, truly efficient techniques for evaluating the Klein integrals that take proper account of the singularities in the integrand were not reported until 1972 [23–25].

One technique for evaluating these integrals accurately is simply to introduce a transformation that removes the singularities. In particular, introduction of the auxiliary variable, $y = \sqrt{v - v'}$, transforms Equations 6.24 and 6.26 into the following forms:

$$r_2(v) - r_1(v) = 4\sqrt{\frac{\hbar^2}{2\mu}} \int_0^{\sqrt{v+1/2}} \left\{ \sqrt{\frac{v-v'}{G_v - G_{v'}}} \right\} dy = 2f \tag{6.29}$$

$$\frac{1}{r_1(v)} - \frac{1}{r_2(v)} = 4\sqrt{\frac{2\mu}{\hbar^2}} \int_0^{\sqrt{v+1/2}} \left\{ B_{v'} \sqrt{\frac{v-v'}{G_v - G_{v'}}} \right\} dy = 2g \tag{6.30}$$

As is illustrated by Figure 6.3d, the integrands in these expressions are smooth and well-behaved and have no singularities(!), so a very modest amount of computational effort will be able to yield turning points converged to machine precision. A particularly convenient procedure is to apply a simple N–point Gauss–Legendre quadrature procedure to the whole interval, and then bisect that interval and apply the same procedure to both halves. At each such stage of subdivision, the error will decrease by a factor of $1/2^{N-2}$ [13]; for $N = 12$ this means an error reduction by three orders of magnitude at each stage of bisection.

Remember that while experimental data are only associated with integer values of v, in order to evaluate these integrals the vibrational energy and rotational constant expressions G_v and B_v must be continuous functions of v. Moreover, as illustrated by Figure 6.2b, the quantization integral of Equation 6.13 may be evaluated for *any* energy E (or G_v), independent of whether or not it corresponds to an integer value of v. Thus, we are free to solve the RKR equations and evaluate turning points for any chosen mesh of integer or noninteger v values. This is quite important, because solving the Schrödinger equation numerically requires an interpolation procedure to provide a mesh of accurate potential function values at distances which will not correspond to the calculated turning points, and if one was restricted to turning points at integer v, such interpolations would often not be highly accurate, in spite of the fact that the calculated turning points would be smooth to machine precision.

Two other practical considerations intrude upon the use of RKR potentials. One is the perhaps obvious, but sometimes overlooked point that calculated turning points cannot really be trusted beyond the vibrational range of the experimental data used to determine the G_v and B_v functions. This restriction is partially lifted if "near-dissociation expansions" of the type described in Section 6.2.3 are used to represent G_v and B_v. However, use of the resulting potential to generate reliable Schrödinger equation solutions would still require smoothly attached functions for extrapolating inward and outward beyond the range of the calculated turning points.

The second practical concern arises from the fact that shortcomings of the experimentally derived functions characterizing G_v and B_v will give rise to errors in calculated RKR turning points. Since the repulsive inner wall of a potential function is very steep, especially at high energies, such errors often manifest themselves as nonphysical behavior of the inner wall of the potential. For example, rather than have a (negative) slope and positive curvature, which vary slowly with energy, this inner wall might pass through an inflection point and take on negative curvature, or it may turn outward with increasing energy, with the slope becoming positive. In practice, the experimental G_v function is usually defined with greater relative accuracy than is the B_v function. However, whatever the source of the problem, a modest degree of inappropriate behavior of either the G_v or B_v function can give rise to nonphysical behavior of the inner wall of the potential, since the expected monotonic increase in slope with energy will greatly amplify the effect of even very small errors in the f and/or g integrals. Thus, one should always examine the behavior of the inner wall of any calculated RKR potential, and if the slope deviates from smooth behavior with positive curvature, it should be smoothed or replaced with a physically sensible extrapolating function.

Although small relative errors in the f or g integral can make the curvature or slope of the high-energy inner wall change in an unacceptable nonphysical manner, the rapid growth of the f integral with increasing G_v means that the width of the potential $[r_2(v) - r_1(v)]$ as a function of energy may still be relatively well-defined by Equation 6.24 or 6.29, even when the directly calculated inner potential wall is unreliable. In this case, combining this directly calculated well-width function with a reasonable extrapolated inner potential wall would yield a "best" estimate of the upper portion of the potential (a procedure first introduced by R. D. Verma [26]). Similarly, even in the complete absence of rotational data, a combination of the well-width information yielded by the calculated f integrals with an inner wall defined by a model such as a Morse potential can give a realistic overall potential function [27]. A "black box" computer code (accompanied by a manual) for performing RKR calculations, which allows the use of a variety of possible expressions for G_v and B_v and takes account of the practical concerns described above, is available on the Internet [28].

Finally, remember that the manipulations of Equation 6.13 to obtain the RKR equations (6.24) and (6.26) [or (6.29) and (6.29)] are mathematically exact! In other words, within the first-order semiclassical or Wentzel–Kramers–Brillouin (WKB) approximation [12], this method yields a unique potential energy function that *exactly* reflects the input functions representing the v-dependence of the vibrational energy G_v and inertial rotational constant B_v. A nagging weakness, however, is the fact that the quantization condition of Equation 6.13 is *not* exact, so quantum mechanical properties of an RKR potential will not agree precisely with the input G_v and B_v data used to generate that potential.

Table 6.1 illustrates this point for four species for which accurate and extensive G_v and B_v functions are available from the literature. Those functions were used to

TABLE 6.1

Root Mean Square Errors in Vibrational Level Spacings and Rotational Constants Calculated from Rydberg–Klein–Rees Potentials for Selected Molecules (Energies in cm^{-1})

Molecule	μ	$\mathfrak{D}_e$	v_{max}	$\dfrac{G_{v_{max}}}{\mathfrak{D}_e}$	Error$\{\Delta G_{v+\frac{1}{2}}\}$	Percent Error$\{B_v\}$
BeH	0.906	17590	9	0.895	0.527	0.031
N$_2$	7.002	79845	20	0.529	0.052	0.0026
Ca$_2$	19.981	1102	25	0.916	0.00079	0.0021
Rb$_2$	42.456	3993	85	0.916	0.00017	0.0013

generate RKR potentials, and an exact quantum procedure [3] was used to calculate the associated vibrational level spacings and inertial rotational constants. The last two columns of Table 6.1 show the root mean square differences between those calculated properties and the G_v and B_v functions used to generate the original RKR potential. In each case, the range considered was truncated at $G_{v_{max}}$, which is the smaller of the upper end to the range of the experimental data used to determine the G_v and B_v functions, or the point at which the onset of irregular behavior of the inner-wall turning points (see page 172) required smoothing and inward extrapolation to be applied.

These results show that errors in RKR potentials due to neglect of the second term in Equation 6.17 are largest for species with small reduced mass. For a hydride, they are quite significant, but their importance drops rapidly with increasing reduced mass, and for $\mu \geq 20$ u (Ca$_2$ and Rb$_2$) the vibrational spacing discrepancies are smaller than typical experimental vibrational energy uncertainties. However, such discrepancies add up, and even for these "heavy" species, the accumulated error in the vibrational energy can be significant. Overall, although the situation is less satisfactory for light molecules, the first-order semiclassical nature of the RKR procedure has only a modest negative effect on the quality of the resulting potential, or of quantities calculated from it. At the same time, the fact that RKR potentials are defined as sets of many-digit turning points that often need to be smoothed on the inner wall, and always need extrapolation functions attached at their inner and outer ends, are persistent inconveniences. These problems are resolved, however, by use of the methodology described in Section 6.3.

In conclusion, we note that the RKR method itself provides no new information about equilibrium structures beyond that implicit in the first-order semiclassical result of Equation 6.8. Although the higher-order quantization condition of Equation 6.17 is not amenable to the exact inversion procedures described above, it has been suggested that better-than-first-order results could be obtained simply by replacing the lower bound on the integrals of Equations 6.24 and 6.26 by $v_{min} = -1/2 - \delta v_{min}$ from Equation 6.18 [29]. Unfortunately, tests analogous to those of Table 6.1 show that while this does give somewhat better results near the potential minimum, at higher v the discrepancies are larger than those yielded by the usual first-order method.

6.2.3 NEAR-DISSOCIATION THEORY

The preceding discussion shows that the RKR method can give quite an accurate potential spanning the range of vibrational energies for which experimental data are available. However, it offers no advice regarding how to address the question posed by Figure 6.4a: how to estimate the number, energies, and other properties of levels lying above the highest one observed, or even to estimate the distance from that last observed level to the dissociation limit.

Figure 6.4b illustrates a graphical means of addressing this question, which was introduced by Birge and Sponer in 1926 [30], and remained the method of choice for most of the next half century. In a Birge–Sponer plot, the vibrational level spacings $\Delta G_{v+\frac{1}{2}} \equiv G_{v+1} - G_v$ are plotted versus the vibrational quantum number, with the points placed at half-integer values of the abscissa. In this diagram, the numerical $\Delta G_{v+\frac{1}{2}}$ value is equal to the area of the narrow vertical rectangle whose upper edge is centered at that point. As a result, the sum of the areas of the six rectangles shown there is equal to the sum of the six $\Delta G_{v+\frac{1}{2}}$ values, which of course is the distance from level $v = 0$ to level $v = 6$.

It is immediately clear that the area under a smooth curve through these points from $v = 0$ to 6 is a very good approximation to that energy difference. Birge and Sponer then pointed out that if this curve was extrapolated to cut the v axis, the area under the curve in the extrapolation region would be a very good approximation to the distance from the highest observed level to the dissociation limit. Moreover, the points where that extrapolated curve crossed half-integer v value gives predicted vibrational spacings for unobserved levels, all the way to the limit. If these predictions were correct, an RKR potential could then be calculated for the whole well.

The only problem with Birge–Sponer plots is the uncertainty regarding how to perform the extrapolation, a problem which remained an open question for 44 years.

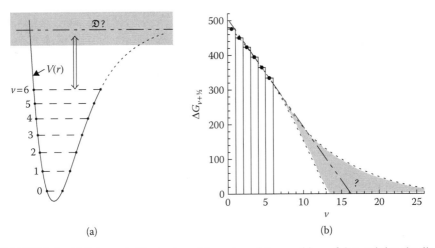

(a)   (b)

FIGURE 6.4 (a) Schematic illustration of the extrapolation problem of determining the dissociation limit $\mathfrak{D}$. (b) A Birge–Sponer plot in which the shaded area illustrates the uncertainty associated with conventional vibrational extrapolation.

The dash–dot–dot line in Figure 6.4b shows a linear extrapolation through the last two experimental points, while the dotted curves bounding the shaded region are plausible alternative extrapolations with negative versus positive curvature. The ratio of the area of the shaded region relative to the overall area under the curve in the extrapolation region is then an indication of the relative uncertainty in the distance from the last observed level to the dissociation limit. Unfortunately, it is clear that this uncertainty could be 50–100%!

A solution to this problem was finally reported in 1970 [31,32]. It was based on the realization that another type of potential for which an explicit analytic expression for the vibrational level energies may be obtained from Equation 6.13 is the attractive inverse-power function $V(r) = \mathfrak{D} - C_n/r^n$ whose form matches the limiting long-range behavior of all intermolecular interactions. As in the RKR method, the derivation is remarkably straightforward.

Since we are interested in the *distribution* of vibrational levels near dissociation, the derivation begins by taking the derivative of Equation 6.13 with respect to the vibrational level energy to obtain the following expression for the density of states at energy G_v (for $J = 0$):

$$\frac{dv}{dG_v} = \frac{1}{2\pi}\sqrt{\frac{2\mu}{\hbar^2}} \int_{r_1(v)}^{r_2(v)} \frac{dr}{[G_v - V(r)]^{1/2}} \tag{6.31}$$

Let us now consider the nature of the integrand appearing in Equation 6.31. For a model Lennard–Jones (12, 6) potential function

$$V_{LJ}(r) = \frac{C_{12}}{r^{12}} - \frac{C_6}{r^6} + \mathfrak{D}_e = \mathfrak{D}_e\left[\left(\frac{r_e}{r}\right)^6 - 1\right]^2 \tag{6.32}$$

which supports 24 vibrational levels, Figure 6.5b shows a plot of that potential and indicates the positions of the energies and turning points of selected levels. Figure 6.5a then shows the nature of the integrand in Equation 6.31 for those four levels; note that while the integrand goes to infinity at both turning points, the area under the curve is always finite. It is immediately clear that for the higher vibrational levels, the area under the curve—and hence the value of the integral—is increasingly dominated by the nature of the integrand (i.e., of the potential) in the long-range region near the outer turning point.

From the early days of quantum mechanics, scientists have known that at long range all atomic and molecular interaction potentials become a sum of inverse-power terms,

$$V(r) \simeq \mathfrak{D} - \sum_{m=n} \frac{C_m}{r^m} \underset{\{\text{very large } r\}}{\Rightarrow} \mathfrak{D} - \frac{C_n}{r^n} \tag{6.33}$$

in which the powers m and coefficients C_m are determined by the nature of the interacting atoms. (A brief summary of the rules governing the terms that appear in this sum for a given case is presented in the appendix to this chapter.) This suggests that for levels whose outer turning points lie at sufficiently large r for the leading (C_n/r^n)

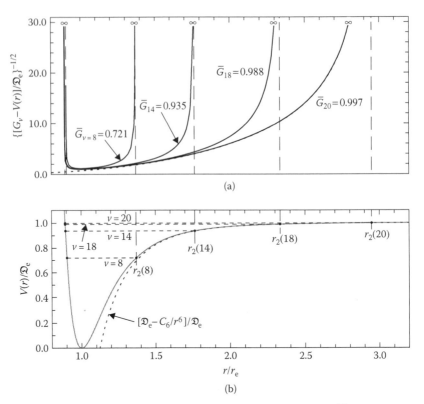

FIGURE 6.5 (a) Integrand of Equation 6.31 for selected levels with $\overline{G}_v \equiv G_v/\mathfrak{D}_e$. (b) A 23-level LJ(12,6) potential with selected level energies and turning points labeled.

term to dominate the interaction, it would be a reasonable approximation to replace $V(r)$ in Equation 6.31 by the simple function $V(r) \approx \mathfrak{D} - C_n/r^n$ to obtain

$$\frac{dv}{dG_v} \approx \frac{1}{2\pi}\sqrt{\frac{2\mu}{\hbar^2}} \int_{r_1(v)}^{r_2(v)} \frac{dr}{\left[G_v - \left(\mathfrak{D} - \left(\dfrac{C_n}{r^n}\right)\right)\right]^{1/2}} \qquad (6.34)$$

Making the substitution $y = r/r_2(v)$ and noting that $[G_v - V(r_2(v))] = 0$, and hence that $[\mathfrak{D} - G_v] = C_n/[r_2(v)]^n$, Equation 6.34 becomes

$$\frac{dv}{dG_v} \approx \frac{1}{2\pi}\sqrt{\frac{2\mu}{\hbar^2}} \frac{(C_n)^{1/n}}{[\mathfrak{D} - G_v]^{(n+2)/2n}} \int_{r_1/r_2}^{1} \frac{dy}{(y^{-n} - 1)^{1/2}} \qquad (6.35)$$

The dotted curve in Figure 6.5a shows what happens to the exact integrand of Equation 6.31 for level $v = 20$ if the actual potential is replaced by the single inverse-power term $\mathfrak{D} - C_6/r^6$. It is immediately clear that both the effect of this substitution on the value of this integrand and the effect of replacing the lower bound of the integral in Equation 6.35 by zero will be very small and will become increasingly negligible for higher vibrational levels (here, $v = 21$–23). Making use of the mathematical identity

$$\int_0^1 \frac{dy}{(y^{-n}-1)^{1/2}} = \frac{\pi^{1/2}}{n} \frac{\Gamma(\frac{1}{2}+\frac{1}{n})}{\Gamma(1+\frac{1}{n})} \tag{6.36}$$

and inverting the resulting expression then gives the following basic near-dissociation theory (NDT) result:

$$\frac{dG_v}{dv} = \left\{ \frac{2n\sqrt{\pi}}{(C_n)^{1/n}} \sqrt{\frac{\hbar^2}{2\mu} \frac{\Gamma(1+\frac{1}{n})}{\Gamma(\frac{1}{2}+\frac{1}{n})}} \right\} [\mathfrak{D}-G_v]^{(n+2)/2n} = K_n [\mathfrak{D}-G_v]^{(n+2)/2n} \tag{6.37}$$

It is usually more convenient to work with the integrated form of this equation; this is the central result that

$$G_v = \mathfrak{D} - X_0(n)(v_\mathfrak{D} - v)^{2n/(n-2)} \tag{6.38}$$

in which $X_0(n) = \left[\frac{(n-2)}{2n} K_n \right]^{2n/(n-2)}$. For $n > 2$, the integration constant $v_\mathfrak{D}$ is the noninteger effective vibrational index associated with the dissociation limit—the intercept of the correctly extrapolated Birge–Sponer plot for the given system—and its integer part $\bar{v}_\mathfrak{D}$ is the index of the highest vibrational level supported by the given potential. For $n = 1$, this expression becomes the Bohr eigenvalue formula for the levels of a Coulomb potential, and $v_\mathfrak{D}(n = 1) = -(1 + \delta)$, where δ is the Rydberg quantum defect. An attractive $n = 2$ long-range potential is not physically possible for a diatomic molecule, but integration of Equation 6.37 for that case gives essentially the same exponential eigenvalue expression known from quantum mechanics. In order to express this result in a practical form, it is convenient to take the first derivative of Equation 6.38 to obtain

$$\frac{dG_v}{dv} = \left[\left(\frac{2n}{n-2} \right) X_0(n) \right] (v_\mathfrak{D} - v)^{(n+2)/(n-2)} \tag{6.39}$$

Since the vibrational level energies and level spacings are the actual physical observables, we then use the fact that $\left[\frac{dG_{v'}}{dv'} \right]_{v'=v+1/2} \approx \Delta G_{v+1/2}$ and rearrange Equations 6.37, 6.38, and 6.39 to give the following expressions:

$$(\Delta G_{v+1/2})^{2n/(n+2)} = [K_n]^{2n/(n+2)} (\mathfrak{D} - G_{v+1/2}) \tag{6.40}$$

$$(\mathfrak{D} - G_v)^{(n-2)/2n} = [X_0(n)]^{(n-2)/2n} (v_\mathfrak{D} - v) \tag{6.41}$$

$$(\Delta G_{v+1/2})^{(n-2)/(n+2)} = \left[\left(\tfrac{2n}{n-2}\right)X_0(n)\right]^{(n-2)/(n+2)}\left(v_{\mathfrak{D}} - v - \frac{1}{2}\right) \qquad (6.42)$$

Thus, NDT predicts that if the observable quantities on the left-hand side of these equations are plotted versus the vibrational mid-point energy $G_{v+1/2} \approx \frac{1}{2}(G_{v+1} + G_v)$ (for Equation 6.40) or the vibrational quantum number v (for the others), for levels lying close to dissociation those plots should be precisely linear, with slopes defined by the constants K_n or $X_0(n)$ (i.e., by μ, n and C_n), while the intercept determines either the energy at the dissociation limit $\mathfrak{D}$ or the vibrational intercept $v_{\mathfrak{D}}$. Plots of this type, sometimes called Le Roy-Bernstein plots, are often used to illustrate applications of NDT.

NDT expressions analogous to Equation 6.38 have been reported for a number of other properties, such as expectation values of the kinetic energy or of powers of the internuclear distance, and for values of the rotational constants B_v, D_v, H_v, ... While it has little direct import for the present discussion, it is interesting to note the algebraic structure of the latter, as it explains the reason for the subscript on the symbol $X_0(n)$ appearing in Equation 6.38. In particular,

$$B_v = X_1(n)\,(v_{\mathfrak{D}} - v)^{\frac{2n}{n-2}-2} \qquad (6.43)$$

$$D_v = -X_2(n)\,(v_{\mathfrak{D}} - v)^{\frac{2n}{n-2}-4} \qquad (6.44)$$

$$H_v = X_3(n)\,(v_{\mathfrak{D}} - v)^{\frac{2n}{n-2}-6} \qquad (6.45)$$

$$L_v = X_4(n)\,(v_{\mathfrak{D}} - v)^{\frac{2n}{n-2}-8} \qquad (6.46)$$

...

where $X_k(n) = \dfrac{\overline{X}_k(n)}{[\mu^n(C_n)^2]^{1/(n-2)}}$ and the $\overline{X}_k(n)$ are known numerical factors [33,34].

One type of application of these results is summarized in Figure 6.6. It illustrates an NDT treatment of data for the ground electronic state of the very weakly bound Van der Waals molecule Ar$_2$, which was first observed in 1970 [35]. The square points are the experimental vibrational level spacings and the dash–dot–dot line is the conventional linear Birge–Sponer (B-S) extrapolation (left-hand ordinate axis) reported by the experimentalists, the shaded area defining their estimate of the distance from the highest observed level ($v = 4$) to the dissociation limit. This approach clearly predicts that $v = 5$ is the highest bound level of this molecule.

As with all molecular states formed from atoms in electronic S states, $n = 6$ for the ground electronic state of Ar$_2$ (see the Appendix). The round points in Figure 6.6 then show exactly those same experimental data plotted (against the right-hand axis) in the manner suggested by Equation 6.42. Since the data for the lowest bound levels are not expected to obey our NDT equation, one would not trust a simple linear fit

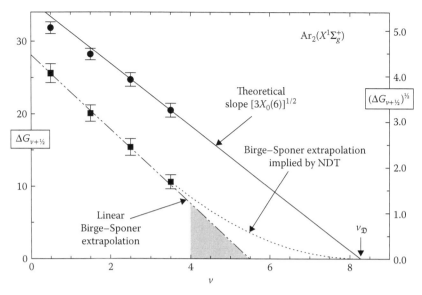

FIGURE 6.6 Illustrative application of near-dissociation theory to data for Ar_2. Left axis: square points, dash–dot–dot line and dotted curve. Right axis: round points and solid line. (Adapted from Le Roy, R. J. 1972. *J. Chem. Phys.* 57, 573–4.)

to these data to provide a reliable extrapolation. However, an accurate value of the C_6 coefficient for this species was available from theory, so the expected limiting slope of this plot could be predicted from the resulting known value of the $X_0(n)$ coefficient.

The solid line on this plot shows the NDT prediction of the extrapolation obtained when a line with this theoretical slope passes through the experimental datum for $v = 3$. The fact that the second-last point also lies on this line while those for the two larger level spacings only gradually deviate from it attests to the validity of this extrapolation. The value of $v_{\mathfrak{D}} = 8.27$ implied by this NDT extrapolation shows that this molecule actually has 50% more bound levels than was implied by the linear B-S type extrapolation, and comparison of the shaded area with the area under the dotted curve in the extrapolation region shows that the estimate of the distance from the highest observed level to dissociation yielded by the traditional B-S extrapolation was too small by a factor of more than two [36].

A second type of application of NDT is the use of Equation 6.41 in the analysis of photoassociation spectroscopy (PAS) data, for which the measured observable is the binding energy $[\mathfrak{D} - G_v]$. The $1^1\Sigma_u^+$ state of Yb_2 dissociates to yield one $^1S_0$ atom and one $^1P_1$ atom, a case for which $n = 3$ (see the Appendix). Hence, Equation 6.41 shows that for levels lying near dissociation, a plot of $[\mathfrak{D} - G_v]^{1/6}$ is expected to be linear with a slope of $[X_0(3)]^{1/6}$ determined by the value of the C_3 coefficient for this state and the intercept by its $v_{\mathfrak{D}}$ value. Figure 6.7 shows a plot of this type based on the recent results of Takahashi et al. [37]. The precise linearity of the points in Figure 6.7 over a range of almost 80 vibrational levels is a very strong endorsement of the validity of Equations 6.38 through 6.42, and it illustrates the fact that NDT provides the most reliable method

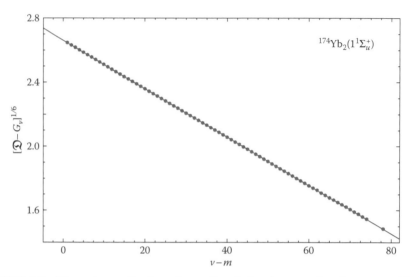

FIGURE 6.7 Illustrative application of near-dissociation theory to data for a state of Yb$_2$ for which $n = 3$. The unspecified integer m indicates that the absolute vibrational assignment is not known.

known for experimentally determining values of long-range C_n potential function coefficients.

The two cases considered above are both situations in which experimental data are available for levels lying sufficiently close to dissociation that NDT may be expected to be valid there. However, in the much more common situations in which this is not true, NDT still offers a valuable way of obtaining optimal estimates of the distance from the highest observed levels to dissociation, and of the number and energies of unobserved levels. For this purpose we introduce the use of near-dissociation expansions (NDEs), expressions that combine the limiting functional behavior of Equation 6.38 with empirical expansions that account for deviations from that limiting behavior. Most work of this type has involved the use of rational polynomials in the variable $(v_{\mathfrak{D}} - v)$ as follows:

$$G_v = \mathfrak{D} - X_0(n)\,(v_{\mathfrak{D}} - v)^{2n/(n-2)}\left[\frac{L}{M}\right]^s \tag{6.47}$$

where the power s is set at either $s = 1$ (to yield "outer" expansions) or $s = 2n/(n-2)$ (to yield "inner" expansions), and

$$\left[\frac{L}{M}\right] = \frac{1 + \displaystyle\sum_{i=1}^{L} p_{t+i}(v_{\mathfrak{D}} - v)^{t+i}}{1 + \displaystyle\sum_{j=1}^{M} q_{t+j}(v_{\mathfrak{D}} - v)^{t+j}} \tag{6.48}$$

while the value of t is determined by the theoretically known form of the leading correction to the limiting behavior of Equation 6.38 [38].

The fundamental *ansatz* underlying the use of NDEs is that fitting experimental data to expressions (such as Equation 6.47) that incorporate the correct theoretically known limiting near-dissociation behavior, will yield more realistic estimates of the physically significant extrapolation parameters $\mathfrak{D}$ and $v_{\mathfrak{D}}$ than could otherwise be obtained. In effect, it replaces *blind empirical extrapolation* using Dunham-type polynomials by *interpolation* between experimental data for levels in the lower part of the potential well and the exactly known functional behavior at the limit. Moreover, such expressions often provide more compact representations of the data than do conventional power series in $\left(v + 1/2\right)$.

Figure 6.8 summarizes the results of performing NDE fits to experimental data for the $A^2\Pi$ state of MgAr$^+$ [39]. Since this species is a molecular ion, the (inverse) power of the leading term in its long-range potential is $n = 4$, and since at least one of its dissociation fragments is in an S state, the power of the second term is $m = 6$ (see the Appendix). For this case, the theory shows that the power t in Equation 6.48 should be $t = 2$ [38]. Theory also tells us that for *any* molecular ion the value of the C_4 coefficient in atomic units is $\alpha/2$, where α is the polarizability of the neutral dissociation fragment, so the value of the limiting NDT coefficient $X_0(4)$ is readily obtained. Moreover, a good theoretical estimate of the C_6 coefficient could be generated for this state, so a realistic value of the leading-deviation coefficient p_2 (for fixed $q_2 = 0$) could also obtained [39].

The dotted line in Figure 6.8 shows the limiting slope $[4X_0(4)]^{1/3}$ defined by the known C_4 coefficient, while the dot–dash curve labeled "linear B-S extrapolation" shows the extrapolation behavior implied by a linear Birge–Sponer plot. The cluster of seven dashed curves shows the results of NDE fits for different $\{L, M, s\}$ in which the C_4 coefficient was held fixed at the theoretical value and two $\{p_i, q_j\}$ parameters were

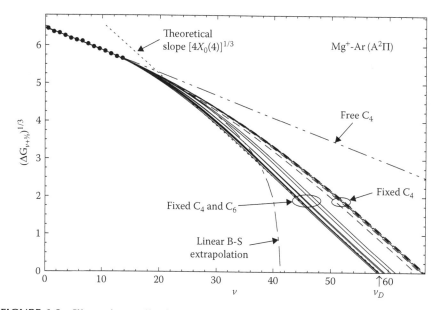

FIGURE 6.8 Illustrative application of near-dissociation expansions fitting to data for the $A^2\Pi$ state of MgAr$^+$.

allowed to vary, with no C_6-based constraint being applied to the p_2 value. The cluster of nine solid curves then shows the results of fits in which both $X_0(C_4)$ and $p_2(C_4, C_6)$ were fixed at the theoretical values, and again two $\{p_i, q_j\}$ parameters were allowed to vary (as well as $v_\mathfrak{D}$ and $\mathfrak{D}$). The quality of fit for all these cases was essentially the same. It is clear that for a given type of model (i.e., only $X_0(4)$ fixed, vs. $X_0(4)$ and p_2 fixed), the NDE models corresponding to different choices of $\{L, M, s\}$ are in reasonably good agreement with one another. However, the difference between the extrapolation behavior for these two classes of models shows that when better theoretical constraints are applied, significantly better extrapolation behavior is attained. For reference, the dash–dot–dot curve labeled "Free C_4" shows that in the absence either of a realistic value of the leading long-range C_n coefficient or of data for levels lying near dissociation, NDE fits can give quite unrealistic extrapolations and should not be trusted.

A final point raised by this example is the question of model-dependence, which is an ever-present, but usually ignored, problem in scientific data analysis. While all of the nine models corresponding to "Fixed C_4 and C_6" give fits to the data of equivalent quality, they all extrapolate slightly differently, and the associated values of the physically interesting parameters $\mathfrak{D}$ and $v_\mathfrak{D}$ differ by substantially more than the parameter uncertainty associated with any individual fit. In cases such as this there is no possibility of selecting a unique "best" model, since there is no physical basis for choosing one set of $\{L, M, s\}$ values over another. The best one can do is to consider as wide a range of models as possible and then average the resulting values of the physically interesting parameters and estimate their uncertainties based on both the variance about their mean and the uncertainties in the individual values. A practical scheme for carrying this out, which was introduced in reference [39], led to the value of $v_\mathfrak{D} = 58.4(\pm 1.2)$ indicated by the pointer at the bottom of Figure 6.8.

On completion of a study such as that illustrated by the results shown in Figure 6.8, one would then choose a representative "optimal" NDE function for the vibrational energies and use it in an RKR calculation to generate a potential spanning essentially the entire potential energy well. Analyses of this type have been performed for a number of molecular systems. In general, fits of vibrational energies and rotational constants to NDEs tended to be somewhat more compact than conventional Dunham polynomials—fewer parameters being required to yield a given quality of fit. However, the interparameter correlation increases rapidly with the number of free $\{p_i, q_j\}$ parameters, and it becomes increasingly difficult to obtain sufficiently realistic preliminary estimates of those parameters for the nonlinear fit to be stable.

Tellinghuisen and Ashmore addressed this problem by introducing "mixed representations" for G_v and B_v, in which conventional Dunham polynomials are used at low v and NDEs at high v, with a switching function merging the two domains [40,41]. Such representations certainly work, and they have been implemented in standard data analysis [41] and RKR programs [28]. However, the complexity of these mixed representations makes them somewhat inconvenient to use, and (to date) neither pure NDEs nor these mixed representations have been widely adopted. Indeed, in recent years the whole approach of attempting to provide global descriptions of molecular vibrational-rotational energies using expansions in terms of vibration-rotation quantum numbers is increasingly being supplanted by the DPF approach described in Section 6.3.

6.2.4 CONCLUSIONS REGARDING SEMICLASSICAL METHODS

Since 1932, fits to the Dunham eigenvalue expression of Equation 6.16 have been a central tool in empirical analyses of diatomic molecule spectroscopic data, and since the early 1960s, use of the resulting G_v and B_v expressions in the RKR procedure has been a ubiquitous technique for determining diatomic potential energy functions. Replacing those expressions by NDEs or the mixed representation functions described on page 190 offered a way of addressing a primary weakness of the Dunham polynomial description—its inability to provide realistic extrapolation behavior. However, the undesirable complication of the latter and the inconvenience of having to perform nonlinear least-squares fits, which require realistic trial parameters, seems to have discouraged the widespread use of these two approaches. Moreover, the following more general shortcomings limit the utility and accuracy of determining potential functions in this way:

- The RKR method is a first-order semiclassical procedure that lacks full quantum mechanical accuracy, a problem that is most serious for species of a small reduced mass (see Table 6.1).
- It is inconvenient to work with a potential defined by a large array of many-digit turning points that have to be interpolated over and extrapolated beyond to yield the type of smooth uniform mesh of function values required for use in practical calculations. This step also introduces the specter of *interpolation noise*—uncertainties in calculated properties associated with the choice of a particular interpolation scheme—a problem that is usually simply ignored.
- The RKR method of Section 6.2.2 is based on the fact that determination of the potential function requires only knowledge of G_v and B_v. However, the discussion in Section 6.1 pointed out that the exact quantum mechanical values of diatomic molecule CDCs $\{D_v, H_v, L_v, \ldots\}$—the derivatives of the energy with respect to $[J(J + 1)]$ evaluated at $J = 0$—can be calculated from any given potential energy function. Thus, CDCs are not independent parameters, but are implicitly determined by the G_v and B_v functions. In spite of this, in most published data analyses, the centrifugal distortion expansion coefficients (Dunham $Y_{\ell,m}$ coefficients with $m \geq 2$) have been treated as independent, free parameters. As a result, errors in the fitted CDCs introduce small compensatory errors into the B_v functions used to define the potential. In the last twenty-five years it has become increasingly common to address this problem by performing self-consistent data analyses in which CDC constants calculated from a preliminary RKR potential would be held fixed in a new fit to the global data set to obtain improved G_v and B_v functions, and hence a better RKR potential. Iteration of this procedure would generally converge quickly. However, its use complicates the process of empirical data analysis.
- Combined-isotopologue data analysis fits can be performed using versions of Equation 6.16, which include atomic mass-dependent terms to account for BOB effects and the breakdown of the simple reduced-mass scaling implied by the first-order semiclassical quantization condition [42]. However, there

is no simple way of distinguishing between these two types of corrections, so BOB contributions to the interaction potential cannot be readily determined in this way.

In view of these concerns, there is clearly a need for the type of exact quantum mechanical data analysis procedure yielding global analytic potential energy functions which is described in Section 6.3. Nonetheless, the semiclassical methods described in Section 6.2 remain valuable for a number of reasons. One of these is simply the fact that they are friendly, familiar, and fairly easy to apply. A more fundamental reason, however, is the fact that the methods described in Section 6.3 always involve nonlinear least-squares fits that require realistic initial trial parameters if the fit is to be at all stable (see Chapter 2), and this traditional methodology provides an excellent way of generating such trial parameters (see Section 6.3.3). Moreover, the use of NDT remains the best method known for extrapolating beyond observed vibrational data to determine bond dissociation energies, as well as for determining experimental values of the leading long-range inverse-power C_n coefficient. As a result, it remains a central tool in the interpretation and analysis of PAS measurements and other types of data for levels lying very near dissociation. Thus, these semiclassical methods will remain essential tools in a spectroscopist's arsenal for the foreseeable future.

6.3   QUANTUM-MECHANICAL DIRECT-POTENTIAL-FIT METHODS

6.3.1   OVERVIEW AND BACKGROUND

In recent years it has become increasingly common to analyze diatomic molecule spectroscopic data by performing *direct potential fits* (DPFs), in which observed transition energies are compared with eigenvalue differences calculated from an effective radial Schrödinger equation based on some parameterized analytic potential energy function, and a least-squares fit is used to optimize the parameters defining that potential. The effective radial Hamiltonian may also include radial strength functions that characterize atomic mass-dependent adiabatic and nonadiabatic BOB functions, and also (if appropriate) radial strength functions that account for splittings due to angular momentum coupling in electronic states with nonzero electronic angular momentum. Because the well depth and equilibrium bond length are usually central parameters of the potential function model, these equilibrium properties are determined directly from the fit.

The DPF approach was originally introduced for the treatment of atom-diatom Van der Waals molecules, for which the lack of any well-defined structure precluded the effective use of traditional methods of analysis [43]. However, over the past two decades it has been increasingly widely used for diatomic data analyses. The essence of the method is as follows: The upper and lower levels of any observed spectroscopic transition are eigenvalues of Equation 6.1 for the appropriate effective potential energy function. As discussed in Section 6.1, for any given potential, this equation can be solved readily using standard methods to yield the eigenvalue $E_{v,J}$ and eigenfunction $\psi_{v,J}(r)$ of any given vibration-rotation level $\{v, J\}$. Moreover, the Hellmann–Feynman theorem shows that the partial derivative of that eigenvalue with respect to any given potential function parameter p_j may be calculated using the expression

$$\frac{\partial E_{v,J}}{\partial p_j} = \left\langle \psi_{v,J}(r) \left| \frac{\partial V(r)}{\partial p_j} \right| \psi_{v,J}(r) \right\rangle \qquad (6.49)$$

The difference between such derivatives for the upper and lower level of each observed transition is then the partial derivative of that datum with respect to that parameter required by the least-squares fitting procedure.

One challenge of this approach is the fact that the data set often consists of many thousands or tens of thousands of individual transitions involving a wide range of vibration-rotation levels, so being able to solve Equation 6.1 efficiently is a matter of some importance. It is always necessary for the Schrödinger-solver subroutine to start from a realistic initial trial energy for each level of interest, and the more accurate that initial estimate, the smaller the time required for obtaining the desired solution. One way of addressing this challenge is as follows: Prior to beginning the fit, the data set would be surveyed to determine the highest observed vibrational level for each electronic state considered. Then, at the beginning of each cycle of the nonlinear fit to optimize the potential function parameters, an automatic procedure would locate each of those (pure) vibrational levels. A particularly efficient way of doing this would make use of the semiclassical energy derivatives computed using Equation 6.31 to generate an estimate of the distance from a given level to the next. Once the pure vibrational levels are known, Hutson's method (references [6,7], see Section 6.1) may be used to generate values of the first few rotational constants for each vibrational level. As the fitting procedure considers the data one at a time, these stored band constants can be used in Equation 6.4 to generate a good initial estimate of the required initial trial energy for the level in question. This combined quantum or semiclassical procedure is quite efficient.

Two other key problems associated with DPF treatment of diatomic molecule data are what potential function form to use, and how to obtain the realistic initial trial parameters required by the nonlinear least-squares fitting procedure. These topics are discussed in Section 6.3.2 and 6.3.3.

6.3.2 POTENTIAL FUNCTION FORMS

A central challenge of the DPF method has been the problem of developing an optimum analytic potential function form. Ideally, such a function should satisfy the following criteria:

- It should be flexible enough to represent very extensive, high-resolution data sets to the full degree of experimental accuracy.
- It should be robust and well-behaved, with no spurious extrapolation behavior outside the region to which the experimental data are most sensitive.

- It should be smooth and continuous everywhere.
- It should incorporate the correct theoretically known limiting behavior at large distances.
- It should be compact and portable—that is, be defined by a relatively modest number of parameters.

Devising a potential function form that satisfies all these criteria has been a nontrivial problem, and work on developing new and better forms (potentiology) remains an active area of research. Sections 6.3.2.1 through 6.3.2.4 describe and compare four families of potential function forms that have been used in diatomic DPF analyses.

6.3.2.1 Polynomial Potential Function Forms

The oldest type of potential function form used in DPF data analyses is a simple polynomial expansion in a radial-coordinate such as the Dunham variable $\xi_{Dun} = (r - r_e)/r_e$ (see Equation 6.14). Dunham expansions themselves have the obvious shortcoming that $V_{Dun}(r) \to +\infty$ or $-\infty$ as $r \to \infty$, the sign depending on the sign of the last nonzero polynomial coefficient. However, this singularity problem is resolved if the Dunham radial variable is replaced by an alternative such as that proposed by Ogilvie and Tipping [44],

$$\xi_{OT} = 2\left(\frac{r - r_e}{r + r_e}\right) \tag{6.50}$$

This quantity has the nice property that it approaches finite values at the two limits $r \to 0$ and $r \to \infty$, so a potential defined as a power series in this variable will have no singularities. Moreover, the simplicity of such power-series forms makes it very easy to generate expressions for the partial derivatives of the potential required for calculating partial derivatives of the observables using Equation 6.49. However, the resulting potential functions may still behave unphysically outside the range of the data used to determine them.

In principle, polynomial potentials may be constrained to approach a known asymptote with some theoretically specified inverse-power long-range behavior [45]. However, the expressions required to impose this behavior are quite complex and require the inclusion of multiple additional polynomial coefficients. For example, requiring a potential defined as a polynomial in the coordinate ξ_{OT} to approach an asymptote with a specified $V(r) \sim \mathfrak{D} - C_6/r^6$ limiting behavior would require the potential-function polynomial to have seven additional high-order terms beyond those required to represent the experimental data. Unfortunately, the resulting functions have a tendency to be somewhat unstable and to display spurious oscillatory behavior in the interval between the data region and the limiting long-range region. Thus, a simple polynomial in variables such as ξ_{OT} is not a viable way of describing an overall potential energy function.

The problem of imposing a constraint on the long-range behavior of a polynomial potential is somewhat reduced if one uses a radial variable of the type proposed by Šurkus [46],

$$\xi_{\text{Šur}}^{(p)} = \frac{r^p - (r_e)^p}{r^p + (r_e)^p} \equiv y_p^{\text{eq}}(r) \tag{6.51}$$

At large distances $\xi_{\text{Šur}}^{(p)} \simeq 1 - 2\left(\dfrac{r_e}{r}\right)^p + \cdots$, so if p is set equal to the power (n) of the leading inverse-power term in the long-range potential, it is a fairly straightforward matter to constrain such a potential to have the desired limiting $\mathfrak{D} - C_n/r^n$ behavior. However, when this power has a moderately large value such as $p = 6$, this variable is relatively "stiff" (see Section 6.3.2.2), so a polynomial function using it will have limited flexibility. Moreover, it would be impossible to constrain such a polynomial to mimic a more sophisticated long-range behavior such as $V(r) \simeq \mathfrak{D} - \dfrac{C_6}{r^6} - \dfrac{C_8}{r^8}$. Hence, potentials defined as polynomials in a Šurkus variable with large p are of limited use.

A practical way to circumvent the problem of the bad long-range behavior of polynomial potentials is simply to attach the desired long-range inverse-power-sum tail smoothly to the fitted polynomial potential at some point near the outer end of the data-sensitive range of r. This has been the approach used by a group at the University of Hannover in a large number of very careful studies of alkali metal and alkaline earth diatomics, for many of which the data span almost the entire potential energy well. That work represents the potential function within the data range by a polynomial in the variable $\xi_{\text{Han}} = (r - r_m)/(r + br_m)$ in which r_m is a fixed distance located near r_e, and b is a fitted constant [47]. However, no polynomial is reliable outside the range of the data to which it is fitted, so to get a useful overall potential it is always necessary to attach both some simple repulsive analytic function at the inner end of the data region and the desired inverse-power-sum at the outer end.

Although these extrapolation functions are parameterized so that they attach to the polynomial smoothly, the precise point of attachment is an *ad hoc* choice. Moreover, the whole attachment procedure relies on the fact that both the polynomial potential and its first derivative are physically correct at the extreme ends of the data-sensitive region, a sometimes questionable assumption. This type of potential form also has three other worrisome shortcomings:

1. The polynomials required to account fully for the data often have very high orders—orders between 20 and 40 being common for extensive experimental data sets—and fits to polynomials of such high order tend to be very highly correlated and sometimes have difficulty converging.
2. While the independent variable typically spans the range $\xi_{\text{Han}} \in (-0.1, 0.6)$ and the potential function on that domain spans an energy range of order 10^4 cm^{-1}, the higher-order polynomial coefficients are usually of oscillating sign and have magnitudes as much as four to eight orders of magnitude larger than the range of the function being fitted. This is a signature for a marginally stable model.
3. In most published analyses using this form, all the (many) potential coefficients reported are listed to 18 significant digits. This means that quadruple

precision arithmetic would be needed to reproduce these functions on most computers, a point that can make such potentials somewhat inconvenient for others to use. Moreover, if the parameters of any model describing data (such as energy level spacing) known to 6 to 8 significant digits truly requires its parameters to be specified to 18 significant digits, there must be something wrong with the model.

In summary, although a number of very extensive, high-resolution data sets have been fitted accurately using polynomial potentials, such functions are not an optimum way of summarizing what we know about a molecule.

6.3.2.2 The Expanded Morse Oscillator Potential Form and the Importance of the Definition of the Expansion Variable

An important development in potential function modeling was the demonstration by Coxon and Hajigeorgiou that a Morse potential with a distance-dependent exponent coefficient was a compact and flexible function, which could provide a very accurate representation of a potential energy well [48–50]. A potential well typically spans an energy range of 10^3–10^5 cm^{-1} and needs to be known to an accuracy of $\leq 10^{-3}$ cm^{-1} if it is to explain high-resolution experimental data. Coxon and Hajigeorgiou had the insight to realize that the $\mathfrak{D}_e$ value and the algebraic structure of the Morse function would account for the bulk of that change, while modest variations of the exponent coefficient would allow precise changes in the potential function shape to be defined by a relatively modest number of parameters. In their early work, the Morse function exponent coefficient was represented by a simple power series in r. However, that proved to be inappropriate, since it meant that as $r \to \infty$ the exponent coefficient polynomial would approach either $+\infty$ or $-\infty$ (depending on the mathematical sign of the highest-order expansion coefficient). Such singularities are removed if the power series expansion variable is replaced by a quantity such as $\xi_{OT}(r)$ (see Equation 6.50), but problems remain, since the value of the expansion variable at the outer end of the data region remained a long way from its limiting value [51]. However, a more robust model is obtained if a version of the Šurkus variable of Equation 6.51 with $p \gtrsim 3$ is used as the exponent expansion variable [52–54].

The resulting model is called an expanded Morse oscillator (EMO) potential, and is written as follows:

$$V_{EMO}(r) = \mathfrak{D}_e (1 - e^{-\beta(r)\cdot(r-r_e)})^2 \tag{6.52}$$

in which

$$\beta(r) = \beta_{EMO}(r) = \sum_{i=0}^{N} \beta_i (y_p^{ref}(r))^i \tag{6.53}$$

with

$$y_p^{ref}(r) = y_p(r; r_{ref}) = \frac{r^p - (r_{ref})^p}{r^p + (r_{ref})^p} \tag{6.54}$$

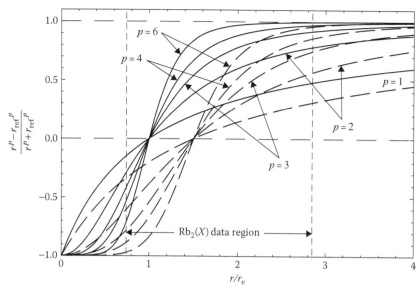

FIGURE 6.9 Plot of the radial variables $y_p^{eq}(r) = y_p(r; r_{ref} = r_e)$ (solid curves) and $y_p^{ref}(r) = y_p(r; r_{ref} = 1.5 r_e)$ (dashed curves) for various p, and the data-sensitive region for ground-state Rb_2.

In all currently published work using this model, the reference distance in the expansion variable was set as $r_{ref} \equiv r_e$. However, recent work using the related Morse/Long-Range (MLR) model (described in Section 6.3.2.3) shows that fixing the exponent-expansion-variable reference distance r_{ref} at a value larger than r_e, typically in the range $1.2 r_e$–$1.5 r_e$, allows accurate fitted potentials to be obtained that require a substantially smaller number of β_i expansion coefficients that would otherwise be needed [55,56].

One other key feature of this model is the power p in the definition of the radial expansion variable. For $p = 1$, the values of $y_p^{ref}(r)$ at the inner and outer ends of the data region are a long way from their limiting values of -1 and $+1$, respectively. As a result, a moderately high-order polynomial function of that variable will have a large probability of behaving badly (e.g., showing oscillatory behavior) on the intervals between the data region and the $y_p = \pm 1$ limits. As p increases, however, the mapping of $y_p(r)$ onto r places the values of $y_p(r)$ at the inner and outer ends of the data region ever closer to the limiting values of ± 1, and hence removes the possibility of such misbehavior.

The importance of being able to set $p > 1$ and $r_{ref} > r_e$ is illustrated by Figure 6.9, which shows how $y_p(r; r_{ref})$ depends on p and r for the two cases $r_{ref} = r_e$ (solid curves) and $r_{ref} = 1.5 r_e$ (dashed curves). The vertical broken lines on this plot are the inner and outer ends of the data-sensitive region* for the ground $X^1\Sigma_g^+$ state of Rb_2, as

---

* Defined here as the inner and outer classical turning points of the highest vibrational level involved in the experimental data set.

considered in the analysis of reference [56]. For $p = 1$ the variable $y_1(r; r_e)$ at the outer end of the data region is barely half way to its limiting value of $+1$, while $y_1(r; r_{ref} = 1.5 \ r_e) \approx 0.33$ there. Thus, any function of those variables whose coefficients are defined by its behavior within the data region would have ample opportunity to behave unphysically between the end of the data region and the limit where $r \to \infty$. In contrast, for higher values of p the variable $y_p(r)$ at the outer end of the data region is relatively flat and close to its upper bound, so an exponent-coefficient function $\beta(r)$ defined as a polynomial in that variable would change very little at larger distances. This bodes well for stable extrapolation behavior of the associated potential.

The solid curves in Figure 6.9 show that the fact that $y_p(r; r_e)$ is close to its upper limit at the outer end of the data region does not mean that the same is true at the inner end; hence, for example, setting $p \geq 3$ does not suffice to ensure sensible inward extrapolation. However, the dashed curves in Figure 6.9 show that choosing a value of r_{ref} somewhat larger than r_e makes the range of y_p^{ref} more symmetric at the two ends of the data range. Indeed, if r_{ref} is set at the geometric mean of the inner and outer bounds, $r_{ref} = \sqrt{r_{inner} \times r_{outer}}$, the range of $y_p(r; r_{ref})$ will be precisely symmetric on that domain; that is, $y_p(r_{inner}; r_{ref}) = -y_p(r_{outer}; r_{ref})$ independent of the choice of p. When combined with an appropriate choice of $p > 1$, this will ensure stable extrapolation beyond both the inner and the outer ends of the data region.

The EMO potential form has been used in a number of empirical data analyses, yielding accurate analytic potentials which fully reproduce the experimental data considered, giving very good estimates of the equilibrium distance r_e, and (if $\mathfrak{D}_e$ is a fitting parameter) it also can give a realistic estimate of the well depth. Indeed, this potential function form satisfies all of the criteria itemized in Section 6.3.1 except one: its fundamental exponential-type nature means that it cannot incorporate the inverse-power-sum behavior characteristic of all long-range intermolecular potentials. For molecular states with simple single-well potentials for which no realistic estimates of the leading long-range inverse-power C_m coefficients are known, the EMO form is arguably the best model potential function available. However, if the values of those coefficients are known, it is always better to incorporate the theoretically predicted inverse-power behavior into the potential using the type of model potential described below.

6.3.2.3 The Morse/Long-Range (MLR) Potential Form

The potential function form described in this section has the same basic algebraic structure as the Morse-type potential discussed in Section 6.3.2.2, but it incorporates two key differences. The first is the replacement of the exponent factor $(r - r_e)$ of Equation 6.52 by the variable $y_p^{eq}(r)$ of Equation 6.51; the second is the introduction of a preexponential factor to incorporate the desired long-range behavior. The resulting function is

$$V_{MLR}(r) = \mathfrak{D}_e \left(1 - \frac{u_{LR}(r)}{u_{LR}(r_e)} e^{-\beta(r) \cdot y_p^{eq}(r)} \right)^2 \tag{6.55}$$

The fact that $y_p^{eq}(r_e) = 0$ and the preexponential factor equals one at $r = r_e$ ensures that this function retains the Morse-type property of having its minimum at r_e and a well depth of $\mathfrak{D}_e$. Since the exponent coefficient function $\beta(r)$ is written as a (constrained) polynomial in $y_p^{\text{ref}}(r)$, the exponent in Equation 6.55 will approach a finite value as $r \rightarrow \infty$. If this limiting value of the exponent is defined as

$$\beta_\infty \equiv \lim_{r \rightarrow \infty} \{\beta(r) \cdot y_p^{\text{eq}}(r)\} = \lim_{r \rightarrow \infty} \{\beta(r)\} = \ln\left(\frac{2\mathfrak{D}_e}{u_{\text{LR}}(r_e)}\right) \tag{6.56}$$

then the limiting long-range behavior of the MLR function will be

$$V_{\text{MLR}}(r) \simeq \mathfrak{D}_e - u_{\text{LR}}(r) + \mathcal{O}\left(\frac{u_{\text{LR}}^2}{4\mathfrak{D}_e}\right) \tag{6.57}$$

Thus, if $u_{\text{LR}}(r)$ is defined as the appropriate sum of inverse-power terms

$$u_{\text{LR}} = \frac{C_{m_1}}{r^{m_1}} + \frac{C_{m_2}}{r^{m_2}} + \cdots \tag{6.58}$$

the long-range tail of the MLR potential will have the correct theoretically predicted inverse-power long-range form.

A simple way to constrain the polynomial expression for $\beta(r)$ to approach the limiting value β_∞ as $r \rightarrow \infty$ is to write it in the form

$$\beta(r) = \beta_{\text{MLR}}(r) = y_p^{\text{ref}}(r)\,\beta_\infty + \left[1 - y_p^{\text{ref}}(r)\right]\sum_{i=0}\beta_i\left[y_q^{\text{ref}}(r)\right]^i \tag{6.59}$$

However, in order to prevent the leading contributions to the asymptotic expansion of the exponential term in Equation 6.55 from modifying the long-range behavior specified by Equation 6.58, the power p in Equations 6.55 and 6.59 must satisfy the condition $p > m_{\text{last}} - m_{\text{first}}$, where m_{first} and m_{last} are, respectively, the powers of the first and last terms included in the chosen definition of $u_{\text{LR}}(r)$ [57]. For states formed from ground-state atoms, the leading terms contributing to the long-range potential often correspond to $m = 6$, 8, and 10; if these three terms define $u_{\text{LR}}(r)$, it would be necessary to set $p > 4$. Note, however, that there are no restrictions on the value of q, and giving it a somewhat smaller value than p often yields good fits with lower-order polynomials than would otherwise be required [55,56].

The basic form of the MLR potential seen in Equation 6.55 is remarkably simple, and the fact that the leading terms in the long-range potential are explicitly incorporated within its algebraic form rather than being a separate attached function is a great improvement over other models. The fact that the empirical function that determines the details of the potential function shape appears in an exponent, together with the use of an expansion variable $y_p^{\text{ref}}(r)$ centered at a distance $r_{\text{ref}} > r_e$, make this form particularly flexible, and allow accurate fits to be obtained with a relatively modest number of expansion parameters. Moreover, the physically interesting quantities r_e, $\mathfrak{D}_e$, and C_n are explicit parameters of the MLR model, which may be varied in the fit, while for other potential forms the determination of the two latter parameters depends partly on where and how the long-range tail is attached to the polynomial spanning the data region.

Illustrations of the importance of the parameterization details described above are provided by the results of DPF analyses of extensive high-resolution data sets for the ground electronic states of MgH and Ca_2 summarized in Table 6.2. In a recently published study, analysis for MgH using a simple version of the MLR form, which fixed $r_{ref} = r_e$ and $p = q = 4$, an exponent polynomial of order 18 was required to give a good fit, and the resulting higher-order β_i expansion coefficients had alternating mathematical signs and magnitudes of order 10^6–10^7 [58]. The fact that polynomial coefficients of this magnitude are required to represent a function with the range $-2.8 \lesssim \beta(r) \lesssim -0.5$ on the data-sensitive domain $-0.3 \lesssim y_p^{eq}(r) \lesssim 0.6$ is a clear signature of a marginally stable model. In contrast, use of the extended MLR model described above with $r_{ref} = 2.3\text{Å} \approx 1.33 r_e$ and $\{p, q\} = \{5, 4\}$ yields the same quality of fit with an exponent polynomial order of only 14, and the magnitudes of the resulting β_i coefficients ranged from ~0.05 to 28, values much more consonant with the domain and range of the function being fitted.

In the second example considered in Table 6.2, we see that use of a more sophisticated MLR potential with $r_{ref} > r_e$ and $q < p$ again allows the data to be fully explained by a much more compact potential than had been obtained using a basic MLR function in which $r_{ref} = r_e$ and $q = p$ [56,57]. The first Ca_2 entry in Table 6.2 shows that a fit to the same data set using a Hannover polynomial potential of the type described in Section 6.3.2.1 requires many more parameters than either MLR model. Thus, the type of generalized MLR function described here clearly yields a much more compact and robust model than do either the early version of the MLR form or the polynomial potentials discussed in Section 6.3.2.1. As a result, we may expect fits using this form to yield more reliable fitted values of physically significant parameters such as $\mathfrak{D}_e$, the equilibrium bond length r_e, and any fitted C_m coefficients.

The MLR potential function form described in this section is being used in an increasing number of practical data analyses, and is arguably the best model for

---

TABLE 6.2
Results of Fits Performed Using Different Potential Function Models; $\overline{dd}$ is the Relative (Normalized by the Data Uncertainties) Root-Mean-Square Difference between Simulated and Experimental Transition Energies

Species	Model	$u_{LR}(r)$	No of Parameters			
			Polynomial	Fitted	Total	$\overline{dd}$
MgH $(X\,^1\Sigma^+)$	"Basic" MLR	C_6, C_8	19	21	24	0.78
	Full MLR	C_6, C_8, C_{10}	15	18	22	0.78
$Ca_2\left(X\,^1\Sigma_g^+\right)$	Hannover polynomial	C_6, C_8, C_{10}	21	25	31	0.69
	"Basic" MLR	C_6, C_8	12	15	18	0.637
	Full MLR	C_6, C_8, C_{10}	7	12	15	0.627

Source: Le Roy, R. J., N. Dattani, J. A. Coxon, A. J. Ross, P. Crozet, and C. Linton. 2009. *J Chem Phys* 131:204309/1–17.

single-well potentials developed to date. However, it appears unlikely that the exponent-polynomial version of this function that is described earlier in Section 6.3.2.3 will prove able to accurately describe double-minimum or shelf-state potentials. Fortunately, a completely different type of potential function model has been developed, which seems ideally suited for dealing with such cases.

6.3.2.4   The Spline-Pointwise Potential Form

A spline-pointwise potential (SPP) is an analytic potential defined as a cubic spline function passing through a chosen grid of points, with the energies of the spline points being the parameters varied in the DPF procedure. This novel approach was first introduced in the late 1980s by Tiemann and Wolf, who applied it to the analysis of data for a number of systems with shallow wells and quasibound levels supported by rotationless potential energy barriers [59–62]. While remarkable for its time, as originally implemented, fits using this approach tended to be somewhat unstable and required very careful monitoring. As a result, it saw little further use until the turn of the century, when Pashov et al. rediscovered this model and showed that use of singular-value decomposition at the core of the least-squares procedure allowed fits using this form to be stable and routine [63–65].

A cubic spline function is normally thought of as a set of cubic polynomials with a distinct cubic spanning the interval between each pair of adjacent points, with the coefficients being constrained to impose continuity and smoothness at each internal grid point. For DPF applications, however, Pashov et al. showed that it was more convenient to write such a spline function as a linear combination of basis functions associated with the N specified mesh points $\{r_i\}$,

$$V_{\mathrm{SPP}}(r) = \sum_{i=1}^{N} V(r_i)\, S_i^N(r) \tag{6.60}$$

in which the potential function values $V(r_i)$ at the grid points r_i are the parameters to be varied in the fits. They also chose to use *natural* cubic splines, so defined by the fact that the second derivative of the function equals zero at the first and last grid points. This is a very simple function to use, since the partial derivative functions required by the least-squares procedure are the parameter-independent functions $S_i^N(r)$.

SPP functions have been used in successful DPF data analyses for regular single-well potentials, for double-well potentials, and for states whose potential function has a single well with a rotationless barrier protruding above the asymptote. However, one shortcoming is the fact that there is no natural way to extrapolate such a function outside the data region at small or large r. In particular, the fact that natural splines are used means that the resulting functions will always have zero curvature at the first and last grid points, and some *ad hoc* procedure has to be used to attach a sensible steep extrapolating function at the inner end of the data-sensitive region and the requisite inverse-power-sum tail at the outer end. This makes it difficult for fits using this model to yield accurate estimates of the length of the extrapolation to the dissociation limit, or to determine experimental values of long-range potential coefficients.

Another concern is the fact that a relatively large number (typically ~50) of potential parameters (i.e., grid-point function values) is required to define a high-quality potential, so ~100 many-digit numbers must be precisely transcribed by anyone wishing to use such a potential. Moreover, while physically interesting parameters such as the equilibrium bond length r_e and well depth D_e may be determined from such potentials, they are not explicit parameters of the model, and hence it is difficult to obtain realistic estimates of their uncertainties.

For the reasons listed above, SPP function is not the best type of model to use for ordinary single-well potentials. However, for potentials with double minima or "shelf" behavior, they are an unparalleled success, since no other functional form can give a smooth, flexible function that can handle the relatively abrupt changes of character associated with such cases [64,65]. This form has also been used in successful DPF treatments of molecular states with a rotationless potential energy barrier [66]. Thus, in spite of shortcomings associated with their extrapolation behavior, SPP potentials will remain an essential component of a spectroscopist's toolkit for the foreseeable future. Documented computer programs for using DPF fits to determine accurate analytic potentials from fits to spectroscopic data using the recommended potential function forms described above are publicly available [63,67].

6.3.3 INITIAL TRIAL PARAMETERS FOR DIRECT POTENTIAL FITS

Vibration-rotation level energies and level energy differences are not linear functions of potential function parameters. Thus, DPF data analysis procedures are always based on nonlinear least-squares fits, and they always require realistic initial trial values of the parameters defining the chosen potential function form. This is another situation in which the traditional data analysis methods described in Section 6.2 prove to be of enduring value, since an RKR potential can always provide a good first-order estimate of the potential, and fitting the analytic potential form of interest to RKR points can give good estimates of the required initial trial parameter values. Such preliminary potentials may also be obtained by ab initio methods, but they are less accurate than RKR potentials, so the latter are preferred when available.

Both polynomial potentials and SPP potentials are linear functions of the relevant expansion parameters, so fitting those forms to a preliminary set of RKR or ab initio turning points is a very straightforward matter. Moreover, the Morse-type algebraic structure of EMO and MLR potentials means that it is also a relatively straightforward matter to obtain preliminary estimate of their exponent expansion parameters β_i. For example, the MLR function of Equation 6.55 may be rearranged to give

$$\beta(r) \cdot y_p^{\text{eq}}(r) = -\ln\left\{\frac{u_{\text{LR}}(r_e)}{u_{\text{LR}}(r)}\left(1 \pm \sqrt{\frac{V_{\text{MLR}}(r) - V_{\text{MLR}}(r_e)}{\mathfrak{D}_e}}\right)\right\} \qquad (6.61)$$

Given some plausible initial estimates of r_e and $\mathfrak{D}_e$, that initial set of turning points defines the values of the right-hand side of Equation 6.61. Since Equation 6.59 shows that $\beta_{\text{MLR}}(r)$ is a linear function of the expansion parameters β_i, it is a straightforward matter to perform a linear least-squares fit to determine initial trial β_i values. A similar rearrangement of the expression for an EMO potential Equation 6.52 yields an expression analogous to Equation 6.61 for determining initial estimates of the exponent expansion coefficients for that case. For either case, the initial set of realistic expansion parameters yielded by this linearized treatment the allows one to perform a proper nonlinear least-squares fit to the turning point energies to obtain an optimized set of trial β_i values, and this procedure can also optimize the assumed r_e and $\mathfrak{D}_e$ values.

In addition to serving as a means of providing realistic initial trial parameters for DPF data analyses, such fits of given sets of turning points to Equations 6.52 or 6.55 can serve as a very efficient way of presenting the results of ab initio calculations of potential energy functions. A computer program for performing turning-point fits of this type is available (with a manual) on the Internet [68].

6.4 BORN–OPPENHEIMER BREAKDOWN EFFECTS

The discussion in Section 6.3 focused on the problem of determining an analytic potential energy function that accurately explains all the available discrete spectroscopic data for a given molecular state in terms of eigenvalues of the radial Schrödinger equation of Equation 6.1. For a heavy molecule such as Rb_2, a single potential energy function usually would suffice to explain the results for all isotopic forms of that species. However, for species of small reduced mass, BOB effects give rise both to differences between the effective potentials for different isotopologues and to atomic mass-dependent corrections to the simple centrifugal potential $(\hbar^2/2\mu r^2)\,[J(J+1)]$ of Equation 6.2. This has import with respect to the determination of equilibrium structures, since most observed transitions involve excited rotational levels, and it is the extrapolation of their properties to $J = 0$ that allows us to determine equilibrium structures and properties.

In practice, BOB effects often may be accounted for with ordinary DPF methodology simply by introducing atomic mass-dependent terms into the effective potential energy function of Equation 6.2. Most such work reported to date has been based on the effective radial Schrödinger equation derived by Watson [69,70], in which atomic mass-dependent nonadiabatic contributions to the kinetic energy operator are incorporated both into an effective adiabatic contribution to the electronic potential energy function and into a nonadiabatic BOB contribution to the effective centrifugal potential of the rotating molecule.

Following the conventions of references [52,71], the resulting effective radial Schrödinger equation for isotopologue α of molecule A–B in a singlet electronic state may be written as

$$\left\{-\frac{\hbar^2}{2\mu_\alpha}\frac{d^2}{dr^2}+\left[V_{\mathrm{ad}}^{(1)}(r)+\Delta V_{\mathrm{ad}}^{(\alpha)}(r)\right]+\frac{[J(J+1)]\hbar^2}{2\mu_\alpha r^2}[1+g^{(\alpha)}(r)]\right\}\psi_{v,J}(r)=E_{v,J}\psi_{v,J}(r)$$

$$(6.62)$$

Here, $V_{\mathrm{ad}}^{(1)}(r)$ is the total electronic internuclear potential for a selected reference isotopologue (labeled $\alpha = 1$), $\Delta V_{\mathrm{ad}}^{(\alpha)}(r)$ is the *difference* between the effective adiabatic potentials for isotopologue α and for the reference species ($\alpha = 1$), and $g^{(\alpha)}(r)$ is the nonadiabatic centrifugal potential correction function for isotopologue α.

Both $\Delta V_{\mathrm{ad}}^{(\alpha)}(r)$ and $g^{(\alpha)}(r)$ are written as a sum of two terms, one for each component atom, whose magnitudes are inversely proportional to the mass of the particular atomic isotope [69–72],

$$\Delta V_{\mathrm{ad}}^{(\alpha)}(r) = \frac{\Delta M_{\mathrm{A}}^{(\alpha)}}{M_{\mathrm{A}}^{(\alpha)}} \tilde{S}_{\mathrm{ad}}^{\mathrm{A}}(r) + \frac{\Delta M_{\mathrm{B}}^{(\alpha)}}{M_{\mathrm{B}}^{(\alpha)}} \tilde{S}_{\mathrm{ad}}^{\mathrm{B}}(r) \tag{6.63}$$

$$g^{(\alpha)}(r) = \frac{M_{\mathrm{A}}^{(1)}}{M_{\mathrm{A}}^{(\alpha)}} \tilde{R}_{\mathrm{na}}^{\mathrm{A}}(r) + \frac{M_{\mathrm{B}}^{(1)}}{M_{\mathrm{B}}^{(\alpha)}} \tilde{R}_{\mathrm{na}}^{\mathrm{B}}(r) \tag{6.64}$$

in which $\Delta M_{\mathrm{A}}^{(\alpha)} = M_{\mathrm{A}}^{(\alpha)} - M_{\mathrm{A}}^{(1)}$ is the difference between the atomic mass of atom A in isotopologue α and in the reference isotopologue ($\alpha = 1$).

For a given isotopologue α, Equation 6.62 effectively has the same form as Equation 6.1, so the full machinery of DPF data analysis described in Section 6.3 can be applied. The only difference is that in addition to the parameterized analytic potential energy function $V_{\mathrm{ad}}^{(1)}(r)$, the fit must simultaneously consider parameters defining the adiabatic and nonadiabatic radial strength functions $\tilde{S}_{\mathrm{ad}}^{\mathrm{A/B}}(r)$ and $\tilde{R}_{\mathrm{na}}^{\mathrm{A/B}}(r)$. This is a very straightforward matter and raises no significant practical problems. Indeed, the fact that these contributions to the radial Hamiltonian are relatively small means that no effort need be devoted to obtaining initial trial values of the parameters defining these BOB radial strength functions. As a result, simultaneous fits to data sets for multiple isotopologues to determine both an analytic potential function and BOB radial strength functions have been routine since the early 1990s [49].

As in the case of the potential energy function, some thought must be given to the analytic form of the radial strength functions $\tilde{S}_{\mathrm{ad}}^{\mathrm{A/B}}(r)$ and $\tilde{R}_{\mathrm{na}}^{\mathrm{A/B}}(r)$ [52]. For one thing, $\tilde{S}_{\mathrm{ad}}^{\mathrm{A/B}}(r)$ must have the same limiting long-range inverse-power behavior as the potential function itself, since different isotopic forms of a given molecular species are normally* expected to have the same limiting long-range functional behavior. A second point is that for an electronic state that dissociates to yield an atom in an excited electronic state, the limiting asymptotic value of $\tilde{S}_{\mathrm{ad}}^{\mathrm{A/B}}(r)$ must correlate with any difference in the associated atomic energy level spacing. For the case of a molecular state of species A–B which dissociates to yield both atoms in excited electronic states, A* + B*, the overall adiabatic correction to the potential for isotopologue α must approach a limiting value equal to the sum of the associated atomic isotope shifts, as follows:

$$\lim_{r \to \infty} \Delta V_{\mathrm{ad}}^{(\alpha)}(r) = \delta E^{(\alpha)}(\mathrm{A}^*) + \delta E^{(\alpha)}(\mathrm{B}^*) \tag{6.65}$$

---

* An example of a special exception to this rule is the $A^1\Sigma_u^+$ state of $^{6,7}\mathrm{Li}_2$ discussed in reference [55], Le Roy, R. J., N. Dattani, J. A. Coxon, A. J. Ross, P. Crozet, and C. Linton. 2009. *J. Chem. Phys.* 131:204309/1–17.

where $\delta E^{(\alpha)}(A^*)$ is the difference between the $A \rightarrow A^*$ atomic excitation energy of the isotope of that atom in molecular isotologue α versus the corresponding value for the reference isotopologue. Note that this expression assumes that the absolute zero of energy for a given molecular species is set at the limit for dissociation to yield two ground state atoms; analogous (but more complicated) constraints would arise for other choices of the reference energy.

The above constraints on the adiabatic radial strength functions $\tilde{S}_{\mathrm{ad}}^{A/B}(r)$ are incorporated into the expression:

$$\tilde{S}_{\mathrm{ad}}^{A}(r) = y_{p_{\mathrm{ad}}}^{\mathrm{eq}}(r)\, u_\infty^A + \left[1 - y_{p_{\mathrm{ad}}}^{\mathrm{eq}}(r)\right] \sum_{i=0} u_i^A \left(y_{q_{\mathrm{ad}}}^{\mathrm{eq}}(r)\right)^i \tag{6.66}$$

where $p_{\mathrm{ad}} = m_1$ is the power of the leading inverse-power term in the long-range potential for the state in question, and

$$u_\infty^A = \delta E^{(\alpha)}(A^*) \left(\frac{M_A^{(\alpha)}}{\Delta M_A^{(\alpha)}}\right) \tag{6.67}$$

There is no physical constraint on the value of the integer q_{ad} defining the power series expansion variable in Equation 6.66, but experience shows that it should be larger than 1 (say $q_{\mathrm{ad}} \geq 3$) to prevent the resulting function from having implausible extrema in the interval between the data region and the asymptotic limit [52]. If the associated radial expansion variables are expressed relative to r_e [i.e., solely in terms of $y_{p/q}^{\mathrm{eq}}(r)$], this form also allows the difference in isotopic dissociation energies to be written as

$$\delta \mathfrak{D}_e^{(\alpha)} = \mathfrak{D}_e^{(\alpha)} - \mathfrak{D}_e^{(1)} = \frac{\Delta M_A^{(\alpha)}}{M_A^{(\alpha)}}\left(u_\infty^A - u_0^A\right) + \frac{\Delta M_B^{(\alpha)}}{M_B^{(\alpha)}}\left(u_\infty^B - u_0^B\right) \tag{6.68}$$

Most of the considerations discussed above also apply to the nonadiabatic centrifugal radial strength functions $\tilde{R}_{\mathrm{ad}}^{A/B}(r)$, so the same type or analytic function may be used to represent them,

$$\tilde{R}_{\mathrm{na}}^{A}(r) = y_{p_{\mathrm{na}}}^{\mathrm{eq}}(r)\, t_\infty^A + \left[1 - y_{p_{\mathrm{na}}}^{\mathrm{eq}}(r)\right] \sum_{i=1} t_i^A \left(y_{p_{\mathrm{na}}}^{\mathrm{eq}}(r)\right)^i \tag{6.69}$$

In this case, however, there is no constraint on the limiting long-range form of the function, so there is no point in using different powers to define the radial variables in the summation and the other factors. Moreover, physical arguments indicate that for neutral molecules $t_\infty^{A/B} = 0$, while for a molecular ion which dissociates to yield (say) $A^{+Q} + B$, it is defined in terms of Watson's charge-modified reduced mass [69] and the conventional reduced mass of the dissociation products [52]. Note that the power-series summation in Equation 6.69 has no constant term because the derivation of Equation 6.62 gave rise to an indeterminacy which is best accounted for by assuming that $g^{(\alpha)}(r = r_e) = 0$ [69,70].

The constraint that $g^{(\alpha)}(r_e) = 0$ (i.e., $t_0^{A/B} = 0$) means the isotopologue dependence of the equilibrium bond length is defined by the value of the first radial derivative of the adiabatic correction function $\Delta V_{ad}^{(\alpha)}(r)$ at $r = r_e$. In terms of the parameterization presented above, this means that

$$\delta r_e^{(\alpha)} = r_e^{(\alpha)} - r_e^{(1)} = \frac{\Delta M_A^{(\alpha)}}{M_A^{(\alpha)}} \frac{\tilde{S}_{ad}^A(r_e)'}{\tilde{k}} + \frac{\Delta M_B^{(\alpha)}}{M_B^{(\alpha)}} \frac{\tilde{S}_{ad}^B(r_e)'}{\tilde{k}} \tag{6.70}$$

where $\tilde{k}$ is the harmonic force constant of the potential function at its minimum and

$$\tilde{S}_{ad}^A(r_e)' \equiv \left(\frac{d\tilde{S}_{ad}^A(r)}{dr} \right)_{r=r_e} = \frac{\left(u_\infty^A - u_0^A\right)p_{ad}^A + u_1^A q_{ad}^A}{2r_e} \tag{6.71}$$

Finally, we note that while it is beyond the scope of the present discussion, straightforward extensions of Equation 6.62 have been developed that take account of the *e/f* Λ-doubling splittings that occur for singlet states with nonzero integer electronic orbital angular momentum [53], and the spin splittings of rotational levels in $^2\Sigma$ states [58,67]. Treatments of all these BOB effects are incorporated into the publicly available DPF data analysis program DPotFit [67].

6.5   CONCLUSION

This chapter has shown that the analysis of diatomic molecule spectroscopic data can yield both very accurate equilibrium properties, including bond lengths and well depths, and accurate overall potential energy functions. The accuracy of bond lengths determined in this way will match that of the experimental B_v values for the lowest observed vibrational levels, with typical uncertainties of order 10^{-4}–10^{-6} Å, while the uncertainties in the associated dissociation energies are no more than a few percent of the binding energy of the highest observed vibrational level.

The standard methods described herein often also allow us to accurately resolve isotopic differences in both equilibrium parameters and the overall potential curves, which are due to BOB. However, we must raise one cautionary note about the isotopologue dependence of equilibrium bond lengths. The discussion in Section 6.4 attributes all such effects to the slope and curvature of the effective adiabatic correction potential $\Delta V_{ad}^{(\alpha)}(r)$ appearing in Equation 6.62, because we have adopted the Watson convention [69,70] of defining the equilibrium value of the nonadiabatic centrifugal potential correction function $g^{(\alpha)}(r_e)$ to be precisely zero. As we pointed out in reference [69], this is an *ad hoc* assumption that was introduced because of a fundamental indeterminacy associated with removing the atomic mass-dependent nonadiabatic contribution to the kinetic energy operator in order to obtain the working Hamiltonian of Equation 6.62. Unfortunately, it is impossible to improve on this description using the information contained in transition energy information alone, and since the effects of this approximation may be expected to be very small, it suffices for almost all practical purposes.

We have also seen here that traditional parameter-fit analyses of experimental data based on Equations 6.4 and 6.16 have been at least partially superceded by the DPF methods of Section 6.3. However, both those traditional methods and the associated semiclassical methods of Section 6.2 remain of enduring value, both because of the physical insight they offer, and because of the importance of their practical use in providing the realistic initial trial parameters required by DPF methods. Within the DPF methodology, it is also clear that use of an appropriate analytic potential function form is of central importance for obtaining an optimal compact, flexible, and accurate potential function which extrapolates realistically at both large and small distances. We have seen that three keys to these objectives are: (i) the step of placing most of the flexibility of the potential function in a parameterized exponent coefficient, rather than in linear factors, (ii) the choice of an optimal definition for the radial expansion variable, and (iii) the incorporation of theoretically known limiting long-range behavior within the overall functional form, rather than as a separate attached function.

While the present discussion has been focused on diatomic molecules, the DPF method was originally introduced as a way of describing three-dimensional atom-diatom systems [43]. The type of potential energy function description used herein has already been proved useful for atom–molecule and molecule–molecule Van der Waals systems [73,74] and in due course we may expect to see this type of approach applied to more "normal" polyatomic molecules.

APPENDIX: WHAT TERMS CONTRIBUTE TO A LONG-RANGE POTENTIAL?

If two atoms lie sufficiently far apart that their electron clouds overlap negligibly, their interaction energy may be expanded as the simple inverse-power sum of Equation 6.33. The nature of the atomic species to which a given molecular state dissociates determines which powers contribute to this sum, and also sometimes determines their sign (more complete discussions may be found in references [75–78]).

An $m = 1$ term will arise only for an ion-pair state that dissociates to yield two atoms with permanent charges. In this case, the interaction coefficient is $C_1 = - Z_a Z_b e^2/4\pi\varepsilon_0$, in which Z_a and Z_b are the ($\pm$ integer) number of charges on atoms a and b, respectively.

An $m = 2$ term arises classically from the interaction between a permanent charge and a permanent dipole moment. Although no atom possesses a permanent dipole moment, an electronically excited one-electron atom such as excited H or He^+ may behave as if it does, since the presence of the interaction partner can cause a mixing of degenerate states of different symmetry to yield a hybrid atomic orbital that is effectively dipolar. If its interaction partner is an ion, it will contribute an $m = 2$ term to Equation 6.33.

An $m = 3$ term arises classically from the interaction between two permanent dipole moments. This could occur in the interaction of two electronically excited one-electron atoms, each of which is in a dipolar hybrid state. However, a much broader range of cases involves the interaction between a pair of atoms of the same species in different atomic states between which electric dipole transitions are allowed. In this case, the resonance mixing of the wave functions for two equivalent atoms whose total orbital angular momentum quantum numbers differ by one (i.e., S with P, or P with D) effectively makes them act as if they both had permanent dipole moments, and an r^{-3} interaction energy arises.

Another type of r^{-3} term can arise from the first-order interaction between an ion and a particle with a permanent quadrupole moment (e.g., with a P-state atom). For this case $C_3 \propto Z_a e Q_b$, where $Z_a e$ is the charge on the ion and Q_b the permanent quadrupole moment on its interaction partner.

An $m = 4$ term could arise in first order from the interaction of an ion with a particle having a permanent octupole moment (e.g., a D-state atom), or between a particle with (or acting as if it had) a permanent dipole moment and a species with a permanent quadrupole moment. In both of these cases, the associated C_4 interaction coefficients would be proportional to the product of the two charge moments with a factor defined by the symmetry of the particular molecular state.

A more common type of r^{-4} potential term arises as the second-order charge-induced dipole interaction between an ion and the electron distribution of its interaction partner. For this case $C_4 = \left(\dfrac{Z_a^2 e^2}{4\pi\varepsilon_0} \right) \dfrac{\alpha_d^b}{2}$, in which $Z_a e$ is the charge on the ion (atom a) and α_d^b the dipole polarizability of particle b. This is usually the leading long-range term for molecular ions.

An $m = 5$ term arises from the classical electrostatic interaction of two permanent quadrupole moments. Thus, it will contribute to the long-range potential whenever *neither* of the interacting atoms is in an S state. As with all first-order interactions, the associated C_5 coefficient is proportional to the product of the associated permanent moments with a factor depending on the symmetry of the particular molecular state, or more particularly, as the product of an electronic state symmetry factor times $\langle r_e^2 \rangle_a \langle r_e^2 \rangle_b$, where $\langle r_e^2 \rangle_\alpha$ is the expectation value of square of the electron radius in the unfilled valence shell of atom α.

Dispersion energy terms with $m = 6, 8, 10, \ldots$, arise in second-order perturbation theory and contribute to *all* interactions between atomic systems (except when one particle is a bare nucleus). For the case of uncharged atoms, at least one of which is in an S state, these are the leading (longest-range) contributions to Equation 6.33. Since they arise in second-order perturbation theory, for a pair of ground-state atoms the associated potential coefficients are always attractive.

EXERCISES

6.1. Making use of the fact that the semiclassical value for the expectation value of the property $f(r)$ is

$$\langle f(r) \rangle = \frac{\displaystyle\int_0^\infty \frac{f(r)\,dr}{[G_v - V(r)]^{1/2}}}{\displaystyle\int_0^\infty \frac{dr}{[G_v - V(r)]^{1/2}}} \tag{6.72}$$

use the machinery of Section 6.2.3 to determine an explicit NDT expression for the expectation value of $f(r) = r^2$.

6.2. Show that a polynomial potential expanded in terms of the Šurkus variable of Equation 6.51 with $p = 4$ cannot give a $1/r^6$ term.

6.3. Derive the geometric mean rule for r_{ref} given in the second-to-last paragraph of Section 6.3.2.2.

6.4. Considering the limiting long-range behavior of Equation 6.59, show why it is necessary to set $p > m_{last} - m_{first}$ in the definition of an MLR potential.

6.5. Determine the limiting long-range behavior of Equation 6.66 and show that it is $\propto r^{-m1}$.

6.6. For a potential whose long-range tail has the form $V(r) \simeq \mathfrak{D} - Ae^{-br}$, derive the analog of Equation 6.37.

6.7. What is the limiting *short-range* functional behavior of an MLR potential energy function?

REFERENCES

1. Le Roy, R. J., and R. B. Bernstein. 1971. *J Chem Phys* 54:5114–26.
2. Le Roy, R. J., and W. -K. Liu. 1978. *J Chem Phys* 69:3622–31.
3. Le Roy, R. J. 2007. Level 8.0: *A Computer Program for Solving the Radial Schrôdinger Equation for Bound and Quasibound Levels.* University of Waterloo Chemical Physics Research Report CP-663, August 1, 2010 http://leroy.uwaterloo.ca/programs/
4. Cooley, J. W. 1961. *Math Comput* 15:363–74.
5. Herzberg, G. 1950. *Spectra of Diatomic Molecules.* New York, Van Nostrand.
6. Hutson, J. M. 1981. *J Phys B At Mol Phys* 14:851–7.
7. Hutson, J. M. 1981. *QCPE Bulletin* 2, no. 2, Program # \$435, Quantum Chemistry Program Exchange, Indiana University, Bloomington, Indiana.
8. Tellinghuisen, J. 1987. *J Mol Spectrosc* 122:455–61.
9. Kilpatrick, J. E., and M. F. Kilpatrick. 1951. *J Chem Phys* 19:930–3.
10. Kilpatrick, J. E. 1959. *J Chem Phys* 30:801–5.
11. Child, M. S. 1980. *Semiclassical Methods in Molecular Scattering and Spectroscopy.* Dordrecht: D. Reidel.
12. Child, M. S. 1991. *Semiclassical Mechanics with Molecular Applications.* Oxford: Clarendon Press.
13. Abramowitz, M., and I. A. Stegun. 1970. *Handbook of Mathematical Functions.* New York: Dover.
14. Dunham, J. L. 1932. *Phys Rev* 41:713–20 and 41:721–31.
15. Fröman, N. 1980. In *Semiclassical Methods in Molecular Scattering and Spectroscopy,* ed. M. S. Child, Series C - Mathematical and Physical Sciences; Vol. 53, 1–44. Dordrecht: Reidel.
16. Kirschner, S. M., and R. J. Le Roy. 1978. *J Chem Phys* 68:3139–48.
17. Paulsson, R., F. Karlsson, and R. J. Le Roy. 1983. *J Chem Phys* 79:4346–54.
18. Schwartz, C., and R. J. Le Roy. 1987. *J Mol Spectrosc* 121:420–39.
19. Oldenberg, O. 1929. *Z Physik* 56:563–75.
20. Rydberg, R. 1931. *Z Physik* 73:376–85.
21. Klein, O. 1932. *Z Physik* 76:226–35.
22. Rees, A. L. G. 1947. *Proc Phys Soc Lond* 59:998–1008.
23. Dickinson, A. S. 1972. *J Mol Spectrosc* 44:183–8.

24. Fleming, H. E., and K. N. Rao. 1972. *J Mol Spectrosc* 44:189–93.
25. Tellinghuisen, J. 1972. *J Mol Spectrosc* 44:194–6.
26. Verma, R. D. 1960. *J Chem Phys* 32:738–49.
27. Tellinghuisen, J., and S. D. Henderson. 1982. *Chem Phys Lett* 91:447–51.
28. Le Roy, R. J. 2003. RKR1 2.0: *A Computer Program Implementing the First-Order RKR Method for Determining Diatomic Molecule Potential Energy Curves.* University of Waterloo Chemical Physics Research Report CP-657, August 1, 2010 http://leroy.uwaterloo.ca/programs
29. Kaiser, E. W. 1970. *J Chem Phys* 53:1686–703.
30. Birge, R. T., and H. Sponer. 1926. *Phys Rev* 28:259–83.
31. Le Roy, R. J., and R. B. Bernstein. 1970. *Chem Phys Lett* 5:42.
32. Le Roy, R. J., and R. B. Bernstein. 1970. *J Chem Phys* 52:3869–79.
33. Le Roy, R. J. 1972. *Can J Phys* 50:953–9.
34. Le Roy, R. J. 1980. *In Semiclassical Methods in Molecular Scattering and Spectroscopy*, ed. M. S. Child, Series C - Mathematical and Physical Sciences; Vol. 53, 109–26. Dordrecht: Reidel.
35. Tanaka, Y., and K. Yoshino. 1970. *J Chem Phys* 53:2012–30.
36. Le Roy, R. J. 1972. *J Chem Phys* 57:573–4.
37. Takasu, Y., K. Komori, K. Honda, M. Kumakura, T. Yabuzaki, and Y. Takahashi. 2004. *Phys Rev Lett* 93:123202–5.
38. Le Roy, R. J. 1980. *J Chem Phys* 73:6003–12.
39. Le Roy, R. J. 1994. *J Chem Phys* 101:10217–28.
40. Tellinghuisen, J., and J. G. Ashmore. 1986. *Chem Phys Lett* 102:10–6.
41. Ashmore, J. G., and J. Tellinghuisen. 1986. *J Mol Spectrosc* 119:68–82 August 1, 2010.
42. Le Roy, R. J. 2005. *DParFit 3.3: A Computer Program for Fitting Multi-Isotopologue Diatomic Molecule Spectra.* University of Waterloo Chemical Physics Research Report CP-660 http://leroy.uwaterloo.ca/programs/
43. Le Roy, R. J., and J. van Kranendonk. 1974. *J Chem Phys* 61:4750–69.
44. Ogilvie, J. F. 1981. *Proc R Soc Lond A* 378:287–300.
45. Ogilvie, J. F. 1988. *J Chem Phys* 88:2804–8.
46. Šurkus, A. A., R. J. Rakauskas, and A. B. Bolotin. 1984. *Chem Phys Lett* 105:291–4.
47. Samuelis, C., E. Tiesinga, T. Laue, M. Elbs, H. Knöckel, and E. Tiemann. 2000. *Phys Rev A* 63:012710/1–11.
48. Coxon, J. A., and P. G. Hajigeorgiou. 1990. *J Mol Spectrosc* 139:84–106.
49. Coxon, J. A., and P. G. Hajigeorgiou. 1991. *J Mol Spectrosc* 150:1–27.
50. Coxon, J. A., and P. G. Hajigeorgiou. 1992. *Can J Phys* 70:40–54.
51. Hajigeorgiou, P. G., and R. J. Le Roy. 2000. *J Chem Phys* 112:3949–57.
52. Le Roy, R. J., and Y. Huang. 2002. *J Mol Struct (Theochem)* 591:175–87.
53. Huang, Y., and R. J. Le Roy. 2003. *J Chem Phys* 119:7398–416; erratum ibid. 2007, 126:169904.
54. Le Roy, R. J., D. R. T. Appadoo, K. Anderson, A. Shayesteh, I. E. Gordon, and P. F. Bernath. 2005. *J Chem Phys* 123:204304/1–12.
55. Le Roy, R. J., N. Dattani, J. A. Coxon, A. J. Ross, P. Crozet, and C. Linton. 2009. *J Chem Phys* 131:204309/1–17.
56. Le Roy, R. J., J. Tao, C. Haugen, R. D. E. Henderson, and Y. -R. Wu. 2010. unpublished work.
57. Le Roy, R. J., and R. D. Henderson. 2007. *E Mol Phys* 105:663–77.
58. Shayesteh, A., R. D. E. Henderson, R. J. Le Roy, and P. F. Bernath. 2007. *J Phys Chem A* 111:12495–505.
59. Tiemann, E. 1987. *Z Phys D: At Mol Clusters* 5:77–82.
60. Wolf, U., and E. Tiemann. 1987. *Chem Phys Lett* 133:116–20.

61. Wolf, U., and E. Tiemann. 1987. *Chem Phys Lett* 139:191–5.
62. Tiemann, E. 1988. *Mol Phys* 65:359–75.
63. Pashov, A., W. Jastrzcebski, and P. Kowalczyk. 2000. *Comp Phys Comm* 128:622–34.
64. Pashov, A., W. Jastrzcebski, and P. Kowalczyk. 2000. *J Chem Phys* 113:6624–8.
65. Pashov, A., W. Jastrzcebski, W. Jasniecki, V. Bednarska, and P. Kowalczyk. 2000. *J Mol Spectrosc* 203:264–7.
66. Bouloufa, N., P. Cacciani, R. Vetter, A. Yiannopoulou, F. Martin, and A. J. Ross. 2001. *J Chem Phys* 114:8445–58.
67. Le Roy, R. J., J. Seto, and Y. Huang. 2007. DPotFit 1.2: *A Computer Program for fitting Diatomic Molecule Spectra to Potential Energy Functions.* University of Waterloo Chemical Physics Research Report CP-664, August 1, 2010. http://leroy.uwaterloo.ca/programs/
68. Le Roy, R. J. 2009. BetaFIT 2.0: *A Computer Program to Fit Potential Function Points to Selected Analytic Functions.* University of Waterloo Chemical Physics Research Report CP-665, http://leroy.uwaterloo.ca/programs/ (accessed August 1, 2010).
69. Watson, J. K. G. 1980. *J Mol Spectrosc* 80:411–21.
70. Watson, J. K. G. 2004. *J Mol Spectrosc* 223:39–50.
71. Le Roy, R. J. 1999. *J Mol Spectrosc* 194:189–96.
72. Ogilvie, J. F. 1994. *J Phys B: At Mol Opt Phys* 27:47–61.
73. Li, H., and R. J. Le Roy. 2008. *Phys Chem Chem Phys* 10:4128–38.
74. Li, H., P. -N. Roy, and R. J. Le Roy. 2010. *J Chem Phys* 132:214309/1–14.
75. Margenau, H. 1939. *Rev Mod Phys* 11:1–35.
76. Hirschfelder, J. O., C. F. Curtiss, and R. B. Bird. 1964. *Molecular Theory of Gases and Liquids.* New York: Wiley.
77. Hirschfelder, J. O., and W. J. Meath. 1967. In *Intermolecular Force*, ed. J. O. Hirschfelder, Advances in Chemical Physics; Vol. 12, 3–106. New York: Interscience.
78. Maitland, G. C., M. Rigby, E. B. Smith, and W. A. Wakeham. 1981. *Intermolecular Forces—Their Origin and Determination.* Oxford: Oxford University Press.

7 Other Spectroscopic Sources of Molecular Properties

Intermolecular Complexes as Examples

Anthony C. Legon and Jean Demaison

CONTENTS

7.1 INTRODUCTION

Quantum chemical methods are well adapted to calculate the equilibrium as well as the temperature-dependent rotationally and vibrationally averaged structures of small molecules, including molecular complexes. For large molecules, it is not always possible to use high-level wave function methods of electronic structure theory with large enough Gaussian basis sets. In these cases, ab initio results are not always reliable. Moreover, in the particular case of weakly interacting complexes, the convergence of the basis set is extremely slow, which limits the usefulness of quantum chemical methods to relatively small complexes. As for the experiments, in most cases the information supplied by the moments of inertia is not sufficient to determine a complete equilibrium or even a complete effective structure (see Chapter 5). There are, however, a few complementary sources of information that can be used for at least partial structure analysis. The most important are line intensity alternations, measurement of electric dipole moment components, inertial defects, variation of the rotational constants with the vibrational quantum number, hyperfine coupling constants, and empirical bond length–frequency relationships, which are discussed in this chapter.

If a molecule has equivalent nuclei, the nuclear spin statistics will affect the population of the molecular states and an intensity alternation of the spectral lines will be observed, which is helpful in determining the symmetry of the molecule. The measurement of the components of the electric dipole moment provides further useful information, which may be used to confirm the symmetry of the molecule or to identify the correct conformer. When a molecule is suspected of being planar, the inertial defect is generally used to prove the planarity. However, we will show that this criterion has to be treated with caution because a small inertial defect may be found for a nonplanar molecule. Conversely, a large inertial defect (in absolute value) may be compatible with a planar molecule. When there is a low-lying vibrational state, the variation of the rotational constants with the vibrational quantum number indicates whether the potential function has a double minimum and, thus, whether the molecule has a plane of symmetry.

The nuclear quadrupole coupling constants are informative about the structure. However, the effects of zero-point vibration affect these constants. More importantly, the bond axis does not necessarily coincide with a principal axis of the nuclear quadrupole tensor. For these reasons, the nuclear quadrupole coupling constants rarely provide useful information in an accurate structure determination. Nevertheless, there is an important exception: they are extremely helpful in establishing the approximate structure of a complex. There are other hyperfine constants, which should be useful, namely the spin–spin coupling constants. Although these constants are quite small and difficult to determine with accuracy, the increasing resolution of the spectrometers may give rise to future applications. Finally, when the stretching vibrational mode of a given bond is isolated, there is a relationship between the bond length and the corresponding stretching frequency.

As usual in molecular spectroscopy, the operators are written in boldface.

7.2 NUCLEAR SPIN STATISTICAL WEIGHTS

When applicable, this is a powerful method to determine the symmetry of a molecule in a given vibrational state. Note that the symmetry of the equilibrium state may be different from the symmetry of this vibrational state. For instance, for molecules with a small double minimum potential (see Section 7.5) it may happen that the ground state is effectively planar, whereas the equilibrium state is nonplanar. This will only be discussed briefly here. Details may be found in books on spectroscopy [1–4].

If there are equivalent nuclei in a molecule, the nuclear spin statistics will affect the population of molecular states and hence the relative intensities of spectral lines. These so-called intensity alternations may be used to determine the symmetry of the molecule.

For a nondegenerate quantum mechanical system, the wave function of a molecule, excluding translation, may be approximately written as a product of functions,

$$\Psi_{\text{total}} = \Psi_e \Psi_v \Psi_r \Psi_{\text{ns}} \tag{7.1}$$

where ψ_e represents the electronic wave function, ψ_v the vibrational function, ψ_r the rotational wave function, and ψ_{ns} the nuclear spin function. According to the Pauli principle, for the exchange (permutation operation $\mathbf{P}$) of two identical nuclei, 1 and 2, that are fermions (have half-integer spins), ψ_{total} must be antisymmetric (i.e., changed in sign: $\mathbf{P}_{12}\psi_{\text{total}} = -\psi_{\text{total}}$). On the other hand, if the nuclei are bosons (have integer spins), ψ_{total} must be symmetric (i.e., unchanged in sign: $\mathbf{P}_{12}\psi_{\text{total}} = \psi_{\text{total}}$). We assume that ψ_e and ψ_v are totally symmetric because we are mainly interested in spectra in the ground electronic and vibrational states of closed shell molecules. In states where ψ_e and ψ_v are not totally symmetric, their contributions must also be taken into account.

The symmetry of the rotational wave function is determined by the principal inertial axis about which rotation occurs. The rotational wave function is unchanged (even) or is changed in sign (odd) by a rotation of π around a principal axis. As there are three principal inertial axes (a, b, and c), there are three possible rotations. A function that is symmetric with respect to all the three rotations is designated an A function. The other functions are designated B functions: B_a, B_b, and B_c, where the subscript designates the principal axis with respect to which ψ_r is symmetric.

If a molecule (without inversion motion) has two equivalent nuclei, it has a symmetry axis, which is a principal axis and a rotation of π about this axis interchanges the position of the two equivalent nuclei. If the spin of the nuclei is I, there are $2I + 1$ values of $m_I = I, I - 1, I - 2, \ldots, -I$, where m_I is the projection of I on an axis fixed in space. For two nuclei, there are $(2I + 1)^2$ such combinations. For the $2I + 1$ cases for which $m_I = m_I'$ (where the prime denotes the second nucleus), the total function is symmetric. If $m_I \neq m_I'$, the total function is neither symmetric nor antisymmetric, but it is possible to form equal numbers of symmetric and antisymmetric combinations. If the spin function of nucleus i is $\sigma_m(i)$, the combinations are $\sigma_m(1)\sigma_{m'}(2) \pm \sigma_m(2)\sigma_{m'}(1)$ with the plus sign for the symmetric combinations and the minus sign for the antisymmetric ones.

This number of symmetric and antisymmetric combinations is $[(2I+1)^2 - (2I+1)]/2$. Thus, the total number of symmetric functions is

$$S = (2I+1)(I+1) \tag{7.2a}$$

and the total number of antisymmetric functions is

$$A = (2I + 1)I \tag{7.2b}$$

If the two nuclei obey the Bose–Einstein statistics, only those rotational and nuclear spin functions whose product is symmetric can occur, while if the two nuclei obey the Fermi–Dirac statistics, only antisymmetric combinations are allowed.

EXAMPLE 7.1A: CASE OF TWO IDENTICAL NUCLEI OF SPIN ZERO

When the spin of the two identical nuclei is zero (Bosons),* the nuclear spin functions are all symmetric and only symmetric rotational states are populated and half the levels are missing or, more exactly, they are not populated because $A = 0$. Typical examples are $N^{16}O_2$ and $S^{16}O_2$ because $I(^{16}O) = 0$.

EXAMPLE 7.1B: CASE OF WATER

Consider the water molecule H_2O, an asymmetric top. There are $(2I + 1)^2 = 4$ nuclear spin functions: $\alpha_1\alpha_2$, $\beta_1\beta_2$, $(\alpha_1\beta_2 + \beta_1\alpha_2)/\sqrt{2}$ and $(\alpha_1\beta_2 - \beta_1\alpha_2)/\sqrt{2}$, where α is for the spin $+1/2$ and β corresponds to the spin $-1/2$. The first three functions are symmetric with respect to the exchange of the two protons and the last one is antisymmetric. As the Hs are fermions (with spin $1/2$), the symmetric spin functions must combine with antisymmetric rotational states. The symmetry axis being the b axis, the levels $K_aK_c = $ eo, oe are antisymmetric (see Table 7.1). The antisymmetric spin functions must combine with the symmetric rotational states ($K_aK_c = $ ee, oo). Thus, from Equation 7.2, the antisymmetric rotational states have a relative statistical weight of 3, compared to a weight of 1 for the symmetric rotational states.

EXAMPLE 7.1C: CASE OF DEUTERATED WATER

D_2O, $I(D) = 1$. There are nine nuclear spin functions. Six of these are symmetric and three antisymmetric (see Equation 7.2). Deuterium is a boson, and the total wave function must be symmetric. The symmetric spin functions are paired with the symmetric rotational wave functions, and the symmetric rotational levels have a 2:1 weight advantage over the antisymmetric levels. On the other hand, in HDO, there are no equivalent nuclei and the statistical weight of all rotational levels is 1.

TABLE 7.1
Symmetry Properties of Wave Functions of Asymmetric Tops

Symmetry Species According to the D_2-Group	$K_aK_c{}^a$	Rotation of π About		
		a	b	c
A	ee	+	+	+
B_a	eo	+	−	−
B_b	oo	−	+	−
B_c	oe	−	−	+

[a] Parity of K_aK_c (e = even, o = odd). The first index refers to the rotation of π about the principal axis a and the second index to the rotation of π about the principal axis c.

---

* A boson is a nucleus of integer spin, whereas the spin of a fermion is half-integer.

EXAMPLE 7.2: CASE OF DIFLUOROMETHANE

If there is more than one pair of equivalent nuclei, the symmetry property of each pair has to be taken into account. Consider the example of CH_2F_2, which has a twofold symmetry axis. Interchange of either the two hydrogen nuclei or the two fluorine nuclei, all of which are fermions, must change the sign of the total wave function. Therefore, the simultaneous exchange of both pairs of nuclei will leave the total wave function unchanged. If I_H is the spin of H, there are $(2I_H + 1)^2$ spin functions, with $S_H = (I_H + 1)(2I_H + 1)$ symmetric functions and $A_H = I_H(2I_H + 1)$ antisymmetric functions (see Equation 7.2). Likewise, if I_F is the spin of F, there are $(2I_F + 1)^2$ spin functions, with $S_F = (I_F + 1)(2I_F + 1)$ symmetric functions and $A_F = I_F(2I_F + 1)$ antisymmetric functions. The product of hydrogen and fluorine functions will give $S_H S_F + A_H A_F$ symmetric and $S_H A_F + S_F A_H$ antisymmetric total spin functions. Hence, the number of symmetric functions will be

$$S = (2I_H + 1)(2I_F + 1)(2I_H I_F + I_H + I_F + 1) \tag{7.3a}$$

and the total number of antisymmetric spin functions will be

$$A = (2I_H + 1)(2I_F + 1)(2I_H I_F + I_H + I_F) \tag{7.3b}$$

In this particular case (total wave function invariant), the symmetric total spin functions must be paired with symmetric rotational wave functions and the antisymmetric total spin functions with the antisymmetric rotational wave functions.

This reasoning may be easily generalized. For a molecule with n pairs of equivalent nuclei, the number of symmetric spin functions is

$$S = \frac{1}{2}\left[\prod_{i=1}^{n}(2I_i + 1)\right]\left[\prod_{i=1}^{n}(2I_i + 1) + 1\right] \tag{7.4a}$$

and the number of antisymmetric functions is

$$A = \frac{1}{2}\left[\prod_{i=1}^{n}(2I_i + 1)\right]\left[\prod_{i=1}^{n}(2I_i + 1) - 1\right] \tag{7.4b}$$

Statistical weights arising from nuclear spins are given for a number of cases in Table 7.2.

When the molecular point group is C_{3v} or higher, the same reasoning applies but the degeneracy of the rotational wave functions has to be taken into account. This is treated in textbooks on rotational spectroscopy [1–4]. The most important case is for molecules of C_{3v} symmetry with no resolvable inversion splitting, such as CH_3CN, which have three equivalent off-axis nuclei with spin I. The statistical weights are for each K level

$$\frac{1}{3}(2I+1)(4I^2 + 4I + 3) \quad \text{if} \quad K = 3n \tag{7.5a}$$

$$\frac{1}{3}(2I+1)(4I^2 + 4I) \quad \text{if} \quad K \neq 3n \tag{7.5b}$$

For degenerate vibrational states K must be replaced by $K - \ell$. More examples are presented in Appendix VII.1

TABLE 7.2

Relative Statistical Weights Due to Spins of Equivalent Nuclei for Asymmetric Tops

Molecule	Identical Nuclei	Spin	ψ_t^a	Sym. Axis	Symmetric Levels	Statistical Weight	Antisymmetric Levels	Statistical Weight
H_2O	2 Fermions	1/2	–	b	ee, oo	1	eo, oe	3
D_2O	2 Bosons	1	+	b	ee, oo	6	eo, oe	3
H_2CO	2 Fermions	1/2	–	a	ee, eo	1	oe, oo	3
D_2CO	2 Bosons	1	+	a	ee, eo	6	oe, oo	3
NO_2, SO_2	2 Bosons	0	+	b	ee, oo	1	eo, oe	0
c-C_2H_4O	4 Fermions	1/2	+	b	ee, oo	10	eo, oe	6
CH_2F_2, $H_2C = CF_2$	4 Fermions	1/2	+	a	ee, eo	10	oe, oo	6
CD_2F_2, $D_2C = CF_2$	2 Fermions 2 Bosons	1/2, 1	–	a	ee, eo	15	oe, oo	21
CH_2Cl_2	4 Fermions	1/2, 3/2	+	b	ee, oo	36	eo, oe	28

[a] Symmetry of the total wave function in the ground state.

7.3   ELECTRIC DIPOLE MOMENTS

The x component of the permanent molecular electric dipole moment vector is defined by

$$\mu_x = |e| \sum_\alpha Z_\alpha x_\alpha - |e| \left\langle 0 \left| \sum_i x_i \right| 0 \right\rangle \tag{7.6}$$

where Z_α is the atomic number of the αth nucleus, x_α and x_i are the coordinates of the nuclei and the electrons, respectively, and $<0|0>$ indicates the average ground-state value. The magnitudes of the components μ_x along the principal axes ($x = a, b, c$) are usually determined by the analysis of the Stark effect in the rotational spectrum.

The Stark effect is lifting the M degeneracy of rotational energy levels, and hence transitions, by a uniform static electric field of strength E_Z applied along a space-fixed direction Z. The frequency shift $\Delta\nu$ of a given so-called Stark component of a transition from the zero-field frequency is usually proportional to either $\mu_g E_Z$ (with $g = a, b, c$) or $(\mu_g E_Z)^2$ and can be measured very precisely. Hence, accurate values of the components μ_x can be obtained. The sign and direction of the electric dipole moment are not directly determined from the Stark effect but can be derived by measuring the dipole moment components of several isotopologues and assuming that the magnitude does not vary upon isotopic substitution. The sign can also be obtained from the Zeeman effect [3].

Also note that the intensity of a given rotational transition in absorption spectroscopy for unsaturated one-photon transitions is proportional to the square of one component of the electric dipole moment: $\propto \mu_g^2$ the value of $g = a, b,$ or c depending on the selection rules. In Fourier transform microwave spectroscopy, the intensity is proportional to the component of the dipole moment: $\propto \mu_g$.

Accurate equilibrium electric dipole moments can be calculated ab initio [5]. They can also be predicted—more or less accurately—on the basis of the vector sum of bond dipole moments [6,7] as follows:

$$\boldsymbol{\mu} = \sum_k \boldsymbol{\mu}_k \tag{7.7}$$

where the $\boldsymbol{\mu}_k$ are bond dipole moments and are assumed transferable from one molecule to another. However, the electric dipole moment is determined by the charge distribution over the entire molecule. Therefore, bond moments are not exactly additive because of the change in bond character from molecule to molecule. Nevertheless, empirical bond moments have been used successfully to predict the structure of many molecules.

The electric dipole moment vector is extremely useful for obtaining information on the symmetry of the molecule: if one component is zero, it indicates the existence of a symmetry plane; if only one component is different from zero, it indicates the existence of a symmetry axis.

Another important application of the electric dipole moment is to identify the observed conformer when there are several possible conformers. An example is discussed in Appendix VII.2.

EXAMPLE 7.3: IDENTIFICATION OF THE THREE ISOMERS OF CHLOROTOLUENE, $CH_3C_6H_4Cl$

The electric dipole moments of ortho-, meta-, and para-chlorotoluenes have been measured. We assume that only the bonds $C–CH_3$ and $C–Cl$ are polar with bond moments μ_1 and μ_2. μ_1 is obtained from toluene: $\mu_1 = \mu[C_6H_5CH_3] = 0.39$ D and μ_2 from chlorobenzene, $\mu_2 = \mu[C_6H_5Cl] = 1.56$ D [6]. In chlorotoluenes, the angle between μ_1 and μ_2 is θ, which is approximately 60° for ortho, 120° for meta, and 180° for para. The total moment μ in chlorotoluene is thus given by

$$\mu^2 = \mu_1^2 + \mu_2^2 - 2\mu_1\mu_2 \cos\theta \tag{7.8}$$

The sign of μ_2 is obvious, Cl being much more electronegative than C. For μ_1, it is more difficult to make a prediction. Trying both signs, it appears that μ_1 and μ_2 are of opposite signs, CH_3 being at the positive end of the bond, for only then is the good agreement illustrated in Table 7.3 [55] obtained.

EXAMPLE 7.4: STRUCTURE OF Ar·HCl

It is assumed that the HCl part in ArHCl is identical in charge distribution and internuclear distance to free HCl [8].

$$\mu_a(\text{ArHCl}) = \mu(\text{HCl})\cos\theta \tag{7.9}$$

where θ is the angle between H-Cl and the a principal inertial axis. As $\mu(\text{HCl}) = 1.1086$ D and $\mu_a(\text{ArHCl}) = 0.81144$ D, we derive $\theta = 42.95°$.

We can slightly increase the accuracy of the result (by 5%) by taking into account the dipole moment induced in Ar by HCl. The experimental dipole moment is

TABLE 7.3
Electric Dipole Moments (D) of Chlorotoluenes

	μ_{calc}	μ_{exp}
Ortho	1.41	1.43
Meta	1.80	1.77
Para	1.95	1.94

Source: From McClellan, A. L. 1963. *Tables of Experimental Dipole Moments.* London: W. H. Freeman.

TABLE 7.4
Electric Dipole Moment and Induced Moment (D) For Hydrogen-Bonded Dimers

Dimer	μ_a	μ_{ind}[a]	Ref.	Dimer	μ_a	μ_{ind}	Ref.
$(HF)_2$	2.98865	0.60	56	$CO_2 \cdot HF$	2.2465	0.60	56
$(H_2O)_2$	2.6429	0.46	56	$SCO \cdot HF$	3.208	0.94	56
$H_2O \cdot HF$	4.073	0.68	56	$OC \cdot HCl$	1.5178	0.388	57
$HF \cdot HCl$	2.4095	0.14	56	$OC \cdot DCl$	1.5623	0.418	57
$H_2S \cdot HF$	2.6239	0.78	56	$OC \cdot BF_3$	0.592	0.482	57
$HCN \cdot HF$	5.612	0.80	56	$CO_2 \cdot HCl$	1.4509	0.446	57
$(CH_2)_2O \cdot HF$	3.85	0.99	56				

Sources: Legon, A. C., and D. J. Millen. 1986. *Chem Rev* 86:635–57. Altman, R. S., M. D. Marshall, and W. Klemperer. 1983. *J Chem Phys* 79:52–6.

[a] $\mu_{ind} = \mu_{dimer} - \mu_A \cos(\theta_A) - \mu_B \cos(\theta_B)$, where θ_X is the angle between the a-principal axis and the symmetry axis of monomer X (X = A, B).

found to be larger than the value calculated from vector addition of the free monomer moments. These enhancements are electrostatically induced as follows:

$$\mu_{ind} = \mu_{ind}^A + \mu_{ind}^B$$
$$= \alpha_A E_B + \alpha_B E_A \tag{7.10}$$

α_i is the polarizability tensor of monomer i and E_j the electric field at that monomer caused by the other substituent (see Appendix VII.2). μ_{ind} is particularly large for hydrogen-bonded molecules; therefore, the simple vector addition of the bond moments fails. See Table 7.4 for a few typical examples. Furthermore, in such cases, Equation 7.10 is sufficiently accurate only if higher-order induced molecular electric multipole moments are involved but this significantly increases the complexity of the method, except when the symmetry of the molecule permits simplification of the equations.

7.4 INERTIAL DEFECT

7.4.1 Definition

For a given vibrational state, the inertial defect Δ is defined as

$$\Delta = I_c - I_a - I_b \tag{7.11}$$

where I_g $(g = a, b, c)$ are the principal moments of inertia of the molecule. For a hypothetical rigid molecule, it may be expressed as

$$\Delta = -2\sum_i m_i c_i^2 \tag{7.12}$$

where m_i and c_i are the mass and the coordinate along the c axis of the ith atom, respectively. If the equilibrium moments of inertia I_ξ^e are used and if the molecule is planar (the principal axes a and b lying in the plane of the molecule), $c_i^e = 0\ \forall i$. Thus,

$$\Delta_e = I_c^e - I_a^e - I_b^e \equiv 0 \tag{7.13}$$

However, if the moments of inertia corresponding to the ground vibrational state are used,

$$\Delta_0 = I_c^0 - I_a^0 - I_b^0 \neq 0 \tag{7.14}$$

mainly because of the zero-point vibrations.

This property was first pointed out experimentally by Mecke [9] for H_2O and was explained subsequently by Darling and Dennison [10]. A general theory for asymmetric tops was later developed by Oka and Morino [11]. From the relationships between ground state and equilibrium rotational constants (see Section 4.3), it is easy to see that the inertial defect is the sum of a vibrational, an electronic, and a centrifugal term,

$$\Delta_0 = \Delta_{vib} + \Delta_{elec} + \Delta_{cent} \tag{7.15}$$

In this sum, Δ_{vib} is the dominant contribution. The value of Δ_{cent} (usually the smallest contribution) depends on the centrifugal distortion corrections applied to the rotational constants. It is generally assumed that the Wilson and Howard rotational constants are used (see Section 4.3.3). In this case

$$\Delta_{cent} = -\hbar^4 \tau_{abab} \left(\frac{3}{4} \frac{I_c}{C} + \frac{I_a}{2A} + \frac{I_b}{2B} \right) \tag{7.16}$$

Likewise, Δ_{elec} is easily calculated as (see Section 4.3.2)

$$\Delta_{elec} = -\frac{m_e}{M_p}(I_c g_{cc} - I_a g_{aa} - I_b g_{bb}) \tag{7.17}$$

where m_e and M_p are the masses of the electron and the proton, respectively; g_{aa}, g_{bb}, and g_{cc}, are the rotational magnetic moment g tensor elements. Δ_{vib} may be calculated using the expressions of the α constants obtained by a standard perturbation calculation (see Section 3.2.2). This was performed by Oka et al. [11,12]. They have shown that

$$\Delta_{vib} = \sum_s \left(\upsilon_s + \frac{1}{2}\right)\Delta_s$$

$$= \sum_s \frac{h}{\pi^2 c}\left(\upsilon_s + \frac{1}{2}\right)\sum_{s'\neq s}\frac{\omega_{s'}^2}{\omega_s(\omega_s^2 - \omega_{s'}^2)}\left[\left(\zeta_{ss'}^{(a)}\right)^2 + \left(\zeta_{ss'}^{(b)}\right)^2 - \left(\zeta_{ss'}^{(c)}\right)^2\right] \tag{7.18}$$

$$+ \sum_t \frac{h}{\pi^2 c}\left(\upsilon_t + \frac{1}{2}\right)\frac{3}{2\omega_t}$$

where ω_s and $\omega_{s'}$ are harmonic wave numbers of the fundamental vibrations, the ζ's are Coriolis coupling constants, and the last term in the expression is summed only over out-of-plane vibrations. With ω in cm^{-1} and Δ in u$\mathring{A}^2$, $h/\pi^2 c = 134.861$.

It is interesting to note that the anharmonic contribution vanishes for a planar molecule. Δ_{vib} may be written as the sum of two terms: $\Delta(i)$ for the contribution of in-plane vibrations and $\Delta(o)$ for that of out-of-plane vibrations. $\Delta(o)$ is negative and $\Delta(i)$ is positive and usually larger, at least for small molecules without low-frequency out-of-plane vibrations. Thus, for "normal" planar molecules, the inertial defect is generally small and positive with magnitude ~0.1 u$\mathring{A}^2$. For weakly bound complexes that are planar, much larger positive values are often found (see Example 7.5). Negative inertial defects become the norm for molecules with 15 or more atoms (see Equation 7.22).

7.4.2 Approximate Methods

Herschbach and Laurie [13] have shown that Δ_{vib} for a planar molecule may be written as the sum of two terms

$$\Delta_{vib} = \Delta_{harm} + \Delta_{Cor} \tag{7.19}$$

The harmonic contribution, Δ_{harm}, is always positive and can be estimated fairly accurately from the quartic centrifugal distortion constants as [14]

$$\Delta_{harm} = 3K\left(\frac{1}{\omega_A} + \frac{1}{\omega_B} + \frac{1}{\omega_{AB}} - \frac{1}{\omega_C}\right) \tag{7.20}$$

where $K = h/8\pi^2 = 16.8576$ uÅ^2/cm^{-1} and the operationally defined wavenumbers (in cm^{-1}) are defined in terms of the centrifugal distortion constants $\tau_{\alpha\beta\gamma\delta}$ by

$$
\omega_A = \sqrt{-\frac{16A^3}{\tau_{aaaa}}} \qquad \omega_B = \sqrt{-\frac{16B^3}{\tau_{bbbb}}}
$$

$$
\omega_C = \sqrt{-\frac{16C^3}{\tau_{cccc}}} \qquad \omega_{AB} = \sqrt{-\frac{16ABC}{\tau_{abab}}}
$$

(7.21)

The Coriolis contribution, Δ_{Cor}, is small for small molecules. For larger molecules, it becomes larger and negative. Watson [14] has proposed a simple empirical formula to estimate this contribution. Finally, the total vibrational inertial defect, where the Coriolis contribution is empirically taken into account, is given by

$$
\Delta_{\text{vib}} = \Delta_{\text{harm}}[1.0292 - 0.0911(N - 3)]
$$

(7.22)

where N is the number of atoms.

A good agreement between the experimental value and the result obtained by applying this formula indicates that the molecule is planar. The breakdown of Watson's formula may be caused by one of the following factors:

- There is a low-frequency out-of-plane vibration.
- The centrifugal distortion constants are not accurate enough.
- The potential is highly anharmonic.
- The molecule is really not planar.

When the low-frequency vibration is well isolated, the value of Δ_{vib} is determined mostly by the wave number of this vibration. When the lowest vibration is an in-plane vibration, Herschbach and Laurie [13] have shown empirically that

$$
\Delta_0 \approx \frac{4K}{\omega_i}
$$

(7.23)

When the lowest vibration is an out-of-plane vibration, the variation in inertial defect for two successive excited states of the out-of-plane vibration is [15]

$$
\Delta_{v_t+1} - \Delta_{v_t} \approx -\frac{4K}{\omega_t}
$$

(7.24)

Following this earlier work, Oka [16] has devised simple semiempirical relations to estimate a negative inertial defect from the low-frequency out-of-plane vibration v_1. For aliphatic molecules, the expression is

$$
\Delta_{\text{vib}} = -\frac{33.715}{v_1} + 0.0186\sqrt{I_c}
$$

(7.25a)

and for aromatic molecules, it is

$$\Delta_{vib} = -\frac{33.715}{v_1} + 0.00803\sqrt{I_c} \qquad (7.25b)$$

In these expressions, v_1 is in cm^{-1} and I_c and Δ_{vib} in $u\mathring{A}^2$. If neither Watson's nor Oka's formulas can be used, Δ_{vib} can still be estimated from Equation 7.18, provided a reliable harmonic force field is available. Ab initio calculations can generally deliver reliable harmonic force fields. However, a word of caution: standard calculations, in which the amplitude of the vibrations is assumed small, are not particularly good at predicting low-frequency vibrations.

7.4.3 APPLICATIONS

EXAMPLE 7.5: PLANARITY OF THE Ar·ClCN COMPLEX

The ground-state inertial defect of Ar·ClCN is found to be very large: $\Delta_0 = 2.75$ $u\mathring{A}^2$ [17]. However, it is possible to determine an experimental harmonic force field from the centrifugal distortion constants assuming that the force constants of ClCN remain unchanged upon complex formation. The vibrational contribution to the inertial defect calculated from this force field is 2.81 $u\mathring{A}^2$ confirming that the complex is planar.

EXAMPLE 7.6: STYRENE, $C_6H_5CH=CH_2$

From the measurement of the electric dipole moment, it might seem that styrene is planar [18]. However, the ground-state inertial defect is also found to be very large (in absolute value) and negative ($\Delta_0 = -0.6958$ $u\mathring{A}^2$) and becomes increasingly more negative with successive excitation of the torsional motion of the ethene group against the benzene ring framework. If the inertial defect is extrapolated back to the hypothetical vibrationless state (as a function of the torsional quantum number), we find that $\Delta_e = +0.0166$ $u\mathring{A}^2$, which is of the magnitude and sign expected for a planar molecule. Furthermore, using the experimental frequency of the C–C torsion, $v = 38.1$ cm^{-1}, we get, from Equation 7.25b, $\Delta_0 = -0.720$ $u\mathring{A}^2$ in good agreement with the experimental value [16].

EXAMPLE 7.7: FORMAMIDE, $HCONH_2$

The ground state inertial defect of formamide is small and positive, $\Delta_0 = 0.00657(2)$ $u\mathring{A}^2$ [19,20]. It first led to the conclusion that formamide is planar in the ground vibrational state [21]. However, the microwave spectra of 10 isotopologues were later measured [22], and it was found that the behavior of the inertial defect upon isotopic substitution is anomalous: its value decreases when a hydrogen atom of the NH_2 group is substituted by a deuterium atom. In some cases, it even becomes negative (see Table 7.5). It was concluded that the molecule is nonplanar with the $H_2N–C$ group forming a shallow pyramid. This is in agreement with the fact that Watson's formula gives a value, $\Delta_0 = 0.0713$ $u\mathring{A}^2$, much larger than the experimental value, indicating that the molecule might be non-planar.

In 1974, a complete substitution (r_s) structure was published [23], and it was found that a quartic-quadratic single minimum potential (see Section 7.5) explained the far-infrared spectrum as well as the behavior of the inertial defect. In 1987, the

TABLE 7.5
Ground State Inertial Defect (uÅ²) for Formamide

	H₂¹⁴NCHO	H₂¹⁴NCDO	c-HD¹⁴NCHO	t-HD¹⁴NCHO	D₂¹⁴NCHO
Δ_0	0.00862	0.0175	0.0029	−0.0154	−0.02069
	H₂¹⁵NCHO	H₂¹⁵NCDO	c-HD¹⁵NCHO	t-HD¹⁵NCHO	D₂¹⁵NCHO
Δ_0	0.00578	0.01748	0.00298	−0.01539	−0.02201

Source: Brown, R. D., P. D. Godfrey, and B. Kleibömer. 1987. *J Mol Spectrosc* 124:34–45.

vibration-rotation spectral data were reanalyzed [24] taking into account the large amplitude motion of the inversion. This confirmed that formamide has a very shallow single minimum potential and is therefore planar in its equilibrium conformation. Further examples are discussed at the end of Section 7.5.

7.4.4 PLANAR MOMENTS OF INERTIA

When the molecule is not planar but has a plane of symmetry, we can generalize the notion of inertial defect by using the planar moments of inertia whose definition is

$$P_x = \frac{1}{2}(-I_x + I_y + I_z) = \sum_i m_i x_i^2 \tag{7.26}$$

where $(x, y, z) = (a, b, c)$, m_i is the mass of atom i and x_i its coordinate along the principal axis x (see also Section 5.2.2).

The quantity

$$\Delta_x = -2P_x \tag{7.27}$$

has been called the pseudoinertial defect. For molecules with a plane of symmetry (assumed to be the a, b plane) and with only two out-of-plane atoms X in forming a symmetry-equivalent pair, we have

$$2P_c = m_X d_{XX}^2 \tag{7.28}$$

where m_X is the mass of atom X and d_{XX} is the distance between the two equivalent X atoms. Although the distance d_{XX} is obviously not constant in different molecules, it often does not vary much and may thus be used to identify pairs of symmetrically equivalent atoms. For instance, for CH_2Y- groups, $X = H$, we have (in uÅ²) [25]

$$3.117 \leq 2P_c \leq 3.490 \tag{7.29}$$

Naturally, the pseudoinertial defect is also affected by a vibrational contribution, which is of the same order of magnitude as the vibrational inertial defect of a planar molecule, the difference being that the anharmonic contribution does not vanish any more.

TABLE 7.6
Planar Moment of Inertia P_b for c-$C_4H_6O \cdot HC \equiv CH$

	P_b (uÅ^2)		P_b (uÅ^2)
C_4H_6O...HCCH	59.4356(2)	$^{13}C_2$...HCCH	60.816(8)
C_4H_6O...DCCD	59.4220(6)	$^{13}C_3$...HCCH	59.872(10)
C_4H_6O...DCCH	59.4396(6)	C_4H_6O...$H^{13}CCH$	59.441(24)
C_4H_6O...HCCD	59.400(12)	C_4H_6O...$HC^{13}CH$	59.439(34)
$[3,4D]C_4H_6O$...HCCH	62.8374(26)		

Source: Cole, G. C., R. A. Hughes, and A. C. Legon. 2005. *J Chem Phys* 122:134311/1–11.

EXAMPLE 7.8: STRUCTURE OF 2,5-DIHYDROFURAN–
ETHYNE, c-$C_4H_6O \cdot HC \equiv CH$

a-type and c-type rotational transitions were observed [26]. It was further checked that b-type transitions are not observable. This indicates that the ac inertial plane is probably a symmetry plane (C_s symmetry). This is confirmed by the analysis of the planar moment P_b, which should be unchanged by isotopic substitution of any atom lying in the ac plane. Furthermore, if the ethyne subunit lies in the ac plane of the complex and if the geometry of 2,5-dihydrofuran is unperturbed by complex formation, the equilibrium value of P_b will be identical to that of free 2,5-dihydrofuran, $P_b = 59.907$ uÅ^2. The values of P_b for nine isotopologues are given in Table 7.6.

P_b is not significantly changed by isotopic substitution of any atom of ethyne, indicating that HCCH lies in the ac plane. On the other hand, isotopic substitution of the atoms of 2,5-dihydrofuran changes the value of P_b indicating that the substituted atoms do lie outside the ac plane. The difference between P_b for $C_4H_6O \cdot HCCH$ and that of free C_4H_6O may be explained by the effect of the zero-point motions, which are different in these two molecules, and by the fact that the geometry of C_4H_6O changes (but probably only very slightly) upon complex formation.

From the internal rotation splittings of a rotational spectrum, we can determine the moment of inertia I_α of the top that is undergoing internal rotation, which is close to the planar moment of inertia. However, it is still difficult to use the top moment of inertia for an accurate structure determination. This point is discussed in Appendix VII.2. The nuclear quadrupole coupling constants may sometimes be used to establish the existence of a symmetry plane (see Appendix VII.6.1).

7.5 TORSIONAL POTENTIAL FUNCTION

When the rotational constants have been determined for excited states of a torsion (or ring puckering) vibration, it is possible to determine the potential function governing the motion and, thus, whether the molecule is planar (or has a plane of symmetry) [27,28]. A zigzag variation behavior of the rotational constants with the torsional quantum number indicates that the potential function has a double minimum and that the molecule is not planar. For almost all the cases investigated, a quartic-quadratic potential function may be used.

The Hamiltonian for a one-dimensional quartic-quadratic oscillator is

$$\mathbf{H} = \frac{\mathbf{P}_x^2}{2\mu} + Ax^4 + Bx^2 \tag{7.30}$$

where x is a generalized vibrational coordinate describing the torsional motion, μ is the reduced mass corresponding to the vibration, and $\mathbf{P}_x$ is the vibrational momentum conjugate to x. A is positive and B may be positive or negative. Odd powers of x in the potential are strictly zero when there is a plane of symmetry. Using the dimensionless coordinate z (see Appendix VII.4), the Hamiltonian may be written as follows:

$$\mathbf{H} = a(\mathbf{P}_z^2 + z^4 + bz^2) \tag{7.31}$$

If $b > 0$, the heavy atom skeleton has a planar equilibrium conformation and therefore a symmetry plane (if b is large, the oscillator is almost harmonic). If $b < 0$, the equilibrium conformation is nonplanar. a and b can be derived from the torsional dependence of the rotational constants. For example, B_v is given by

$$\langle v|B|v \rangle = B^0 + \beta_2 \langle v|z^2|v \rangle + \beta_4 \langle v|z^4|v \rangle \tag{7.32}$$

where the B^0 constants refer to the planar molecule. The experimental determination of the parameters β_2 and β_4 permits the parameters a and b of Equation 7.31 to be obtained.

This method seems to be much safer than the use of the inertial defect. However, interactions between one of the higher vibrational levels of the torsional mode and a nearby level of another vibration are always possible. This will affect the rotational constants and may be misleading.

EXAMPLE 7.9: GEOMETRY OF *SYN*-ACRYLAMIDE, $CH_2=CHCONH_2$

The electric dipole moment of this molecule was measured [29]. When μ_c is fixed at zero, the standard deviation of the fit of the Stark shifts is insignificantly larger (meaning that this component of the dipole moment is likely to be zero. It is in favor of the existence of a plane of symmetry, see Section 7.3). However, this is not enough to conclude that ab is a symmetry plane. The inertial defect is rather large in absolute value, $\Delta_0 = -0.13130(3)$ uÅ², but its sign may be explained as indicating the presence of a low-frequency out-of-plane vibration for the molecule. The lowest vibrational frequency is indeed the torsional frequency of the group CH–C(O). Relative intensity measurements of the rotational transitions give $v = 90(10)$ cm⁻¹. This value is confirmed using the difference between the inertial defect of two consecutive torsionally excited states, Equation 7.24, which gives $v = 85$ cm⁻¹. If the value of 90 cm⁻¹ is put into Equation 7.25a, it gives $\Delta_0 = -0.134$ uÅ² in excellent agreement with the experimental value. Furthermore, the smooth variation of the rotational constants for this torsional vibration ($b > 0$) leads to the conclusion that the heavy atom skeleton is planar. Finally, the inertial defect of the deuterated species, NHD and ND₂, is close to that of the normal species, which is expected for a planar molecule: $\Delta_0(ND_2) = -0.17731(24)$ uÅ²; $\Delta_0(NDH) = -0.1553(4)$ uÅ²; and $\Delta_0(NHD) = -0.1499(5)$ uÅ². In conclusion, acrylamide is likely to possess a planar equilibrium conformation.

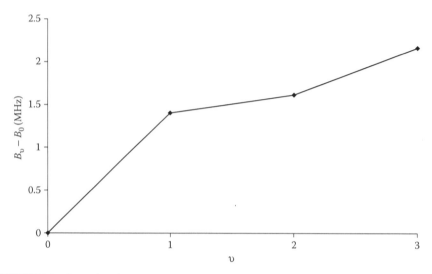

FIGURE 7.1 *B* rotational constant (MHz) of fluorostyrene as a function of the ring-puckering vibrational quantum number υ. (From Villamañan, R. M., J. C. López, and J. L. Alonso. 1989. *J Am Chem Soc* 111:6487–91.)

EXAMPLE 7.10: 2-FLUOROSTYRENE, C$_6$H$_4$FCH=CH$_2$

Because styrene has been found to be planar (see Section 7.4.3), 2-fluorostyrene may be expected to be planar too [30]. However, the ground-state inertial defect is much larger (in absolute value) than that of styrene, $\Delta_0 = -1.215$ uÅ^2 (to be compared with -0.696 uÅ^2 for styrene). Furthermore, a zigzag behavior of the rotational constants as a function of the torsional quantum number is the proof of a double minimum potential and, hence, a nonplanar equilibrium conformation for the molecule. See Figure 7.1 for a plot of $B_υ$ as a function of υ. Finally, the determination of the potential using Equation 7.32 gives $b = -3.35$.

7.6 NUCLEAR HYPERFINE STRUCTURE

Nuclei possess a spin angular momentum **I** of magnitude $\{I(I+1)\}^{1/2}\hbar$, where the quantum number I, whose value is dependent on the composition and structure of the nucleus, is either integral or half integral [3]. The nuclear spin angular momentum I can couple with the rotational angular momentum to produce a hyperfine structure of the rotational transitions.

7.6.1 NUCLEAR QUADRUPOLE INTERACTION

If $I > 1/2$, the nucleus may possess a nonvanishing quadrupole moment Q, which results from a nonspherical charge distribution in the nucleus. The interaction between Q and the electric field gradient of the molecule at the quadrupole nucleus provides a mechanism through which **I** and **J** can couple.

Quantum mechanics limits the number of allowed, discrete orientations of the nuclear spin vector $\mathbf{I}$ with respect to the rotational angular momentum vector $\mathbf{J}$ of the molecular framework. Each allowed orientation of $\mathbf{I}$ and $\mathbf{J}$ corresponds to a different orientation of the nuclear electric quadrupole moment with respect to the electric field gradient and therefore to a different energy of interaction. The Hamiltonian for the interaction is given by the product of two commuting tensors of second rank, namely the nuclear electric quadrupole moment tensor $\mathbf{Q}$ and the electric field gradient tensor $\nabla\mathbf{E}$ (both at the position of the nucleus in question), according to

$$\mathbf{H}_Q = -\frac{1}{6}\mathbf{Q} : \nabla\mathbf{E} \tag{7.33}$$

The most important parameter obtained in the analysis of the nuclear quadrupole hyperfine structure is the electric field gradient, which provides valuable information on the electronic environment of the quadrupolar nucleus. The quadrupole coupling constants form a symmetric tensor whose components are written as

$$\chi_{xy} = eQq_{xy} \tag{7.34}$$

with

$$q_{xy} = \left(\frac{\partial^2 V}{\partial x \partial y}\right) \tag{7.35}$$

eQ is the conventional nuclear electric quadrupole moment and V is the static potential arising from the extranuclear charges. The experimental nuclear quadrupole coupling constants are usually written as a tensor χ relative to the principal inertial axes of the molecule

$$\chi = \begin{pmatrix} \chi_{aa} & \chi_{ab} & \chi_{ac} \\ \chi_{ab} & \chi_{bb} & \chi_{bc} \\ \chi_{ac} & \chi_{bc} & \chi_{cc} \end{pmatrix} \tag{7.36}$$

The Laplace equation implies

$$\chi_{aa} + \chi_{bb} + \chi_{cc} = 0 \tag{7.37}$$

If the molecule is a symmetric top, the off-diagonal terms are zero. If the molecule possesses a symmetry plane, for example, ab, $\chi_{ac} = \chi_{bc} = 0$ but $\chi_{ab} \neq 0$.

7.6.2 SPIN-ROTATION INTERACTION

Many studied molecules have singlet Σ electronic ground states. As all the electrons are paired, their electronic magnetism is canceled. However, the rotation of the molecule slightly excites higher electronic states with nonzero angular momentum, thereby generating a weak magnetic field, which can interact with the nuclear magnetic moments. This magnetic hyperfine interaction, called the *spin-rotation* interaction, is proportional to $\mathbf{I} \cdot \mathbf{J}$ [31]. The spin-rotation Hamiltonian for a nucleus A of spin $\mathbf{I}_A$ may be written as

$$\mathbf{H}_{SR} = -\mathbf{I}_A \mathbf{C}_A \mathbf{J} \tag{7.38}$$

where $\mathbf{C}_A$ is the spin-rotation coupling tensor of rank two, which consists of a sum of nuclear and electronic parts. Note that there are two different definitions; the minus sign is often omitted in Equation 7.38.

7.6.3 SPIN–SPIN INTERACTION

The *spin–spin* interaction between nuclei is a direct magnetic dipole–dipole interaction, which depends on the magnetic dipole moments and masses of the nuclei of spin different from zero. The spin–spin Hamiltonian for the interaction between two nuclei of spin $\mathbf{I}_1$ and $\mathbf{I}_2$ is

$$\mathbf{H}_{SS} = \frac{1}{R^3} \left[\boldsymbol{\mu}_1 \cdot \boldsymbol{\mu}_2 - \frac{3(\boldsymbol{\mu}_1 \cdot \mathbf{R})(\boldsymbol{\mu}_2 \cdot \mathbf{R})}{R^2} \right] \tag{7.39}$$

$\mathbf{R}$ is the vector joining the two nuclei, and $\boldsymbol{\mu}_i$ is the nuclear magnetic moment due to the nuclear angular momentum $\mathbf{I}_i$, which can be expressed as

$$\boldsymbol{\mu}_i = \mu_N g_i \mathbf{I}_i \tag{7.40}$$

where μ_N is the nuclear magneton and g_i the nuclear g value. With these notations, the Hamiltonian may be rewritten as

$$\mathbf{H}_{ss} = \mathbf{I}_1 \mathbf{D} \mathbf{I}_2 \tag{7.41}$$

where $\mathbf{D}$ is the second-rank direct dipolar spin–spin coupling tensor. The scalar spin–spin coupling term is omitted since the electron-coupled spin–spin interaction is often beyond the accuracy actually achievable. However, it should be noted that the contribution of the electron-coupled term was found significant in some molecules, for instance, ClF [32] or InI [33]. The elements of $\mathbf{D}$ are given by [34]

$$D_{ij} = \frac{g_1 g_2 \mu_N^2 (R^2 \delta_{ij} - 3R_i R_j)}{R^5} \text{ with } i, j = x, y, z \tag{7.42}$$

From this equation, it appears that the spin–spin interaction is a powerful tool to determine the distance between two atoms. However, the splitting induced by the spin–spin interaction is extremely small in most cases and the parameters D_{ij} are difficult to determine with the required accuracy. For this reason, they have been determined in very few polyatomic molecules (see Appendix VII.5). However, they have been determined for several complexes, in particular with HF, and they can be used analogously to the quadrupole coupling constants to obtain some information on the structure of a complex (see Section 7.6.5.3 and Equation 7.56).

7.6.4   USE OF THE OFF-DIAGONAL ELEMENTS OF THE QUADRUPOLE COUPLING TENSOR

The diagonal elements of χ are easy to determine. On the other hand, the determination of the off-diagonal elements is a difficult problem, which requires either the finding of a suitable perturbation (which enhances the effects of the off-diagonal elements) or the use of a sub-Doppler technique. The principal axes of the nuclear coupling tensor are generally different from the principal inertial axes, at least for an asymmetric molecule. Diagonalization of χ to obtain χ^P (diagonal tensor in its own principal axes) permits determination of the rotation angles from the inertial axis system to the quadrupole principal axis system. The important point is that it is now often possible to obtain these angles with an excellent precision ($\sim0.01°$) much greater than that given by a standard structure determination using the moments of inertia (typically an order of magnitude less).

When a quadrupolar nucleus is at the end of a bond, this information is extremely useful because it is usually assumed that the electronic density is cylindrically symmetric about the internuclear axis. Actually, a thorough analysis by Kisiel et al. [35] indicates that there is a significant difference $\delta = \angle(CX, a) - \angle(z, a)$, where $X = Cl, Br, I$, between the direction of the symmetry axis z of the field gradient and the direction of the corresponding bond. From a sample of 11 different nuclei, they found $0.06° \leq \delta \leq 1.41°$.

In complexes with a symmetry plane, the determination of the single off-diagonal element of χ provides useful information, as demonstrated in the example of the 2,5-dihydrofuran$\cdot$HCl complex [36]. From the spectroscopic constants, it may be deduced that 2,5-dihydrofuran$\cdot$HCl has probably an equilibrium geometry of C_s symmetry with the principal inertial plane ac being the symmetry plane. This assumption is strengthened by the fact that only a- and c-type transitions are observed and that χ_{ac} is the only determinable off-diagonal element. Hence, $\chi_{ab} = \chi_{bc} = 0$. Let z be the equilibrium direction of the HCl axis in the complex. A good approximation to the equilibrium value of the angle $\angle(a, z) = \theta$ can be obtained from the tensor χ. The y axis is perpendicular to z and coincides with the b axis. The chlorine nuclear quadrupole coupling tensor in the equilibrium conformation is χ_{ij}^e where $i, j = x, y, z$. The C_s symmetry of the complex requires that $\chi_{xy} = \chi_{yz} = 0$. χ_{xz} may be different from zero, but it is probably small. If the electric field gradient is unaffected by the complexation, it should be zero. Thus, it will be neglected. We further assume that the vibrational averaging does not affect these conclusions. The diagonal components $\langle\chi_{ii}\rangle$ ($i = x, y, z$) averaged over the zero-point motion of the HCl subunit in the complex are related to the components of χ according to

$$\chi_{aa} = \langle\chi_{zz}\rangle\cos^2\theta + \langle\chi_{xx}\rangle\sin^2\theta \tag{7.43a}$$

$$\chi_{bb} = \langle\chi_{yy}\rangle \tag{7.43b}$$

$$\chi_{cc} = \langle\chi_{zz}\rangle\sin^2\theta + \langle\chi_{xx}\rangle\cos^2\theta \tag{7.43c}$$

$$\chi_{ac} = \left[\langle\chi_{xx}\rangle - \langle\chi_{zz}\rangle\right]\cos\theta\sin\theta \tag{7.43d}$$

From Equations 7.43a and 7.43c, we get

$$\chi_{aa} - \chi_{cc} = \left[\langle \chi_{zz} \rangle - \langle \chi_{xx} \rangle \right] \cos 2\theta \tag{7.44}$$

Dividing Equation 7.43d by Equation 7.44, we obtain

$$\frac{\chi_{ac}}{\chi_{aa} - \chi_{cc}} = -\frac{1}{2} \tan 2\theta \tag{7.45}$$

Equation 7.45 permits determination of the equilibrium value of θ, the angle between the a-inertial axis and the bond axis HCl. Then, we can calculate the values of $\langle \chi_{ii} \rangle$, $i = x, y, z$. This permits us to check the validity of the assumptions because $\langle \chi_{xx} \rangle$ and $\langle \chi_{yy} \rangle$ should be equal if the z axis coincides with the bond axis. For 2,5-dihydrofuran·HCl, these values are indeed nearly equal: $\langle \chi_{xx} \rangle = 25.31$ MHz and $\langle \chi_{yy} \rangle = 24.90$ MHz. Note that for weakly bound complexes such as 2,5-dihydrofuran·HCl, a knowledge of the equilibrium angle θ allows the position of the HCl subunit to be located with respect to the principal inertial axis system of the complex and therefore (assuming that the HCl subunit is unchanged on complex formation) places the hydrogen atom much more accurately than otherwise possible. In this way, accurate values of the angular deviations of the hydrogen bond from linearity in this and similar complexes of C_s symmetry have been established.

7.6.5 USE OF THE DIAGONAL ELEMENTS OF THE QUADRUPOLE COUPLING TENSOR

7.6.5.1 Principle of the Method

The diagonal elements of χ (usually χ_{aa}) can also be used to get information on the structure of a complex. We discuss the principle of the method by using the example of the Ar·HCl complex [8]. Assume that the HCl part in ArHCl is identical in charge distribution and in geometry to free HCl. The experimental value of χ_{aa}(Cl) in the dimer is the projection of the value of the quadrupole coupling constant χ in free HCl on the instantaneous a axis of the complex averaged over the zero-point motion

$$\chi_{aa} = \frac{1}{2} \langle 3 \cos^2 \theta - 1 \rangle \chi \tag{7.46}$$

with $\theta = \angle(a, \mathrm{HCl})$. Using the experimental value of χ_{aa} and the known value of χ, an average value defined by $\theta_{av} = \cos^{-1} \left\{ \left(2\chi_{aa}/\chi + 1 \right)/3 \right\}^{1/2}$ is deduced. It is determined only within $180°$. The choice between the acute angle and the obtuse angle is made with the help of the moments of inertia and their change upon isotopic substitution.

Because of vibrational averaging, the value of θ will vary significantly from one isotopologue to the other one. This variation can be explained by using the two-dimensional isotropic harmonic oscillator as a model. The average angle is given by [37]

$$\langle \theta^2 \rangle = \frac{\hbar}{2\pi \mu_b \nu_b} \tag{7.47}$$

where

$$\nu_b = \frac{1}{2\pi}\sqrt{\frac{k_b}{\mu_b}} \tag{7.48}$$

is the harmonic bending frequency, k_b is the bending force constant, and the bending reduced mass μ_b is given by

$$\mu_b = \left[\frac{1}{M_{Ar}r_{Ar-H}^2} + \frac{1}{M_{Cl}r_{HCl}^2} + \frac{1}{M_H}\left(\frac{1}{r_{Ar-H}} + \frac{1}{r_{HCl}}\right)^2\right]^{-1} \tag{7.49}$$

The accuracy of the calculation of θ in Equation 7.46 is limited by two assumptions: the charge transfer and the change in geometry upon complexation are neglected. We will now analyze the effects of these two approximations.

7.6.5.2 Variation of the Electric Field Gradient upon Complexation

Different approximate methods may be used depending on the complexity of the molecule. When applicable, the Townes–Dailey method [38] is the simplest. We discuss it using the example of the $Ar \cdot ICl$ complex [39].

The presence of the Ar atom induces a slight change in the electric field gradients at I and Cl because the ICl electric charge distribution induces electric dipole, quadrupole, and higher moments in Ar which in turn produce additional electric field gradients at I and Cl. The transfer of one electron from I to Cl increases the iodine nuclear quadrupole coupling constant in magnitude from $\chi_A(I)$ to $2\chi_A(I)$, where $\chi_A(I)$ is the coupling constant of the free iodine atom. Correspondingly, the chlorine coupling constant decreases from $\chi_A(Cl)$ to zero. Hence, the presence of the Ar atom will induce a fraction δ of an electron to be transferred from I to Cl

$$\chi_{aa}^e(X) = \chi_0(X) \pm \delta\chi_A(X) \tag{7.50}$$

where the positive sign corresponds to $X = I$ and the negative sign to $X = Cl$. Substituting this equation into Equation 7.46 gives

$$\chi_{aa}(X) = \frac{1}{2}[\chi_0(X) \pm \delta \cdot \chi_A(X)]\langle 3\cos^2\theta - 1\rangle \tag{7.51}$$

Using the values of $\chi_A(X)$ usually derived from atomic spectra, Equation 7.51 can be solved with $X = I$, Cl to give δ and the average value of θ. For $Ar \cdot ICl$, $\delta = 5.41(8) \times 10^{-3}\, e$ and $\theta = 5.454(10)°$. A generalization of this approach to determine similarly both the fraction δ_i of an electronic charge transferred between a Lewis base B and a dihalogen XY and also that δ_p transferred from X to Y is set out in [40].

More generally, to calculate the variation of the electric field gradient at an atom X (in a complex B...HX, for example) due to the nearby molecule B, the distributed-point multipole model may be used, where point multipoles are assigned to each atom of B. When the molecule B...HX is axially symmetric, the electric field gradient at X due to the atoms Y_i of B is [41]

$$q = \sum_i q(i) \tag{7.52}$$

where i refers to the atoms Y_i, and $q(i)$ is given by

$$q^0(i) = -\frac{1}{4\pi\varepsilon_0}\left[2\frac{Q_i}{r_i^3}P_0(\cos\alpha) + 6\frac{\mu_i}{r_i^4}P_1(\cos\alpha) + 12\frac{\Theta_i}{r_i^5}P_2(\cos\alpha)\right.$$
$$\left. + 20\frac{\Omega_i}{r_i^6}P_3(\cos\alpha) + \cdots \right] \tag{7.53}$$

where Q_i is the charge located at nucleus i, μ_i is the dipole, Θ_i is the quadrupole, Ω_i is the octupole, r_i is the distance from nucleus i to nucleus X, and $P_n(\cos\alpha)$ are Legendre polynomials, where α is the angle between the a-principal axis of the complex and the symmetry axis of B. Next, the Sternheimer shielding effects have to be taken into account by

$$q(i) = q^0(i)(1 - \xi) \tag{7.54}$$

where ξ is the Sternheimer shielding factor. The Y_i atoms induce an electric quadrupole moment in the electronic charge surrounding the nucleus X, which leads to a contribution to the electric field gradient at X.

The more general case of an asymmetric complex is treated in reference [41]. A third, simpler, and more accurate method is to calculate ab initio the variation of the electric field gradient [42].

7.6.5.3 Determination of the Bond Lengthening from the Nuclear Quadrupole Coupling

From the known hyperfine coupling constants (nuclear quadrupole and spin–spin), we can determine the lengthening of a bond due to the formation of a dimer [41]. We treat the hydrogen-bonded dimers B · HF as an example where the lengthening may be sizeable because HF forms strong hydrogen bonds.

The coupling constants for isolated HF (or DF) are called D_0 for the H–F nuclear spin–spin coupling and χ_0 for the D-nuclear quadrupole coupling constant (in DF). When the bond length r of HF (or DF) is stretched by δr ($\ll r$), the "effective" coupling constants are

$$D_0^{\text{eff}} = D_0 + \frac{dD_0}{dr}\delta r \tag{7.55}$$

and

$$\chi_0^{\text{eff}} = \chi_0 + \frac{d\chi}{dr}\delta r \tag{7.56}$$

dD_0/dr is calculated from the geometry of HF by differentiating Equation 7.42. $d\chi/dr$ can be obtained by scaling the results of an ab initio calculation of χ at different values of r.

TABLE 7.7

Bond Lengthening δr for Various B...HF Dimers

B	δr (Å)	B	δr
Ar	0	$H_2{}^{32}S$	0.010
^{84}Kr	0	$H^{12}C^{15}N$	0.014
^{133}Xe	0	$^{12}CH_3{}^{12}C^{15}N$	0.016
$^{15}N_2$	0.001	$H_2{}^{16}O$	0.015
$^{12}C^{16}O$	0.007		

Source: Legon, A. C., and D. J. Millen. 1986. *Proc R Soc Lond A* 404:89–99. With permission; Legon, A. C., and D. J. Millen. 1986. *Chem Rev* 86:635–57.

The H–F spin–spin coupling constant in $B \cdot HF$ is

$$D_{aa} = \frac{1}{2} D_0^{\text{eff}} \left\langle 3\cos^2 \theta^H - 1 \right\rangle \tag{7.57}$$

where $\theta^H = \angle(a, HF)$. For the D-quadrupole coupling constant, the analogous equation is

$$\chi_{aa}^{\text{corr}} = \frac{1}{2} \chi_0^{\text{eff}} \left\langle 3\cos^2 \theta^D - 1 \right\rangle \tag{7.58}$$

where χ_{aa}^{corr} is the D-nuclear quadrupole coupling constant in the dimer corrected for the contribution of B to the electric field gradient at D (see Section 7.6.5.2). From Equation 7.47, the relationship between θ^H and θ^D is known, and its use in Equations 7.57 and 7.58 permits the calculation of δr. This quantity has been determined for various $B \cdot HF$ dimers and the values are given in Table 7.7.

It is worth noting that an accurate ab initio calculation of the equilibrium bond lengthening δr in $H_2O \cdot HF$ gives 0.0156 Å [43] in perfect agreement with the experimental value. Another application of the nuclear quadrupole coupling constants, namely to establish the existence of a symmetry plane, is discussed in Appendix VII.6.1.

7.7 ISOLATED STRETCHING FREQUENCIES

There are many correlations between structural and other parameters [44]. Although most of them are difficult to use or are inaccurate, there is one relation that may be of great help for a structure determination. The stretching force constants are related to the bond strength, and one can expect a good correlation between the length of a bond and the corresponding diagonal stretching force constant, as shown, for instance, by Badger [45,46]. In principle, such a correlation might be used to determine bond lengths, but it is at least as difficult to accurately determine force constants as bond lengths. There are, however, a few cases in which the bond length versus diagonal force constant correlation is relevant. One is when the molecule is simple enough and this correlation has been, for instance, used to determine the Au-Au and Ag-Ag bond lengths in several molecules [47]. More generally, when a vibrational mode r has characteristic frequency far from the others, r', it may be considered isolated. In other words, the contribution of nondiagonal force constants $f_{rr'}$ is then negligible when the corresponding energy differences $E_r - E_r'$ are large. Thus, in this particular case, a relationship between the bond length and the corresponding stretching vibrational frequency is to be expected. Such a relationship was first pointed out by Bernstein [48] and considerablsy developed by McKean [49] for CH bonds.

X-H ($X = C, N, O$) stretching fundamentals may be isolated because, owing to their small reduced mass, they appear at much higher frequencies than the other fundamentals. However, it remains exceptional because these fundamentals are often perturbed by anharmonic resonances. McKean used selective deuteriation, with all CH bonds but the relevant CH one being deuterated. The CH group of interest therefore decouples from the rest of the molecule in order to make sure that the CH stretching fundamental is not affected. The main difficulty is often to synthesize a molecule where all hydrogen atoms but one have been replaced by deuterium. Furthermore, in some cases the stretching vibration is perturbed by some anharmonic resonance and, except if a detailed vibration-rotation analysis was carried out to provide a deperturbed stretching wavenumber, another method is required to estimate the deperturbed wavenumber.

Another method for determining isolated XH stretching frequencies relies on data from overtone bands, that is, resulting from the multiexcitation of a single vibrational mode. It is well-known that the motions of the individual bonds become increasingly localized upon increasing excitation and thus less and less affected by internal couplings [50]. The band origin of as many overtones of the XH stretch as possible needs to be measured and inserted into a so-called Birge–Sponer plot [51], as follows:

$$v = v\tilde{\omega} - v(v+1)\tilde{\omega}x \tag{7.59}$$

where $\tilde{\omega}$ is the mechanical wavenumber and $\tilde{\omega}x$ is the first anharmonic correction term. This procedure thus also allows the wavenumber of the "unperturbed" fundamental band, $v_{is}(XH)$, to be determined by

$$v_{is} = \tilde{\omega} - 2\tilde{\omega}x \tag{7.60}$$

The major advantage of this second method is to avoid isotopic labeling. However, it does not always avoid the problem of resonance.

The accuracy of the correlation corresponds to a bond-length change of about 0.001 Å for a shift in the isolated CH stretching fundamental wavenumber of 14 cm^{-1} [52]. A least-squares fit of 60 data points results in the following expression (see also Figure 7.2):

$$r_e(CH)[\text{Å}] = 1.3047(29) - 7.311(97) \times 10^{-5} v_{is}(CH)[cm^{-1}] \tag{7.61}$$

with a correlation coefficient $\rho = 0.990$ and a standard deviation $\sigma = 0.001$ Å. For the alkynyl hydrogen $\equiv$C–H, the following slightly different relation should be used:

$$r_e(C_{[sp]}H)[\text{Å}] = 1.558(33) - 14.90(99) \times 10^{-5} v_{is}(CH)[cm^{-1}] \tag{7.62}$$

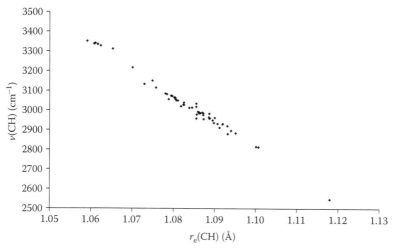

FIGURE 7.2 Correlation between isolated stretching wavenumbers ν(CH) (cm⁻¹) and equilibrium distances r_e(CH) (Å). (From Demaison, J., and H. D. Rudolph. 2008. *J Mol Spectrosc* 248:66–76. With permission.)

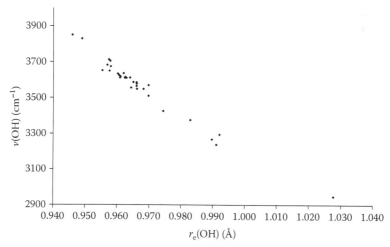

FIGURE 7.3 Correlation between isolated stretching wavenumbers ν(OH) (cm⁻¹) and equilibrium distances r_e(OH) (Å). (From Demaison, J., M. Herman, and J. Liévin. 2007. *Intern Rev Phys Chem* 26:391–420.)

Such a linear relationship was also shown to exist between r(OH) and ν(OH) [53] (see also Figure 7.3). For the OH bond length, a fit of 28 data points results in the following expression:

$$r_e(OH)[\text{Å}] = 1.2261(76) - 7.29(21) \times 10^{-5} \nu_{is}(OH)[\text{cm}^{-1}] \qquad (7.63)$$

with a correlation coefficient $\rho = 0.979$ and a standard deviation $\sigma = 0.0015$ Å. A similar relation has also been found to exist between r(NH) and ν(NH) [54].

REFERENCES

1. Townes, C. H., and A. L. Schawlow. 1975. *Microwave Spectroscopy*. New York: Dover Publications.
2. Kroto, H. W. 1975. *Molecular Rotation Spectra*. New York: Dover Publication.
3. Gordy, W., and R. L. Cook. 1984. *Microwave Molecular Spectra*. New York: Wiley.
4. Bunker, P. R., and P. Jensen. 2004. *Fundamentals of Molecular Symmetry*. Bristol: IOP Publishing.
5. Bak, K. L., J. Gauss, T. Helgaker, P. Jørgensen, and J. Olsen. 2000. *Chem Phys Lett* 319:563–8.
6. Smyth, C. P. 1955. *Dielectric Behavior and Structure*. New York: McGraw-Hill.
7. Gierke, T. D., H. L. Tigelaar, and W. H. Flygare. 1972. *J Am Chem Soc* 94:330–8.
8. Novick, S. E., P. Davies, S. J. Harris, and W. Klemperer. 1973. *J Chem Phys* 59:2273–9.
9. Mecke, R. 1933. Z. *Phys A* 81:313–31.
10. Darling, B. T., and D. M. Dennison. 1940. *Phys Rev* 57:128–39.
11. Oka, T., and Y. Morino. 1961. *J Mol Spectrosc* 6:472–82.
12. Oka, T., and M. -F. Jagod. 1990. *J Mol Spectrosc* 139:313–27.
13. Herschbach, D. R., and V. W. Laurie. 1964. *J Chem Phys* 40:3142–53.
14. Watson, J. K. G. 1993. *J Chem Phys* 98:5302–9.
15. Hanyu, Y., C. O. Britt, and J. E. Boggs. 1966. *J Chem Phys* 45:4725–8.
16. Oka, T. 1995. *J Mol Struct* 352/353, 225–33.
17. Keenan, M. R., D. B. Wozniak, and W. H. Flygare. 1981. *J Chem Phys* 75:631–40.
18. Caminati, W., B. Vogelsanger, and A. Bauder. 1988. *J Mol Spectrosc* 128:384–98.
19. Fogarasi, G., and P. G. Szalay. 1997. *J Phys Chem A* 101:1400–8.
20. Demaison, J., A. G. Császár, I. Kleiner, and H. Møllendal. 2007. *J Phys Chem A* 111:2574–86.
21. Kurland, R. J. 1955. *J Chem Phys* 23:2202–3.
22. Costain, C. C., and J. M. Dowling. 1960. *J Chem Phys* 32:158–65.
23. Hirota, E., R. Sugisaki, C. J. Nielsen, and G. O. Sørensen. 1974. *J Mol Spectrosc* 49:251–67.
24. Brown, R. D., P. D. Godfrey, and B. Kleibömer. 1987. *J Mol Spectrosc* 124:34–45.
25. Demaison, J., G. Wlodarczak, K. Siam, J. D. Ewank, and L. Schäfer. 1988. *Chem Phys* 120:421–8.
26. Cole, G. C., R. A. Hughes, and A. C. Legon. 2005. *J Chem Phys* 122:134311–1–134311–11.
27. Gwinn, W. D., and A. S. Gaylord. 1973. *MTP Int Rev Sci: Phys Chem Ser* Two 3:205–61.
28. Legon, A. C. 1980. *Chem Rev* 80:231–62.
29. Marstokk, K. -M., H. Møllendal, and S. Samdal. 2000. *J Mol Struct* 524:69–85.
30. Villamañan, R. M., J. C. López, and J. L. Alonso. 1989. *J Am Chem Soc* 111:6487–91.
31. Flygare, W. H. 1974. *Chem Rev* 74:653–87.
32. Fabricant, B., and J. S. Muenter. 1977. *J Chem Phys* 66:5274–7.
33. Walker, N. R., S. G. Francis, J. J. Rowlands, and A. C. Legon. 2006. *J Mol Spectrosc* 239:126–9.
34. Read, W. G., and W. H. Flygare. 1982. *J Chem Phys* 76:2238–46.
35. Kisiel, Z., E. Bialkowska-Jaworska, and L. Pszczólkowski. 1998. *J Chem Phys* 109:10263–72.
36. Legon, A. C., and J. C. Thorn. 1994. *Chem Phys Lett* 227:472–9.
37. Balle, T. J., E. J. Campbell, M. R. Keenan, and W. H. Flygare. 1980. *J Chem Phys* 72:922–32.
38. Townes, C. H., and B. P. Dailey. 1949. *J Chem Phys* 17:782–96.
39. Davey, J. B., A. C. Legon, and E. R. Waclawik. 1999. *Chem Phys Lett* 306:133–44.

40. Legon, A. C. 2008. The interaction of dihalogens and hydrogen halides with Lewis bases in the gas phase: An experimental comparison of the halogen bond and the hydrogen bond. In *Halogen Bonding: Fundamentals and Applications*, ed. P. Metrangolo and G. Resnati, in the series Structure and Bonding; Vol. 126, 17–64. Berlin: Springer.
41. Legon, A. C., and D. J. Millen. 1986. *Proc R Soc Lond A* 404:89–99.
42. Cummins, P. L., G. B. Bacskay, and N. S. Hush. 1985. *J Phys Chem* 89:2151–5.
43. Demaison, J., and J. Liévin. 2008. *Mol Phys* 106:1249–56.
44. Mastryukov, V. S., and S. H. Simonsen. 1996. Empirical correlations in structural chemistry. In *Advances in Molecular Structure Research*, ed. M. Hargittai and I. Hargittai. Vol. 2, 163–89. Greenwich, CT: JAI press.
45. Badger, R. M. 1934. *J Chem Phys* 2:128–31.
46. Herschbach, D. R., and V. W. Laurie. 1961. *J Chem Phys* 35:458–63.
47. Perreault, D., M. Drouin, A. Michel, V. M. Miskowski, W. P. Schaefer, and P. D. Harvey. 1992. *Inorg Chem* 31:695–702.
48. Bernstein, H. J. 1962. *Spectrochim Acta* 18:161–70.
49. McKean, D. C. 1978. *Chem Soc Rev* 7:399–422.
50. Henry, B. R. 1987. *Acc Chem Res* 20:429–35.
51. Birge, R. T., and H. Sponer. 1926. *Phys Rev* 28:259–83.
52. Demaison, J., and H. D. Rudolph. 2008. *J Mol Spectrosc* 248:66–76.
53. Demaison, J., M. Herman, and J. Liévin. 2007. *Intern Rev Phys Chem* 26:391–420.
54. Demaison, J., L. Margulès, and J. E. Boggs. 2000. *Chem Phys* 260:65–81.
55. McClellan, A. L. 1963. *Tables of Experimental Dipole Moments*. London: W.H. Freeman.
56. Legon, A. C., and D. Millen. 1986. *Chem Rev* 86:635–57.
57. Altman, R. S., M. D. Marshall, and W. Klemperer. 1983. *J Chem Phys* 79:52–6.

CONTENTS OF APPENDIX VII (ON CD ROM)

8 Structures Averaged over Nuclear Motions

Attila G. Császár

CONTENTS

8.1   POSITION VERSUS DISTANCE AVERAGES

Quantum mechanics tells us that molecules can never rest; in real molecules, nuclei are always subject to vibrations. Thus, the overwhelming problem of molecular structure research is that for polyatomic molecules it is impossible to measure equilibrium structures experimentally. The closest one can get to the equilibrium (r_e-type) structures without significant modeling efforts are the vibrationally averaged structures corresponding to the ground-state vibrations (empirical r_0- or r_z-type structures, see Section 5.3). The differences between the average and equilibrium structures are substantial and may be comparable in magnitude, for example, to structural changes caused by chemical changes, like substitutions in a given family of compounds.

Driven by the need for more and more accurate and detailed spectroscopic and diffraction experiments and the emergence of predictive computational quantum chemistry, structural chemists developed several distance definitions, many of a physical and some of an operational origin. The number of differently averaged structure types is relatively large (see pp. xix–xx), and this has resulted in some confusion about their true meaning, use, and utility.

One can define either vibrationally averaged internuclear distances or distances between vibrationally averaged nuclear positions. Thus, one can determine either "distance" or "position" averages. As shown below, these averages can be converted into each other using different assumptions. Three levels of sophistication must be discussed when moving between different vibrationally averaged and equilibrium structures. First, especially for bonded distances, one can use diatomic paradigms. The most important of these is the Morse oscillator (see Section 8.2.3), the prototypical anharmonic oscillator and an excellent model for most chemical bonds, invented at the dawn of quantum chemistry.* Second, one can turn attention to polyatomic semirigid molecules and use perturbation theory (PT) to give appropriate and approximate formulas for distance averagings and distance conversions. The most common approach is to use PT at first and second orders using harmonic vibrations as the zeroth-order model. Third, one can perform variational nuclear motion computations, determine vibrational-rotational energy levels and wave functions, and use the wave functions to perform the rovibrational averaging. All these routes are discussed in this chapter.

It is clear from the nature of the three routes that understanding averaging cannot be separated from the theory of nuclear motion and of vibration-rotation interactions. The quantum mechanics to be used is greatly simplified by the fact that only stationary states or equilibrium distributions among stationary states are of interest to us. Thus, even if the averages represent thermal averages and not that corresponding to a particular single (ro)vibrational state, it is sufficient to assume a Boltzmann (ro)vibrational distribution.†

In equilibrium structures, distances and positions strictly correspond to each other. When molecules vibrate, which they do along curvilinear paths as shown in Figure 8.1, it seems better to put more emphasis on distance averages compared to position averages as the former ones have more clear physical meaning and are amenable to simple but "exact" variational treatments. Nevertheless, traditionally the position averages received more attention as their perturbative treatment proved slightly simpler.

The r_z structure, where the subscript z stands for *zero-point positional average*, is the most common position average. An r_z structure strictly refers to the average

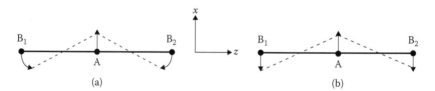

(a)   (b)

FIGURE 8.1 Curvilinear (a) and rectilinear (b) coordinates representing the bending displacement of a linear AB_2 molecule. During pure curvilinear bending $\Delta r(AB_1) = \Delta r(AB_2) = 0$, while during pure rectilinear bending $\Delta z(AB_1) = \Delta z(AB_2) = 0$.

---

* Morse, P. M. 1929. *Phys Rev* 34:57.

† Note that there are modern spectroscopic techniques, for example, those done in a beam (supersonic or not), where the experimental conditions are far from a thermal equilibrium. Nevertheless, these are rarely used to determine parameters of structural interest.

nuclear positions in the ground vibrational state. Thus, by definition, r_z structures should not have temperature dependence. An r_z distance can be considerably different from an r_0 distance since the latter contains the effect of vibrations perpendicular to the bond. The connection between the r_z and r_e (equilibrium internuclear) distances is usually written in the form

$$r_z = r_e + \langle \Delta z \rangle_0 \tag{8.1}$$

where Δz denotes differences between the local Cartesian displacements in z of the two atoms involved in the bond, the z axis of the molecule-fixed frame is aligned along the internuclear direction as in Figure 8.2, $\langle \ \rangle$ means vibrational averaging, and the subscript 0 following $\langle \ \rangle$ refers to the ground vibrational state, in which all the normal modes of the molecule are characterized by vibrational quantum numbers of zero.

Another position average is the so-called r_α distance. The r_α distance corresponds to the nuclear positions averaged at a given temperature T under the assumption of thermal equilibrium; thus, it is the temperature-dependent extension of the r_z structure. Therefore, it is better to denote this temperature-dependent position average as $r_{\alpha,T}$. Similar to Equation 8.1, we can write

$$r_{\alpha,T} = r_e + \langle \Delta z \rangle_T \tag{8.2}$$

The $r_{\alpha,T}$ distance can be converted to r_z by an approximate extrapolation to zero temperature; in other words, $r_{\alpha,0} = r_z$. Note that it is usual in the literature to write $r_{\alpha,0}$ as r_α^0.

The lowest-order "distance" averages are related to $\langle r \rangle, \langle r^{-1} \rangle, \langle r^2 \rangle, \langle r^{-2} \rangle, \langle r^3 \rangle$, and $\langle r^{-3} \rangle$, where $\langle \ \rangle$ means vibrational or preferably vibrational-rotational averaging. The distance averages, in the order they are written above, will be denoted as mean, inverse, rms (where rms stands for root mean square), effective, cubic, and inverse cubic distances, after taking $\langle r^n \rangle^{1/n}$. In general, the averages may represent averages in any (ro)vibrational state or thermal averages over a Boltzmann (ro)vibrational distribution. The mean $\left(\langle r \rangle \right)$ and inverse $\left(\langle r^{-1} \rangle^{-1} \right)$ distances have relevance in gas electron diffraction (GED) experiments.* The effective bond length $\left(\langle r^{-2} \rangle^{-1/2} \right)$ has strong

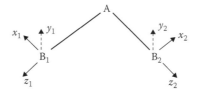

FIGURE 8.2 Local Cartesian coordinates of an AB_2 triatomic fragment.

* For details about the GED technique, please consult textbooks and reviews dealing with this experimental method. See, for example, I. Hargittai and M. Hargittai, Eds., Stereochemical Applications of Gas-Phase Electron Diffraction, Part A: The Electron Diffraction Technique, VCH: Weinheim, 1988.

connection with rotational spectroscopy. The inverse cubic average $\left(\left\langle r^{-3} \right\rangle^{-1/3}\right)$ appears in dipolar coupling constants measured by nuclear magnetic resonance spectroscopy of partially oriented molecules and also in rotational spectroscopy when the hyperfine splitting due to nuclear spin–spin coupling is resolved.

Two distance averages of central importance are $r_{g,T}$ and $r_{a,T}$, as they are usually denoted. The subscript g in $r_{g,T}$ stands for "center of gravity of the probability distribution of the interatomic distance," and $r_{g,T}$ denotes the thermal average value of an r_g-type internuclear distance at temperature T. The working definition to compute $r_{g,T}$ is

$$r_{g,T} = \sum_{v,J_\tau} W_{v,J_\tau}(T)\langle r\rangle_{vJ_\tau} \tag{8.3}$$

where the Boltzmann weight factor is given as

$$W_{v,J_\tau}(T) = \frac{e^{-\frac{E_{vJ_\tau}-E_0}{k_BT}}}{\sum_{vJ_\tau} e^{-\frac{E_{vJ_\tau}-E_0}{k_BT}}} \tag{8.4}$$

In Equation 8.4, the $\exp\left(\dfrac{-\Delta E}{k_B T}\right)$ factor is known as the Boltzmann factor,* and k_B represents the Boltzmann constant,[†] E_0 is the vibrational zero-point energy, E_{vJ_τ} represents the vibrational-rotational energies preferably computed variationally (especially if higher temperatures or large vibrational amplitudes are involved), and the averaging may use the corresponding (ro)vibrational wave functions determinable in a variational nuclear motion computation. Therefore, it is clear that the $r_{g,T}$-type distance average basically means an $\langle r\rangle$ average with weights provided by the Boltzmann factors. In these expressions, v and J_τ stand for the vibrational and rotational labels (approximate or good quantum numbers), respectively. In the most usual case, v is a collective index of normal-mode quantum numbers and J is a "good" quantum number corresponding to the overall rotation of the molecule, and $\tau = K_a - K_c$, where K_a and K_c have their usual meaning, that is, the values of $|K|$ for the prolate and oblate symmetric rotor limits, respectively, with which the particular rotational level of the asymmetric top correlates.[‡]

---

* The distribution embodied in Equation 8.3 is usually called the canonical distribution and was established in 1902 as a general principle of statistical mechanics by Gibbs. Although it does not follow this historical fact, the canonical distribution, applicable to any system in a state of thermal equilibrium with a constant number of particles, is mostly referred to as Boltzmann distribution (or Boltzmann law). We adhere to this general usage in what follows.

[†] For the latest recommended value of k_B, as well as of all fundamental physical constants, see http://www.codata.org

[‡] Those interested in the theory of the overall rotation of (asymmetric top) molecules should consult a textbook on the subject, for example, Kroto, H. W. 1992. *Molecular Rotation Spectra*. New York: Dover.

The connection between $r_{g,T}$, $r_{\alpha,T}$, and r_e can be given in terms of local Cartesian coordinates, see Figure 8.2, as*

$$
\begin{aligned}
r_{g,T} &= r_e + \langle \Delta r \rangle_T \cong \left\langle \left[\left(r_e + \Delta z \right)^2 + \left(\Delta x \right)^2 + \left(\Delta y \right)^2 \right]^{1/2} \right\rangle_T \cong \\
&= r_e + \langle \Delta z \rangle_T + \left(\langle (\Delta x)^2 \rangle_T + \langle (\Delta y)^2 \rangle_T \right) \Big/ (2 r_e) + \cdots \cong \\
&\cong r_{\alpha,T} + \left(\langle (\Delta x)^2 \rangle_T + \langle (\Delta y)^2 \rangle_T \right) \Big/ (2 r_e)
\end{aligned}
\tag{8.5}
$$

The thermal average of an r_a-type internuclear distance is defined similarly to r_g as

$$
r_{a,T} = \frac{\displaystyle\sum_{vJ_\tau} e^{-\frac{E_{vJ_\tau} - E_0}{k_B T}}}{\displaystyle\sum_{vJ_\tau} \langle 1/r \rangle_{vJ_\tau} e^{-\frac{E_{vJ_\tau} - E_0}{k_B T}}}
\tag{8.6}
$$

Thus, an r_a-type distance average basically means an $\langle r^{-1} \rangle$ average and thus could be called an "inverse" distance average. Naturally, the mean (r_g) and inverse (r_a) distance averages can be defined also within any simple model defining energy levels and wave functions, like simple model anharmonic oscillators, or within vibrational perturbation theory carried out to second order (VPT2; see Chapter 3).

Structures built upon distance averages, for example $r_{g,T}$, and position averages, for example, r_z or $r_{\alpha,T}$, have different strengths and weaknesses. As is clear from its definition, the distance between two bonded atoms can be better represented by $r_{g,T}$, because it is a real, physically defined vibrational average, its temperature dependence also follows a simple physical picture, and it is amenable in principle to exact variational nuclear motion treatments. The r_z or $r_{\alpha,T}$ distances are projected, "position" averages, missing certain vibrational effects and thus cannot be obtained directly through variational nuclear motion computations. As to the distances between nonbonded atoms, if they are expressed in $r_{g,T}$ they are slightly inconsistent with the $r_{g,T}$ distances between the corresponding bonded atoms. In other words, the three $r_{g,T}$-type distances and angles of a (nonlinear) triatomic fragment do not satisfy the simple geometrical requirement expressed by the law of cosines.† This has been known for a long time and was thoroughly investigated both for linear and nonlinear molecules.

Among the practitioners of GED experiments, this problem is often referred as the Bastiansen–Morino shrinkage effect. These unwanted $r_{g,T}$ differences (shrinkages) can be quite substantial, about 0.01 Å for distances and 10° for angles, especially if the related vibrations are excited by several quanta. Because r_z or $r_{\alpha,T}$ refer

---

* These relations have been derived originally by Morino, Y., K. Kuchitsu et al.; see, for example, Morino, Y., J. Nakamura, and P. W. Moore. 1962. *J Chem Phys* 36:1050 and Morino, Y., K. Kuchitsu, T. Oka. 1962. *J Chem Phys* 36:1108.

† The law of cosines (also called the cosine rule) simply states that in a triangle $c^2 = a^2 + b^2 - 2ab \cos \gamma$, where c is the side of the triangle opposite of the angle γ, and a and b are the sides that form the angle γ.

to the average nuclear positions, which do satisfy the law of cosines, a bond angle or a nonbonded distance can be better represented by them. This, and the way it allows for the combination of GED and rotational constant (B_z) information, explains the popularity of $r_{\alpha,T}$-type bond angles reported in GED studies.

Finally, a few words about empirical (sometimes operational) distance types discussed extensively in this book but not directly relevant here as they cannot be determined in a direct, quantum mechanical way. The basic reason for this inability is that the direct relation of these distance types to some well-defined physical concept, like that of an equilibrium structure, has never been established. Empirical r_0 distances are determined from effective ground-state rotational constants (like B_0^ξ, where $\xi = a, b, c$ are the inertial axes) obtained from spectroscopic (usually microwave [MW] and millimeterwave [MMW]) measurements (see Section 5.3). An "effective" r_0 structure does not reflect the average structure of the molecule in the ground vibrational state; furthermore, the untreated contributions of the zero-point vibrations to B_0^ξ often cause inconsistencies and anomalies in the r_0 distances determined.

Another empirical structure, the so-called r_s or substitution structure (see Section 5.4), may be close to an equilibrium structure, but not necessarily so, and r_s structures have no clear physical meaning either. The empirical mass-dependence molecular structures, $r_m^{(1)}$ and $r_m^{(2)}$ (Section 5.5), provide another set of estimates to the true equilibrium structures of molecules but their deviation from r_e remains unclear even after the structural analysis (see Chapter 5 for details).

8.2 DIATOMIC PARADIGMS

8.2.1 A Short Treatise on Probability Theory

Probability, a term that one may encounter in everyday life, occupies a central place in quantum mechanics. Unlike when describing the result of throwing a dice, in quantum mechanics we often deal with continuous variables. For example, the distance coordinate r and the displacement coordinate $x = r - r_e$ are both continuous variables. One can talk about the probability of a distance being between r and $r + dr$ or the probability of a displacement being between x and $x + dx$, dr and dx being small. Of course, the probability will be proportional to the actual extent of dr or dx. The probability will also vary along r or x. Thus, one could write the combined probability as the product $p(r)dr$ or $p(x)dx$, where $p(x)$, for example, is some yet unknown function of the continuous variable x that characterizes the mentioned variability. Functions of the type $p(x)$ are called probability density functions as they can be viewed as giving the probability per unit interval of x (and similarly for other functions).

Following a postulate formulated by Born, in quantum mechanics the probability density is given by $|\Psi(r)|^2$; thus, it is the product $|\Psi(r)|^2\,dr$ that gives the probability of finding the distance in the region between r and $r + dr$, where Ψ is the so-called vibrational state (or wave) function of the system. To find out the probability of an internuclear distance r being in the finite interval $[r_A, r_B]$, one needs to evaluate the definite integral $\int_{r_A}^{r_B} |\Psi(r)|^2\,dr$. For problems where it can be said with certainty that the displacement x from equilibrium must be within a given range $\Re$, $\int_{\Re} |\Psi(x)|^2 \, dx = 1$. Actually, all probability densities must satisfy this property as it is simply the mathematical expression that the displacement is somewhere in the range $\Re$.

The probability density function for the vth vibrational state ψ_v of a diatomic molecule is written as

$$P_v(x) = |\psi_v(x)|^2 \tag{8.7}$$

The probability density function for a diatomic molecule in thermal equilibrium at temperature T is then simply

$$P_{v,T}(x) = \sum_v W_v |\psi_v(x)|^2 \tag{8.8}$$

In the theory of probabilities, it is customary to define the following five quantities, all relevant to later discussions: expectation value, variance, covariance, skewness, and moments. For the rest of this subsection we switch, as in fact has been done for Equation 8.8, from continuous to discrete variables as the energies, which govern the averagings are discrete in nature. Thus, we switch from appropriate integrals to appropriate sums.

The expectation value $\langle X \rangle$ of a variable X for a given probability distribution, defined by probabilities $p_i \left(\sum_i p_i = 1 \right)$ for measurement outcomes X_i, is defined as the weighted sum

$$\langle X \rangle = \sum_i p_i X_i \tag{8.9}$$

The variance $(\Delta X)^2$, measuring the deviation from the expectation value (mean), is

$$(\Delta X)^2 = \left\langle \left(X - \langle X \rangle \right)^2 \right\rangle = \sum_i p_i \left(X_i - \langle X \rangle \right)^2 = \langle X^2 \rangle - \langle X \rangle^2 \tag{8.10}$$

The covariance of two (random) variables X and Y is defined as

$$\left\langle \left(X - \langle X \rangle \right)\left(Y - \langle Y \rangle \right) \right\rangle = \langle XY \rangle - \langle X \rangle \langle Y \rangle \tag{8.11}$$

If X and Y are independent (they are then called uncorrelated), the covariance clearly vanishes.

The dimensionless standard coefficient of skewness of the probability distribution curve is defined as

$$A_3 = \frac{\left\langle \left(X - \langle X \rangle \right)^3 \right\rangle}{\left\langle \left(X - \langle X \rangle \right)^2 \right\rangle^{3/2}} \tag{8.12}$$

The quantities $\langle X^n \rangle$ are usually called moments, where n can take both negative and positive values. For $n = 1, 2,$ and 3, one talks about first, second, and third moments, respectively. Thus, A_3 depends on the first three moments, $\langle X \rangle$, $\langle X^2 \rangle$, and $\langle X^3 \rangle$.

8.2.2 THE LINEAR HARMONIC OSCILLATOR MODEL

The simplest description of the vibrations of a diatomic molecule (or in fact of any N-atomic molecule) is provided by the linear harmonic oscillator model. The harmonic potential is given as $V(x) = (1/2) f_{rr} x^2$, where the quadratic force constant f_{rr} is related to the harmonic frequency ω as $f_{rr} = 4\pi^2 c^2 \omega^2 \mu$, μ being the reduced mass of the oscillator and $x = r - r_e$ is the displacement from equilibrium. A quantity of later interest, the relative displacement from equilibrium, sometimes called the Dunham expansion variable,* is defined as $\xi = \dfrac{x}{r_e}$. As all elementary textbooks on quantum mechanics show or even prove, the energy level structure of a linear harmonic oscillator is extremely simple, $E_v = hc\omega(v + 1/2)$, where v is the vibrational quantum number. The corresponding eigenfunctions (harmonic oscillator wave functions) are slightly more complicated and are given in the following mathematical form:

$$\psi_v(y) = A_v H_v(y) \exp\left(-y^2/2\right) \tag{8.13}$$

where $A_v = \left(2^v v! \sqrt{\pi}\right)^{-1/2}$ are normalization factors, the orthogonal polynomials H_vs are solutions to Hermite's differential equation, $H_v'' - 2yH_v' + 2vH_v = 0$, and are thus called Hermite polynomials, and $y = \left(\mu\omega/\hbar\right)^{1/2} x$ is a dimensionless displacement coordinate. For a pictorial representation of the quantum mechanical anharmonic oscillator results see Figure 6.1, with eigenpair results similar to the harmonic picture at low excitation.

Let us see how the linear harmonic oscillator model helps us understand the intricacies of vibrational averaging. To start out easy, we note that within the linear harmonic oscillator model, due to the symmetry of the probability density $\left|\psi_v(x)\right|^2$ about the origin, it holds for all eigenstates that $\langle x \rangle = 0$ and thus for the first moment $\langle r \rangle = r_e$. One has to keep in mind that moments other than the first, like $\langle r^{-2} \rangle$ of importance for rotational spectroscopy as $r_0 = \langle r^{-2} \rangle_0^{-1/2}$ for a diatomic molecule,† are not equal to r_e even when the vibrations are harmonic. In fact, $\langle r^{-2} \rangle$, for example, can be expanded (see also Equation 6.15) as

$$\langle r^{-2} \rangle = \left(r_e + \langle x \rangle\right)^{-2} = r_e^{-2}\left(1 + \langle \xi \rangle\right)^{-2} = r_e^{-2}\left(1 - 2\langle \xi \rangle + 3\langle \xi^2 \rangle - 4\langle \xi^3 \rangle + \cdots\right) \tag{8.14}$$

Thus, the first term in the "effective" vibrationally averaged distance, $\langle r^{-2} \rangle^{-1/2}$, is just r_e, the effect of the extra terms is given by a perturbation calculation and can be expressed as a certain sum of the moments involving ξ. Of course, Equation 8.14 holds not only for the harmonic oscillator but also for all (anharmonic) oscillators, though only for a limited range (see the end of Section 6.2.1 for related realistic bond length estimates for the H_2 molecule).

---

* Dunham, J. L. 1932. *Phys Rev* 41:721.
† As a warning to those who tend to overextend the validity of simple case studies, we note that for polyatomic molecules the r_0 distance definition no longer equals $\langle r^{-2} \rangle^{-1/2}$.

EXERCISE 8.1

For a diatomic molecule, determine the expansions of the inverse (r^{-1}) and inverse cubic (r^{-3}) distances as a function of the equilibrium bond length r_e and the dimensionless relative displacement variable $\xi = \dfrac{(r - r_e)}{r_e}$.

Next, let us work out the first two energy moments, $\langle E \rangle$ and $\langle E^2 \rangle$, of a linear harmonic oscillator in thermal equilibrium at temperature T. Let $P_i = P(E_i)$ be the probability of the oscillator having an energy E_i and introduce the notation $\beta = (k_B T)^{-1}$. Clearly,

$$P_i = Z^{-1} \exp(-\beta E_i) = \exp(-\beta E_i) \left(\sum_{k=1}^{\infty} \exp(-\beta E_k) \right)^{-1} \tag{8.15}$$

where Z is called the partition function,* and its definition is shown in Equation 8.15. In this special case, the degeneracy factors generally present in the probabilities and the partition function have been neglected.

The mean (average) energy $\langle E \rangle$ can be found directly from Z as follows:

$$\langle E \rangle = \sum_i P_i E_i = Z^{-1} \sum_i E_i \exp(-\beta E_i) = -Z^{-1} \frac{\partial Z}{\partial \beta} = -\frac{\partial \ln Z}{\partial \beta} \tag{8.16}$$

Similarly,

$$\langle E^2 \rangle = \sum_i P_i E_i^2 = Z^{-1} \sum_i E_i^2 \exp(-\beta E_i) = Z^{-1} \frac{\partial^2 Z}{\partial \beta^2} \tag{8.17}$$

It is thus clear that the variance of the energy of the harmonic oscillator can be given as $(\Delta E)^2 = \langle E^2 \rangle - \langle E \rangle^2 = \dfrac{\partial^2 (\ln Z)}{\partial \beta^2}$. Note in this respect that an important statement of statistical thermodynamics is that all its important energy quantities can be derived from the partition function Z.

For a single oscillator

$$Z = \sum_k \exp(-\beta E_k) = \exp\left(\frac{-\beta hc\omega}{2} \right) \sum_k \exp(-k\beta hc\omega)$$

$$= \exp\left(\frac{-\beta hc\omega}{2} \right) \sum_k \exp(-\beta hc\omega)^k \tag{8.18}$$

---

* As in several cases in quantum mechanics, the simple translation of the original German word, *Zustandsumme* ("state sum") would have been a more descriptive name than the exclusively used English *terminus technicus* "partition function."

The sum at the end of the expression of the partition function is a geometric series of the well-known form $\sum_{k=0}^{\infty} a^k = (1-a)^{-1}$ $(a < 1)$. Therefore,

$$Z = \frac{\exp\left(\dfrac{-\beta hc\omega}{2}\right)}{1 - \exp(-\beta hc\omega)} = \frac{1}{2}\operatorname{cosech}\left(\frac{\beta hc\omega}{2}\right)^{*} \tag{8.19}$$

Using the previously defined expressions for the moments, one gets, for example,

$$\langle E \rangle = \frac{hc\omega}{2}\coth\left(\frac{\beta hc\omega}{2}\right) = hc\omega\left(\frac{1}{2} + \frac{1}{\exp(\beta hc\omega) - 1}\right) \tag{8.20}$$

Now, let us turn our attention to probability density functions related to the linear harmonic oscillator. The vibrational probability density function of the harmonic oscillator, $P^{HO}(x)$, can be determined from the known eigenfunctions of the analytic solution of the corresponding time-independent Schrödinger equation given in Equation 8.13. For the ground $(v = 0)$ vibrational state, the probability density function is simply

$$P_0^{HO}(r) = \frac{1}{\sqrt{2\pi\langle x^2 \rangle}}\exp\left(-\frac{x^2}{2\langle x^2 \rangle}\right) \tag{8.21}$$

where $\langle x^2 \rangle = \dfrac{h}{8\pi^2\mu c\omega}$, as can be proven for a harmonic oscillator (see Exercise 8.2).

This is, of course, a prototypical Gaussian function. The harmonic oscillator probability density function can be rewritten in a slightly simpler-looking form as

$$P_0^{HO}(r) = \left(\frac{\alpha}{\pi}\right)^{1/2}\exp(-\alpha x^2) \tag{8.22}$$

where $\alpha = \dfrac{4\pi^2\mu c\omega}{h}$.

EXERCISE 8.2

Prove that for a linear harmonic oscillator $\langle x^2 \rangle = \dfrac{h}{8\pi^2\mu c\omega_e}$ where x is the displacement from equilibrium, μ is the reduced mass of the oscillator, and ω_e is the angular frequency.

---

* The hyperbolic cosecant function "cosech" is sometimes written as "csch" and is defined as $\operatorname{cosech} z \equiv (\sinh z)^{-1} = 2/[\exp(z) - \exp(-z)]$. It is related to the hyperbolic cotangent function coth as $\operatorname{cosech} z = \coth\left(\frac{1}{2}z\right) - \coth z$.

8.2.3 THE MORSE POTENTIAL

The simple, one-dimensional Morse potential has the form

$$V^M(x) = D_e \left[1 - \exp\left(-a_3 x\right)\right]^2 \tag{8.23}$$

where D_e is the dissociation energy (the energy difference between the minimum energy $V^M(r_e)$ and the energy at infinite separation) and the Morse asymmetry parameter a_3 characterizes the shape (the skewness) of the Morse curve. Because the Morse potential provides a much improved description of the behavior of bonds upon contraction or elongation over the linear harmonic oscillator model and it still can be solved analytically, those interested in the structures of molecules in the gas phase often employ arguments based on the Morse oscillator approximation to understand and model anharmonic vibrations of bonded atoms in molecules.

To make the Morse potential useful for molecular applications, the asymmetry parameter a_3 for a bonded atom pair in a polyatomic molecule is assumed to be equal to that of the corresponding diatomic molecule. Furthermore, to a good approximation, the asymmetry parameter for an AB bond is nearly equal to the average of the corresponding values of A_2 and B_2. Finally, the rule of thumb is that the a_3 parameter is about 2 Å^{-1} for a single bond and just slightly larger for multiple bonds.

As can easily be checked, the lowest-order force constants corresponding to the leading terms of the expansion of the Morse potential around the equilibrium structure

$$V^M(x) = \frac{1}{2} f_{rr}^M x^2 + \frac{1}{6} f_{rrr}^M x^3 + \cdots \tag{8.24}$$

are $f_{rr}^M = 2a_3^2 D_e$ and $f_{rrr}^M = -6a_3^3 D_e$. Thus, the quadratic and the cubic force constants of a Morse oscillator are related simply as $f_{rrr}^M = -3a_3 f_{rr}^M$. The general, nth-order expression for the force constants of the Morse oscillator is

$$f_n^M \equiv \frac{\partial^n V^M}{\partial x^n} = 2(-1)^n (2^{n-1} - 1) a_3^n D_e \tag{8.25}$$

EXERCISE 8.3

Show that for a Morse-oscillator model of anharmonic vibrations, $V^M(x) = D_e[1 - \exp(-ax)]^2$, where D_e is the dissociation energy (the energy difference between the minimum energy $V^M(r_e)$ and the energy at infinite separation), a is the Morse asymmetry parameter, and $x = r - r_e$, the quadratic and the cubic force constants are given as $f_{rr}^M = 2a^2 D_e$ and $f_{rrr}^M = -6a^3 D_e$, respectively.

EXERCISE 8.4

Determine the general, nth-order expression for the force constants, $f_n^M \equiv \dfrac{\partial^n V^M}{\partial x^n}$, of the Morse oscillator, whose full form is given in Exercise 8.3.

A useful, perturbation-like expression for the truncated Morse potential is therefore

$$V(x) = \frac{1}{2} f_{rr}^M x^2 \left[1 - a_3 x + \left(\frac{7a_3^2}{12} \right) x^2 - \left(\frac{a_3^3}{4} \right) x^3 + \cdots \right] \qquad (8.26)$$

This expression has a strong similarity to the so-called Dunham potential (see Equation 6.14)*

$$V^D(\xi) = a_0 \xi^2 \left[1 + a_1 \xi + a_2 \xi^2 + \cdots \right] \qquad (8.27)$$

This, of course, is because the Dunham potential is just a Taylor-series expansion of the anharmonic potential with some minor changes in the notation. The truncated cubic function already represents well the part of the potential curve that influences the usual stretching displacement of a bond.

In quantum mechanics, Newton's second law of motion may be written as

$$m \frac{d^2 \langle x \rangle}{dt^2} = -\left\langle \frac{\partial V}{\partial x} \right\rangle \qquad (8.28)$$

This result is a particular consequence of Ehrenfest's theorem.† In the cases of interest to us, the mean displacement $\langle x \rangle$ is chosen to be independent of time t; so, it follows that $\left\langle \frac{\partial V}{\partial x} \right\rangle = 0$ and

$$\frac{1}{2} f_{rr}^M \left(2\langle x \rangle - 3a_3 \langle x^2 \rangle + \cdots \right) = 0 \qquad (8.29)$$

EXERCISE 8.5

Show the connection between the quantities $\frac{d}{dt}\langle x \rangle$ and $\frac{d}{dt}\langle p \rangle$, defined by Ehrenfest's theorem, and Equation 8.28.

Neglecting higher-order terms, one immediately gets for the mean displacement the following simple but useful expression relating the first two moments of the displacement via the Morse asymmetry parameter:

$$\langle x \rangle = \frac{3}{2} a_3 \langle x^2 \rangle \qquad (8.30)$$

Similar to Equation 8.30, in the first order of PT, one can obtain an estimate for the mean internuclear distance as $\langle r \rangle = r_e + \frac{3}{2} a_3 \langle x^2 \rangle$. This is probably the simplest estimate that can be used for guessing a vibrationally averaged bond length in a polyatomic molecule; thus, the quantity $\langle x^2 \rangle$ is of central importance in the theory and practice of structure investigations.

---

* Dunham, J. L. 1932. *Phys Rev* 41:721.
† Ehrenfest, P. 1927. *Z Phys* 45:455.

The instantaneous value of r fluctuates about the mean (average) distance $\langle r \rangle$. A widely applied natural measure of this fluctuation is the mean square deviation ("mean of the square of the deviation") from the average, called "variance" in probability theory, defined as $\left\langle \left(r - \langle r \rangle \right)^2 \right\rangle$. As to an application of the variance, one of the structural parameters refined during GED structure analyses is the mean-square amplitude of vibrations $l^2 = \left\langle \left(r - \langle r \rangle \right)^2 \right\rangle$, which equals $\langle r^2 \rangle - \langle r \rangle^2 = \langle x^2 \rangle - \langle x \rangle^2$.

EXERCISE 8.6

Prove the relations $\left\langle \left(r - \langle r \rangle \right)^2 \right\rangle = \langle r^2 \rangle - \langle r \rangle^2 = \langle x^2 \rangle - \langle x \rangle^2$.

In molecular structure research, it became customary to use the following quantity related to the dimensionless standard coefficient of skewness A_3 (Equation 8.12):

$$\tilde{A}_3 = \frac{\left\langle \left(x - \langle x \rangle \right)^3 \right\rangle}{\left\langle \left(x - \langle x \rangle \right)^2 \right\rangle^2} \tag{8.31}$$

The $\tilde{A}_3$ parameter, which may have a unit of Å^{-1}, reduces at 0 K for a Morse oscillator to the Morse asymmetry parameter a_3.

The exact solution for the mean displacement of the Morse oscillator, keeping higher-order terms, is

$$\begin{aligned}
\langle x \rangle &= \frac{3}{2} a_3 \langle x^2 \rangle - \frac{7}{6} a_3^2 \langle x^3 \rangle + \frac{5}{8} a_3^3 \langle x^4 \rangle + \cdots \\
&= \frac{3}{2} a_3 \langle x^2 \rangle - \frac{109}{24} a_3^2 \langle x^2 \rangle^2 + \cdots
\end{aligned} \tag{8.32}$$

There are several comments one can make about the results derived and presented about the Morse oscillator in this subsection. Since one knows the eigenfunctions for the Morse oscillator analytically, $\langle x^2 \rangle$ can be averaged properly over the anharmonic motion. In a PT solution, $\langle x^2 \rangle$ is replaced by the average over the harmonic motion of the quadratic problem, which one can call $\langle x^2 \rangle_h$. We can show that for a Morse oscillator the difference $\langle x^2 \rangle - \langle x^2 \rangle_h$ is $\frac{15}{4} a^2 \langle x^2 \rangle_h$, keeping terms through quartic in x. This difference amounts to an error of a few percent for distances corresponding to bonds.

Inclusion of the terms of order higher than $\langle x^2 \rangle$ decreases the value of the mean displacement, $\langle x \rangle$, again by a few percent. Thus, the common calculation of mean bond lengths, utilizing first-order perturbation correction for cubic potential terms, contains two errors. The first is the replacement of $\langle x^2 \rangle$ by $\langle x^2 \rangle_h$, and the second is the truncation of $V(x)$. These two drastic looking simplifications have opposite signs and are of nearly equal magnitude. Thus, they very nearly cancel for

Morse-like bonds, making the treatment based on the Morse-oscillator model of much higher apparent accuracy than one would expect without considering this cancellation.

If one assumes that the potential energy surface of a polyatomic molecule is expressed in terms of normal coordinates $\mathbf{Q}$ simply as a cubic expansion, $V(\mathbf{Q}) = \frac{1}{2} \sum_s \lambda_s Q_s^2 + \frac{1}{6} \sum_{ijk} \Phi_{ijk} Q_i Q_j Q_k$, one immediately gets an expression of suffi-

cient accuracy, $\langle Q_s \rangle = -\sum_j \Phi_{sjj} / 2\lambda_s \langle Q_j^2 \rangle$. First-order PT leads to the same relation between the displacement and the amplitude, except that there is a small difference between $\langle Q_i^2 \rangle$ and its harmonic counterpart $\langle Q_i^2 \rangle_h$.

8.2.4 PROBABILITY DISTRIBUTION FUNCTION OF INTERNUCLEAR DISTANCES AND RELATED MOMENTS

For the lowest vibrational and rotational level of the Morse oscillator, it is usual to write the probability density function as an extension of Equation 8.22 as follows:

$$P_0^M(r) = \left(\frac{\alpha}{\pi}\right)^{1/2} \left\{ 1 + ax + \frac{a^2 x^2}{2} + \frac{(a^3 + 2a\alpha)x^3}{6} + \cdots \right\} \exp(-\alpha x^2) \qquad (8.33)$$

Like in the Morse case, let us define the probability density function $P(r_{ij})$ of the r_{ij} internuclear distance in a molecule as an extension of the harmonic picture. To the first order of approximation, the probability density function may then be written as (compare to Equation 8.21)

$$P(r_{ij}) = \frac{A}{\sqrt{2\pi \langle x^2 \rangle}} \exp\left(\frac{-x^2}{2\langle x^2 \rangle}\right) \left\{ 1 + c_1 x + c_3 x^3 \right\} \qquad (8.34)$$

In order to calculate the unknown coefficients c_1 and c_3, it is sufficient to evaluate the first three moments, $\langle x \rangle$, $\langle x^2 \rangle$, and $\langle x^3 \rangle$. It is straightforward to show that

$$c_1 = \frac{5\langle x \rangle \langle x^2 \rangle - \langle x^3 \rangle}{2\langle x^2 \rangle^2} \qquad (8.35)$$

and

$$c_3 = \frac{\langle x^3 \rangle - 3\langle x \rangle \langle x^2 \rangle}{6\langle x^2 \rangle^3} \qquad (8.36)$$

EXERCISE 8.7

Determine the coefficients c_1 and c_3 in Equation 8.34.

Similar but more complex probability density functions can be derived or assumed. These have relevance in different areas of structural research but discussion of these is beyond the scope of the present book.

The average nth power of the bond distance r is given, as a function of the relative displacement ξ, by the general expression, obtained by a truncated Taylor-series expansion:

$$\langle r^n \rangle^{1/n} = \kappa r_e = r_e \left(1 + \langle \xi \rangle + \frac{n-1}{2} \left(\langle \xi^2 \rangle - \langle \xi \rangle^2 \right) + \cdots \right) \tag{8.37}$$

It is interesting to observe the occurrence of the mean and the variance in this expression.

One usually retains only the first two or three terms in the κ scale factor. Then, one obtains the following formulas for some of the common distance averages, after neglecting the usually small $\langle \xi \rangle^2$ term:

$$\langle r \rangle = \left(1 + \langle \xi \rangle \right) r_e \tag{8.38a}$$

$$\langle r^2 \rangle^{1/2} = \left(1 + \langle \xi \rangle + \frac{1}{2} \langle \xi^2 \rangle \right) r_e \tag{8.38b}$$

$$\langle r^{-1} \rangle^{-1} = \left(1 + \langle \xi \rangle - \langle \xi^2 \rangle \right) r_e \tag{8.38c}$$

$$\langle r^{-2} \rangle^{-1/2} = \left(1 + \langle \xi \rangle - \frac{3}{2} \langle \xi^2 \rangle \right) r_e \tag{8.38d}$$

$$\langle r^{-3} \rangle^{-1/3} = \left(1 + \langle \xi \rangle - 2 \langle \xi^2 \rangle \right) r_e \tag{8.38e}$$

EXERCISE 8.8

Prove the general validity of Equation 8.38a.

EXERCISE 8.9

Prove the validity of Equation 8.37.

EXERCISE 8.10

Derive the next, higher-order terms not given in Equation 8.37.

For a diatomic molecule with a perturbing cubic potential term $V' = \left(\frac{h\omega_e^2}{4B_e} \right) a_1 \xi^3$ one can show that

$$\langle \xi \rangle = \langle v | \xi | v \rangle = -a_1 \left(\frac{3B_e}{\omega_e} \right) \left(v + \frac{1}{2} \right) \tag{8.39}$$

and

$$\langle \xi^2 \rangle = \langle v | \xi^2 | v \rangle = \left(\frac{2B_e}{\omega_e} \right) \left(v + \frac{1}{2} \right) \tag{8.40}$$

where ω_e and $B_e = \dfrac{h}{(8\pi^2 \mu r_e^2)}$ are the harmonic vibrational frequency and the rotational constant of the diatomic molecule (in the same units), and a_1 is a (dimensionless) cubic anharmonicity constant, usually around -3. Equations 8.39 and 8.40 suggest that $\langle \xi \rangle / \langle \xi^2 \rangle = -3a_1/2$. Thus, invoking the harmonic approximation, $\langle \xi \rangle = 0$, to simplify Equations 8.38 would lead to incorrect results. Furthermore, the same relation suggests that a measurement that would probe $\langle r^{-7} \rangle$ would lead to an experimental determination of the equilibrium bond length.

The deviation of $\langle r \rangle$ from r_e (see Equation 8.38a) is entirely due to vibrational anharmonicity, and $\langle \xi \rangle$ characterizes the shift in the center of gravity of the distance probability distribution function from the equilibrium position. On the other hand, the deviation between the effective and the mean distances (cf. Equations 8.38a and 8.38c) depends only upon the harmonic part of the potential, so in the ground vibrational state $\langle r^{-2} \rangle^{-1/2} - \langle r \rangle = -\left(\dfrac{3}{2} \right) \left(\dfrac{r_e B_e}{\omega_e} \right) < 0$. The higher moments $\langle \xi^2 \rangle$ and $\langle \xi^3 \rangle$ characterize the variance (the mean-square amplitude) and the skewness of the same (distance) probability distribution curve, respectively. Following from the expressions given, the order of the averaged distances, assuming $\langle \xi \rangle > 0$, is $\langle r^3 \rangle^{1/3} > \langle r^2 \rangle^{1/2} > \langle r \rangle > \langle r^{-1} \rangle^{-1} > \langle r^{-2} \rangle^{-1/2} > \langle r^{-3} \rangle^{-1/3} > r_e$. One of the beauties of the general expression truncated after the second-order term is that any of the distance averages (moments) can be derived by knowing the equilibrium value, r_e, the mean value, $\langle \xi \rangle$, and the second moment $\langle \xi^2 \rangle$.

EXERCISE 8.11

Using the approximate relation for the moment of inertia of a diatomic molecule $I_0 \approx I_e \left(1 + \dfrac{\alpha}{2B_e} \right)$ and the facts that $\alpha = -\left(\dfrac{6B_e^2}{\omega_e} \right)(1 + a_1)$ and that a typical value of a_1 is -3, show that $\dfrac{(r_0 - r_e)}{r_0} \approx \dfrac{3B_e}{\omega_e}$.

8.3 THE PERTURBATIONAL ROUTE

The conventional computation of vibrationally averaged molecular properties, including averaging of structural parameters, involves a low-order anharmonic vibrational analysis via PT, employing the rectilinear vibrational normal ($\mathbf{Q}$) or dimensionless normal ($\mathbf{q}$) coordinates for the expansion of the potential. A few comments must be made here on the use of normal coordinates. $\mathbf{Q}$ and $\mathbf{q}$ are intimately related to the so-called Eckart conditions describing the orientation of the molecule-fixed coordinate system. The Eckart conditions eliminate the overall translation and rotation of the molecule in space during the vibrations and lead to a maximum separation of the two types of internal nuclear motion, vibration and rotation. These conditions depend on the masses of the nuclei of the molecule, making the **Q** and **q** coordinates, and thus the potential expansion based on them, mass-dependent. This isotope dependency of the potential is a serious disadvantage of the expansion based on normal coordinates, and thus the use of mass-independent internal coordinates should be advocated whenever possible for the expansion of the potential. When using the **Q** and **q** coordinates, it is not assumed that the molecule exhibits infinitesimally small vibrations, but the basic assumption is that these coordinates are related to the rectilinear Cartesian coordinates expressed in the Eckart frame through a linear transformation (see Section 8.4.1). Of course, convergence of the expansion of the potential does depend on the rigidity of the molecule. Thus, such an analysis is expected to provide rather accurate results for semirigid molecules at relatively low temperatures. For molecules having large-amplitude motions and when highly excited vibrations of semirigid molecules contribute to the averaging, the perturbational route to be described in this section should not be employed and one needs to switch to the variational route to be described in Section 8.4.

A function P of an arbitrary molecular property can be expressed in a Taylor series of normal coordinates Q_r and dimensionless normal coordinates q_r, where

$$q_r = \left(\frac{4\pi^2 c\omega_r}{h}\right)^{1/2} Q_r = \left(\frac{hc}{\hbar^2}\right)^{1/2} \omega_r^{1/2} Q_r, \text{ as}$$

$$P = P_e + \sum_r P_r Q_r + \frac{1}{2}\sum_{r,s} P_{rs} Q_r Q_s + \cdots \qquad (8.41a)$$

and

$$P = P_e + \sum_r \tilde{P}_r q_r + \frac{1}{2}\sum_{r,s} \tilde{P}_{rs} q_r q_s + \cdots \qquad (8.41b)$$

where P_r and P_{rs} (and similarly $\tilde{P}_r$ and $\tilde{P}_{rs}$) refer to the first- and second-order derivatives of $P(\mathbf{Q})$ [and of $P(\mathbf{q})$], respectively, taken at a reference structure, usually chosen to be the equilibrium structure, as indicated by P_e. The expansion is generally truncated after the quadratic term as indicated in Equations 8.41. The vibrationally averaged value of $P(\mathbf{Q})$ or $P(\mathbf{q})$, $\langle P \rangle$, can then be calculated as

$$\langle P \rangle = P_e + \sum_r P_r \langle Q_r \rangle + \frac{1}{2}\sum_{r,s} P_{rs} \langle Q_r Q_s \rangle + \cdots \qquad (8.42a)$$

and

$$\langle P \rangle = P_e + \sum_r \tilde{P}_r \langle q_r \rangle + \frac{1}{2}\sum_{r,s} \tilde{P}_{rs} \langle q_r q_s \rangle + \cdots \qquad (8.42b)$$

To employ Equations 8.42, the average values of the normal and dimensionless normal coordinates and their products need to be known. To obtain these for the usual small-amplitude vibrations of semirigid molecules, it is necessary to expand the vibrational potential energy as a function of the normal normal coordinates, $V(\mathbf{Q})$ and $V(\mathbf{q})$, and for convenience, to terminate the expansions after the cubic term.

$$
\begin{aligned}
V(\mathbf{Q}) &= \frac{1}{2}\sum_r \lambda_r Q_r^2 + \frac{1}{6}\sum_{r,s,t} \Phi_{rst} Q_r Q_s Q_t + \cdots \\
&= \frac{1}{2}\sum_r \lambda_r Q_r^2 + \sum_{r\le s\le t} K_{rst} Q_r Q_s Q_t + \cdots
\end{aligned}
\tag{8.43a}
$$

and

$$
\begin{aligned}
V(\mathbf{q}) &= \frac{1}{2}\sum_r \omega_r q_r^2 + \frac{1}{6}\sum_{r,s,t} \phi_{rst} q_r q_s q_t + \cdots \\
&= \frac{1}{2}\sum_r \omega_r q_r^2 + \sum_{r\le s\le t} k_{rst} q_r q_s q_t + \cdots
\end{aligned}
\tag{8.43b}
$$

In these expressions, ω_r is the vibrational wave number of the rth normal mode, $\omega_r = \frac{\lambda_r^{1/2}}{(2\pi c)}$, the $\{\Phi_{rst}, K_{rst}\}$ and the $\{\phi_{rst}, k_{rst}\}$ pairs, where $K_{rst} = k_{rst}(\gamma_r\gamma_s\gamma_t)^{1/2}hc$ and $\gamma_r = \frac{hc\omega_r}{\hbar^2}$, are cubic normal and cubic dimensionless normal coordinate force constants, respectively, and the cubic force constants Φ_{rst} and K_{rst} (as well as ϕ_{rst} and k_{rst}) are related by simple numerical factors introduced by the difference between the unrestricted and restricted summations indicated in Equations 8.43a and b.

In order to compute $\langle Q_k \rangle$, the first-order anharmonic wave functions perturbed by the cubic potential terms are used. To compute $\langle Q_k Q_l \rangle$, the zeroth-order harmonic wave functions are needed. After some manipulations, the required vibrational averages of the normal coordinates for the ground vibrational state turn out to be

$$
\langle Q_k \rangle = -\frac{\hbar}{4\omega_k^2}\sum_l \frac{\Phi_{kll}}{\omega_l}
\tag{8.44a}
$$

and

$$
\langle Q_k Q_l \rangle = \frac{\hbar}{2\omega_k}\delta_{k,l}
\tag{8.44b}
$$

where $\delta_{k,l}$ stands for the usual Kronecker delta symbol.

The similar expressions for the averages of the dimensionless normal coordinates, obtained after application of the Ehrenfest theorem, for an arbitrary vibrational state v are

$$\langle q_r \rangle_v = -\left(\frac{1}{\omega_r} \right) \left[3k_{rrr} \left(v_r + \frac{1}{2} \right) + \sum_{s \neq r} g_s k_{rss} \left(v_s + \frac{1}{2} \right) \right] \qquad (8.45a)$$

and

$$\langle q_r^2 \rangle_v = v_r + \frac{1}{2} \qquad (8.45b)$$

where the degeneracy factors g_s of the dimensionless normal coordinates have also been introduced.

The effect of temperature on the vibrationally averaged values just given can also be simply considered. Quantum mechanics tells us that within the linear harmonic oscillator model (see Section 8.2.2)

$$\left\langle v_s + \frac{g_s}{2} \right\rangle_T = \left(\frac{g_s}{2} \right) \coth \left(\frac{hc\omega_s}{2k_B T} \right) \qquad (8.46)$$

which leads to straightforward averaging. The resulting formulas for the required temperature-dependent average values in thermal equilibrium at temperature T are

$$\langle Q_k \rangle_T = -\frac{\hbar}{4\omega_k^2} \sum_l \frac{\Phi_{kll}}{\omega_l} \coth \left(\frac{\hbar\omega_l}{2k_B T} \right) \qquad (8.47a)$$

$$\langle Q_k Q_l \rangle_T = \frac{\hbar}{2\omega_k} \delta_{k,l} \coth \left(\frac{\hbar\omega_k}{2k_B T} \right) \qquad (8.47b)$$

$$\langle q_r \rangle_T = -\left(\frac{1}{2\omega_r} \right) \left[3k_{rrr} \coth \left(\frac{hc\omega_r}{2k_B T} \right) + \sum_{s \neq r} g_s k_{rss} \coth \left(\frac{hc\omega_s}{2k_B T} \right) \right] \qquad (8.47c)$$

$$\langle q_r^2 \rangle_T = \frac{1}{2} \coth \left(\frac{hc\omega_r}{2k_B T} \right) \qquad (8.47d)$$

It is useful to remind ourselves that the technique just described includes several approximations. First, the property function P is given as a truncated Taylor series of the normal coordinates. Second, the potential energy of the molecule is given in a usually slowly converging normal coordinate force field representation truncated at a low order (the third order in Equation 8.42). Third, the T-dependence assumed a canonical distribution and the T-averaging used linear harmonic oscillator results.

As emphasized at the beginning of this section, the Cartesian displacement coordinates of the nuclei are related to the normal coordinates by a strictly linear transformation in the molecule-fixed Eckart axis system. Consequently, the mean values $\langle Q_k \rangle$, as defined in Equation 8.44a, specify the displacements of the average nuclear positions from the equilibrium positions. As learned in Section 8.1, the arrangement of the nuclei in the molecule placed at their average positions in thermal equilibrium at temperature T is referred to as the $r_{\alpha,T}$ structure (or as the $r_z \equiv r_{\alpha,0}$ structure if $T = 0$).

It is useful at this point to consider not only the rectilinear Cartesian and normal coordinates, transformed into each other by a linear transformation, but also the curvilinear internal coordinates, which describe more closely the motions of the vibrating and rotating atoms of a molecule (Figure 8.1). The internal coordinates are usually chosen to be either the simple local valence coordinates, R_r (e.g., bond stretching, angle bending, linear angle bending, torsional, and out-of-plane bending), or their linear combinations, S_r. A special set of internal coordinates are the symmetry coordinates, which reflect the point-group symmetry of the molecule and are determined from local valence coordinates by the required symmetry operations. Traditionally, the local valence coordinates are expressed as

$$R_r = \sum_k L_r^k Q_k + \frac{1}{2} \sum_{k,l} L_r^{k,l} Q_k Q_l + \cdots \tag{8.48}$$

where the expansion coefficients are the so-called **L** tensor elements.* The temperature-dependent average value of R_r can then be calculated as

$$\langle R_r \rangle^T = \sum_k L_r^k \langle Q_k \rangle^T + \frac{1}{2} \sum_k L_r^{k,k} \langle Q_k^2 \rangle^T + \cdots \tag{8.49}$$

where previously given expressions can be employed for computing the thermal average values of the normal coordinates and their products.

8.4   VARIATIONAL ROUTE

The perturbative treatments described in Section 8.3 are mostly adequate for semi-rigid molecules. For molecules with large-amplitude motions, they may become inadequate, and thus one should resort to a variational averaging of the structures or in general of any properties of the molecule. Unlike in the past, variational nuclear motion computations, even those employing exact kinetic energy operators, are no longer limited to small, three- and four-atomic systems but can be extended to

---

* For a detailed discussion on **L** tensor elements, see Hoy, A. R., I. M. Mills, and G. Strey. 1972. *Mol Phys* 24:1265 and Allen, W. D., A. G. Császár, V. Szalay, and I. M. Mills. 1996. *Mol Phys* 89:1213.

somewhat larger ones, including those having large amplitude motion over several minima, which can be accessed even at relatively low temperatures and energies. As to the subject of this chapter, variational routes may yield temperature-dependent (ro)vibrationally averaged structural parameters and vibrationally averaged rotational constants.

As in all cases in quantum mechanics, the approximate variational nuclear motion treatment is based on the variation principle. This states that in the class of basis functions satisfying the boundary conditions of the quantum mechanical problem at hand, for any state corresponding to the given Hamiltonian the energy computed using an approximate wave (state) function serves as an upper bound of the exact energy. The five basic steps of any variational computation of energies and wave functions are as follows: (1) selection of an appropriate coordinate system; (2) determination of the corresponding Hamiltonian operator; (3) choice of a suitable set of basis functions; (4) computation of the (nonzero) elements of the Hamiltonian matrix in the given basis; and (5) diagonalization of the Hamiltonian matrix in order to obtain its (desired) eigenpairs (eigenvalues and eigenvectors). Of course, in all the five steps there are a number of possible choices based on mathematical and/or physical convenience, which greatly influence the effectiveness of the variational computation.

The variational technique to deduce rovibrationally averaged properties is based on simple expectation value computations, where the rovibrational wave functions obtai-ned from variational (or nearly variational) nuclear motion computations are employed to determine expectation values of the given molecular property. Computational cost aside, the variational technique has several advantages over perturbative treatments. The function f, which describes a molecular property, can be given as an arbitrary function of the internal coordinates. It is not required to give the form of f in a Taylor series expansion. The variational vibrational computations provide converged energy levels with the corresponding accurate wave functions. These numerically exact wave functions can be employed for expectation value computations providing "exact" vibrationally averaged properties within the accuracy limit of the potential energy surface (PES). It must also be stressed that the variational technique allows computation of properly rovibrationally averaged properties, and artificial separation of the vibrational and rotational degrees of freedom is not necessary.

8.4.1   ECKART–WATSON HAMILTONIAN

The most straightforward and simplest theoretical route is followed when the Eckart frame (Figure 8.3) is used in a variational nuclear motion computation. The choice of the Eckart frame means that the rotation-vibration interaction is zero at the reference structure, and the interaction is very small close to the reference structure. Note that it is impossible to define a frame in which the rotation-vibration interaction vanishes over a finite region of the configuration space. The use of the Eckart frame is one of the best choices if maximal vibrational-rotational separation is to be achieved, the coupling terms are small for lower vibrational excitations of not too wide-amplitude motions. Using the Eckart frame and universally defined rectilinear

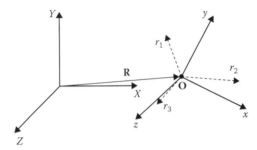

FIGURE 8.3 The Eckart framework: the origin, **O**, of the molecule-fixed x-y-z coordinate system, corresponding to the nuclear center of mass, is located by a vector, **R**, pointing from the origin of the laboratory-fixed X-Y-Z coordinate system to **O**. The $\mathbf{r}_i$ vectors locate each nucleus of the molecule with respect to the x-y-z coordinate system.

internal coordinates (normal coordinates), the rotation-vibration Hamiltonian can be simplified to the Eckart–Watson form*

$$\hat{H}^{\text{rot-vib}} = \frac{1}{2}\sum_{\alpha\beta}(\hat{J}_\alpha - \hat{\pi}_\alpha)\mu_{\alpha\beta}(\hat{J}_\beta - \hat{\pi}_\beta) + \frac{1}{2}\sum_{k=1}^{3N-6}\hat{P}_k^2 - \frac{\hbar^2}{8}\sum_\alpha \mu_{\alpha\alpha} + V \qquad (8.50)$$

with volume element $dQ_1 dQ_2 \ldots dQ_{3N-6}\sin\theta d\varphi d\theta d\chi$, where φ, θ and χ are the Euler angles[†] that describe the overall rotation of the molecule in the Eckart axis system. The third term on the right-hand side of Equation 8.50 is often called the extrapotential term as it involves no derivatives and thus in this sense is similar to the potential energy term V. Rectilinear internal (normal) coordinates are specified as

$$Q_k = \sum_{i=1}^{N}\sum_{\alpha\beta\gamma}\sqrt{m_i}\,l_{i\alpha k}(x_{i\alpha} - c_{i\alpha}), \; k = 1, 2, \ldots, 3N-6 \qquad (8.51)$$

where m_i is the mass associated with the ith nuclei, $c_{i\alpha}$ are the reference coordinates, and $x_{i\alpha}$ are the instantaneous Cartesian coordinates in the Eckart frame. The usage of the Eckart frame and certain orthogonality requirements impose the following conditions on the elements $l_{i\alpha k}$ specifying the actual rectilinear internal coordinates:

$$\sum_{i=1}^{N}\mathbf{l}_{ik}^{\mathrm{T}}\mathbf{l}_{il} = \delta_{kl} \qquad \sum_{i=1}^{N}\sqrt{m_i}\,\mathbf{l}_{ik} = 0 \qquad \sum_{i=1}^{N}\sqrt{m_i}\,\mathbf{c}_i \times \mathbf{l}_{ik} = 0 \qquad (8.52)$$

---

* Many people call the Hamiltonian described in Equation 8.50 the "Watson Hamiltonian," based on his publications Watson, J. K. G. 1968. *Mol Phys* 15:479 and Watson, J. K. G. 1970. *Mol Phys* 19:465. Nevertheless, as a tribute to the seminal contributions of Eckart (see Eckart, C. 1935. *Phys Rev* 47:552) and due to the importance of the choice of Eckart embedding in this Hamiltonian, we follow here those authors who call this Hamiltonian the Eckart–Watson Hamiltonian. In fact, the "Watson Hamiltonian" is the simplest quantum mechanical form of the classical vibrational-rotational Hamiltonian of Eckart.
† The Euler angles are ubiquitous in the description of classical and quantum rotations. Thus, they are treated in detail in most elementary textbooks on classical as well as quantum mechanics.

In Equation 8.51, $\hat{P}_k = -i\hbar\dfrac{\partial}{\partial Q_k}$ ($k = 1, 2, \ldots, 3N{-}6$), $\hat{J}_x, \hat{J}_y,$ and $\hat{J}_z$ are the compo-

nents of the total angular momentum, $\hat{\pi}_\alpha = \displaystyle\sum_{kl=1}^{3N-6} \zeta^\alpha_{kl} Q_k \hat{P}_l$ is the Coriolis coupling oper-

ator (see also Equation 4.13), $\mu_{\alpha\beta} = (\mathbf{I}'^{-1})_{\alpha\beta}$ is the generalized inverse inertia tensor,

$I'_{\alpha\beta} = I_{\alpha\beta} - \displaystyle\sum_{klm=1}^{3N-6} \zeta^\alpha_{km}\zeta^\beta_{lm}Q_kQ_l$ is the generalized inertia tensor, and $\zeta^\alpha_{km} = e_{\alpha\beta\gamma}\displaystyle\sum_{i=1}^{N} l_{i\beta k} l_{i\gamma m}$,

where $e_{\alpha\beta\gamma}$ denotes the Lévi–Civitá symbol defined in Chapter 4. The vibration-only part of the Eckart–Watson operator has the form

$$\hat{H}^{\text{vib}} = \frac{1}{2}\sum_{\alpha\beta}\hat{\pi}_\alpha\mu_{\alpha\beta}\hat{\pi}_\beta + \frac{1}{2}\sum_{k=1}^{3N-6}\hat{P}_k^2 - \frac{\hbar^2}{8}\sum_\alpha\mu_{\alpha\alpha} + V \tag{8.53}$$

8.4.2 Rovibrational Hamiltonians in Internal Coordinates

For floppy, flexible molecules and for those with accessible PES regions exhibiting multiple minima, the choices behind the Eckart–Watson Hamiltonian, namely the Eckart frame of reference and the use of rectilinear (normal) coordinates, do not result in a useful description. For such cases, it is better to use an approach that allows using arbitrarily chosen body-fixed frames and curvilinear internal coordinates. Such Hamiltonians can be developed straightforwardly using the standard theory of vibrations and rotations.

The simplest form of the rovibrational Hamiltonian, $\hat{H}^{\text{rv}}$, for an N-atomic molecule in internal coordinates is the Podolsky form, is

$$\hat{H}^{\text{rv}} = \frac{1}{2}\sum_{kl}^{D+3}\tilde{g}^{-1/4}\hat{p}_k^\dagger G_{kl}\tilde{g}^{1/2}\hat{p}_l\tilde{g}^{-1/4} + V \tag{8.54}$$

where out of the $3N{-}6$ internal coordinates there are $D \le 3N - 6$ active variables $(q_1, q_2, \ldots, q_D)$, $\mathbf{G}$ is the well-known El'yashevich–Wilson $\mathbf{G}$ matrix treated in all textbooks on molecular vibrations (see recommended readings at the end of this

chapter), $\tilde{g} = \det\mathbf{g}$, $\mathbf{g} = \mathbf{G}^{-1}$, and the momenta conjugate to q_k are $\hat{p}_k = -i\hbar\left(\dfrac{\partial}{\partial q_k}\right)$,

$k = 1, 2, \ldots, D$, and $\hat{p}_{D+1} = \hat{J}_x$, $\hat{p}_{D+2} = \hat{J}_y$, $\hat{p}_{D+1} = \hat{J}_z$, the volume element is

$dq_1\, dq_2 \ldots dq_D \sin\theta\, d\theta\, d\phi\, d\chi$, and $\left(\hat{J}_x, \hat{J}_y, \hat{J}_z\right)$ are the components of the operator corresponding to the overall rotation of the molecule. Elements of $\mathbf{G}$ and $\det\mathbf{g}$ are expressed in terms of the internal coordinates and the masses of the nuclei and are not functions of the Euler angles.

In the same formulation, the operator corresponding to the rotationless case, that is, the pure vibrational Hamiltonian can be written as either

$$\hat{H}^{\text{v}} = \frac{1}{2}\sum_{kl}^{D}\tilde{g}^{-1/4}\hat{p}_k^\dagger G_{kl}\tilde{g}^{1/2}\hat{p}_l\tilde{g}^{-1/4} + V \tag{8.55}$$

or

$$\hat{H}^{\mathrm{v}} = \frac{1}{2} \sum_{kl}^{D} \hat{p}_k^{\dagger} G_{kl} \hat{p}_l + U + V \tag{8.56}$$

where U is the so-called extrapotential term, a nonderivative part of the kinetic energy operator, and

$$U = \frac{\hbar^2}{32} \sum_{kl}^{D} \left[\frac{G_{kl}}{\tilde{g}^2} \frac{\partial \tilde{g}}{\partial q_k} \frac{\partial \tilde{g}}{\partial q_l} + 4 \frac{\partial}{\partial q_k} \left(\frac{G_{kl}}{\tilde{g}} \frac{\partial \tilde{g}}{\partial q_l} \right) \right] \tag{8.57}$$

These operators can form the basis for efficient variational nuclear motion computations. When applied, appropriate choices for the internal coordinates and the embedding, that is, attaching the body-fixed frame to the molecule need to be made. Such computations yield the required eigenvalues and rovibrational wave functions for nuclear motion averaging.

8.4.3 VARIATIONAL AVERAGING OF DISTANCES

Variational computation of the averages of different powers of the structural parameters, for example, $\langle r \rangle_{vJ_\tau}$, is achieved as follows. First, the chosen geometric coordinate r has to be given as a function of coordinates used in the variational treatment. Second, one has to compute the expectation values. Vibrationally averaged distances, for example, the mean distance, $\langle r \rangle_v$, correspond to a given vibrational state v and (ro)vibrationally averaged mean distances $\langle r \rangle_{vJ_\tau}$ correspond to the (ro)vibrational state characterized by the labels v and J_τ. Determination of the mean distance requires the computation of the integral $\langle \Psi_{vJ_\tau} | r | \Psi_{vJ_\tau} \rangle$.

Computation of this multidimensional integral becomes especially simple when one works in the so-called discrete variable representation (DVR)* of the (ro)vibrational Hamiltonian, whereby the wave function is known at a set of discrete grid points and thus integration amounts to a simple summation. It is important to emphasize that during (ro)vibrational averaging, one can take advantage of the fact that the internal coordinates do not depend on the Euler angles that describe the overall rotation of the molecule.

The effect of temperature can be taken into account by simple Boltzmann averaging; for an application, see Equations 8.3 and 8.6.

There are several sources of possible errors contaminating the variationally computed nuclear motion averages. First, even if exact kinetic energy operators are

---

* The DVR representation, one of the grid-based representations, was introduced into molecular quantum chemistry in the 1960s. It basically amounts to a useful change in the basis used for the representation of the Hamiltonian and introduces approximations during evaluation of the Hamiltonian matrix elements. For a detailed treatment of the different DVR techniques see Light, J. C., I. P. Hamilton, J. V. Lill. 1985. *J Chem Phys* 82:1400 and Light, J. C., and T. Carrington Jr. 2000. *Adv Chem Phys* 114:263.

employed for the nuclear motion computations, the approximations introduced during construction of the PES-forming part of the Hamiltonian used may result in substantial errors. Second, it is usually insufficient just to use the vibrational Hamiltonian for the nuclear motion averaging; the rotational motion also needs to be taken into account. Third, at elevated temperatures, it may become necessary to compute a very large number of rovibrational energies and wave functions, which may prove prohibitive for some applications and simplifications may need to be introduced. Fourth, in variational computations, the energies converge much faster than the wave functions; thus, more extended basis sets need to be employed during nuclear motion averaging than for the determination of the eigenvalues of the Hamiltonian. Fifth, while the Eckart–Watson Hamiltonian, with which simple computations can be performed, might be suitable for the determination of the lowest eigenpairs, its application may lead to incorrect results for some of the higher-lying (ro)vibrational states; in such cases, it is mandatory to use a Hamiltonian expressed in appropriately chosen internal coordinates.

8.4.4 VIBRATIONALLY AVERAGED ROTATIONAL CONSTANTS

Effective rotational constants, incorporating vibrational averaging, are the principal structural results obtained from fitting appropriate rovibrational Hamiltonians to not only MW and MMW but also to infrared spectroscopic data (see Chapter 5). The average rotational constants A_v, B_v, and C_v, determined experimentally, correspond to the vth vibrational state.

Let us denote the eigenvalues and eigenfunctions of the Eckart–Watson form of $\hat{H}^v$, see Equation 8.53, by E_v^{vib} and ψ_v^{vib}, respectively. By using the vibrational eigenfunctions, effective rotational operators can be produced by averaging the Eckart–Watson form of the exact vibrational-rotational Hamiltonian, $\hat{H}^{rot\text{-}vib}$, for each vibrational state as

$$
\begin{aligned}
\left\langle \hat{H}^{rot\text{-}vib} \right\rangle_v &= E_v^{vib} - \frac{1}{2}\sum_{\alpha\beta} \hat{J}_\alpha \left\langle \mu_{\alpha\beta}\hat{\pi}_\beta \right\rangle_v - \frac{1}{2}\sum_{\alpha\beta} \left\langle \hat{\pi}_\alpha\mu_{\alpha\beta} \right\rangle_v \hat{J}_\beta + \frac{1}{2}\sum_{\alpha\beta} \hat{J}_\alpha \left\langle \mu_{\alpha\beta} \right\rangle_v \hat{J}_\beta \\
&= E_v^{vib} - \sum_{\alpha\beta} \left\langle \hat{\pi}_\alpha\mu_{\alpha\beta} \right\rangle_v \hat{J}_\beta + \frac{1}{2}\sum_{\alpha\beta} \hat{J}_\alpha \left\langle \mu_{\alpha\beta} \right\rangle_v \hat{J}_\beta
\end{aligned}
\tag{8.58}
$$

An effective rotational Hamiltonian used in the evaluation of high-resolution rotation-vibration experiments may have the form of, for instance,

$$
\hat{H}^{rot}_{eff,v} = A_v \hat{J}_x^2 + B_v \hat{J}_y^2 + C_v \hat{J}_z^2 + \sum_{\beta\gamma} T_v^{\beta\gamma} \left(\hat{J}_\beta^2 + \hat{J}_\gamma^2 \right)^2 + K
\tag{8.59}
$$

where A_v, B_v, C_v, and $T_v^{\beta\gamma}$ (β, $\gamma = x, y, z$) are so-called spectroscopic constants corresponding to a given vibrational state v. In order to predict effective spectroscopic constants from variational nuclear motion computations, one should mimic the procedure used by spectroscopists leading to effective rotational Hamiltonians. This topic is still under development and no final recommendations can be given.

Lukka and Kauppi suggested one procedure some time ago.* Within their proposed algorithm one has to (1) start out from an arbitrary (preferably exact) rotation-vibration Hamiltonian; (2) compute vibration-only wave functions; (3) carry out vibrational averaging of the total rotation-vibration Hamiltonian using the computed wave functions; and (4) use a series of numerical contact transformations to convert the effective Hamiltonian to the expected form given in Equation 8.59. This route has never been fully exploited. What one must remember is that the rotational constants, which can straightforwardly be computed variationally, for example, those corresponding to the principal axes system, should not be compared directly with their experimental counterparts as they refer to quantities of different physical origin.

EXAMPLE 8.1: THE WATER MOLECULE

The water molecule was chosen as the molecular model of this section for the following reasons: (1) it is perhaps the only polyatomic and polyelectronic molecule for which unusually accurate semiglobal ab initio (and empirical) adiabatic PESs are available; (2) it is a simple bent triatomic molecule amenable to rigorous treatments both for its electronic and nuclear motions; and (3) it is one of the most important molecules that is also easy to handle experimentally and has been studied in great detail both spectroscopically and by GED, providing critical anchors when comparing the theoretical and experimental results.[†]

Some of the experimental, empirical, and theoretical equilibrium structural parameters (the OX bond lengths, r_e in Å, and the XOX bond angles, θ_e in degrees, where X = H or D) available for the $H_2{}^{16}O$ and $D_2{}^{16}O$ isotopologues of the water molecule are collected in Table 8.1. As Table 8.2 shows, though water is a hard case for most empirical treatments, the more recent empirical and first-principles structural parameters show only a very small scatter.

In order to cover the temperature range usually available experimentally, say between 300 and 1400 K, and thus be able to do proper variational thermal averaging, rovibrational computations need to be performed up to relatively high energies, as determined by the vibrational structure of the molecule. In case of the water molecule, for example, the ab initio database, which needs to be generated for proper quantum mechanical averaging, contains for $H_2{}^{16}O(D_2{}^{16}O)$ some 18,000(24,000) rovibrational energies, the number of vibrational ($J = 0$) levels is 64(61), and the computations had to be performed up to $J = 39(45)$, where J is the rotational quantum number. Representative averaged structural parameters thus determined are collected in Table 8.2. One can easily check the approximate validity of the expressions in Section 8.2.4 using the data available in Table 8.3.

As shown in Table 8.2, vibrational averaging based on variationally computed wave functions yields significantly different results for different moments of r. The average OH distances based on different moments deviate substantially from each other, ranging from 0.996 to 0.977 Å for the (0 0 0) vibrational state. Note also

---

* Lukka, T. J., and E. Kauppi. 1995. *J Chem Phys* 103:6586.
† For full details concerning equilibrium structures of water isotopologues and their rotationally-vibrationally averaged counterparts see Császár, A. G., G. Czakó, T. Furtenbacher, et al. 2005. *J Chem Phys* 122:214305 and Czakó, G., E. Mátyus, and A. G. Császár. 2009. *J Phys Chem A* 113:11665, respectively.

TABLE 8.1
Brief History of the Equilibrium Structure of the $H_2{}^{16}O$ and $D_2{}^{16}O$ Isotopologues of the Water Molecule

| Year | $H_2{}^{16}O$ | | $D_2{}^{16}O$ | | Comment |
	r_e	θ_e	r_e	θ_e	
1932	...	115			a
1945	0.9584	104.45			b
1956	0.9572(3)	104.52(5)	0.9575(3)	104.47(5)	c
1961	0.9561	104.57	0.9570	104.43	d
1997	0.95783	104.509			e
2005	0.95785	104.500	0.95783	104.490	f

Note: Bond length r_e in Å and bond angle θ_e in °.

[a] As reported in Plyler, E. K. 1932. *Phys Rev* 39:77, obtained from the fundamental wave numbers of water assumed to be 5309, 1597, and 3742 cm^{-1} and through the use of an equation derived by Dennison (Dennison, D. M. 1926. *Philos Mag* 1:195).

[b] Based upon careful analysis of results due to Mecke et al. (Mecke, R. 1933. *Z Phys* 81:313; Baumann, W., and R. Mecke. 1933. *Z Phys* 81:445), Darling and Dennison (Darling, B. T., and D. M. Dennison. 1940. *Phys Rev* 57:128), and Nielsen (Nielsen, H. H. 1941. *Phys Rev* 59:565; Nielsen, H. H. 1942. ibid., 62:422), as reported in Herzberg, G. 1945. *Molecular Spectra and Molecular Structure*, Vol. II. Toronto: van Nostrand.

[c] As reported in Benedict, W. S., N. Gailar, and E. K. Plyler. 1956. *J Chem Phys* 24:1139. The differences between the $H_2{}^{16}O$ and $D_2{}^{16}O$ structural parameters reported are about an order of magnitude larger than the well-established first-principles values from 2005.

[d] As reported in Kuchitsu, K., and L. S. Bartell. 1961. *J Chem Phys* 36:2460. The results were obtained from the rotational constants of Benedict et al. from 1956 and the lowest-order vibration-rotation interaction constants determined by Kuchitsu and Bartell.

[e] Characteristic of the fitted empirical PES of Partridge, H., and D. W. Schwenke. 1997. *J Chem Phys* 106:4618.

[f] Mass-dependent (adiabatic) equilibrium values r_e^{ad}, as given in Császár, A. G., G. Czakó, T. Furtenbacher, et al. 2005. *J Chem Phys* 122:214305. The related mass-independent Born–Oppenheimer equilibrium (r_e^{BO}) values are 0.95782 Å and 104.48$_5$°.

that for the (0 1 1) state, the difference between the $<r^{-3}>^{-1/3}$ and the $<r^3>^{1/3}$ values increases to a very substantial 0.03 Å.

In Table 8.3, r_g- and r_a-type distances are given for the OX (X = H or D) distances. It is noteworthy that while the equilibrium OH and OD bond lengths even in the adiabatic approximation are almost the same, they differ from each other only by 0.00002 Å, $\langle r(OH)\rangle_0$ is longer than $\langle r(OD)\rangle_0$ by a substantial 0.00488 Å, where $\langle r\rangle_0$ denotes the vibrational ground state (0 0 0). The change in the OH(OD) bond lengths between room temperature (300 K) and even 1400 K is significant but not large, about 0.004 Å.

Finally, a few words about rotational contributions to effective (averaged) distances. Complete neglect of the rotational contribution to the average T-dependent

TABLE 8.2

Vibrationally Averaged OH Bond Lengths and HOH Bond Angles of H$_2$^{16}O in Different Vibrational States (v) Determined Variationally

	$H_2{}^{16}O$						
v	$<r>$	$<r^2>^{1/2}$	$<r^3>^{1/3}$	$<r^{-1}>^{-1}$	$<r^{-2}>^{-1/2}$	$<r^{-3}>^{-1/3}$	$<\theta>$
(0 0 0)	0.97565	0.97809	0.98052	0.97079	0.96835	0.96592	104.430
(0 1 0)	0.97805	0.98052	0.98300	0.97311	0.97063	0.96816	103.641
(0 2 0)	0.98029	0.98281	0.98534	0.97523	0.97271	0.97019	107.059
(1 0 0)	0.99252	0.99745	1.00235	0.98264	0.97773	0.97287	104.130
(0 0 1)	0.99301	0.99797	1.00291	0.98307	0.97814	0.97325	103.372
(0 3 0)	0.98227	0.98486	0.98745	0.97709	0.97451	0.97192	108.763
(1 1 0)	0.99502	0.99999	1.00493	0.98507	0.98013	0.97524	105.267
(0 1 1)	0.99562	1.00064	1.00563	0.98558	0.98059	0.97566	104.465
(0 4 0)	0.98386	0.98652	0.98917	0.97855	0.97590	0.97325	110.899

Note: Bond length r in Å and bond angle θ in °.

TABLE 8.3

Temperature Dependence of the Average Internuclear (r_g) and Inverse Internuclear (r_a) Structural Parameters (in Å) of the H$_2$^{16}O and D$_2$^{16}O Isotopologues of the Water Molecule

T (K)	r_g(OH)	r_a(OH)	r_g(OD)	r_a(OD)
0	0.97565	0.97079	0.97077	0.96724
200	0.97605	0.97118	0.97116	0.96763
400	0.97646	0.97159	0.97158	0.96805
600	0.97692	0.97204	0.97211	0.96856
800	0.97747	0.97257	0.97279	0.96919
1000	0.97813	0.97319	0.97368	0.96997
1200	0.97895	0.97392	0.97476	0.97090
1400	0.97993	0.97477	0.97600	0.97193

distances, that is, doing constrained vibrational ($J = 0$) averagings does not yield correct bond length increases. The distance corrections due to rotations are substantial, but turn out to be linear (shown in Figure 8.4), as suggested by classical mechanics

$$\left\langle \delta r \right\rangle_{\text{rot}}^{T} = \sigma T \tag{8.60}$$

Thus, the centrifugal distortion correction can be treated perfectly well through a few simple computations, as only the linear factor in front of T needs to be determined.

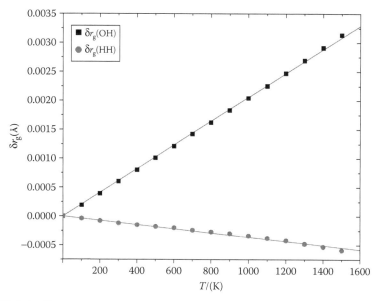

FIGURE 8.4 Temperature dependence of the rotational contributions to the r_g(OH) and r_g(HH) parameters of $H_2^{16}O$. The linear fits, $\delta r_g = \sigma T$, gave σ parameters of 2.06×10^{-6} and -3.6×10^{-7} Å/K for δr_g(OH) and δr_g(HH), respectively.

These factors appear to be isotope-independent and positive and negative for the bonded (OH/OD) and nonbonded (HH/DD) distances, respectively.

REFERENCES AND SUGGESTED READINGS

The elementary theory of vibrations and rotations has been treated in several excellent textbooks. The following are some of the classic and easily accessible sources on vibrations and rotations.

Kroto, H. W. 1992. *Molecular Rotation Spectra*. New York: Dover.
Wilson Jr., E. B., J. C. Decius, and P. C. Cross. 1955. *Molecular Vibrations*. New York: McGraw-Hill.

Major Texts on Molecular Rotations

Gordy, W., and R. L. Cook. 1984. *Microwave Molecular Spectra*. New York: Wiley-Interscience.
Townes, C. H., and A. L. Schawlow. 1955. *Microwave Spectroscopy*. New York: Mc-Graw-Hill.
Wollrab, J. E. 1967. *Rotational Spectra and Molecular Structure*. New York: Academic Press.

Major Texts on Molecular Vibrations

Califano, S. 1976. *Vibrational States*. New York: Wiley.
Papoušek, D., and M. R. Aliev. 1982. *Molecular Vibration-Rotation Spectra*. Amsterdam: Elsevier.

Major Reviews on Nuclear Averaging

Kuchitsu, K. 1992. The potential energy surface and the meaning of internuclear distances. In *Accurate Molecular Structures*, ed. Domenicano, A., I. Hargittai. Oxford, UK: Oxford University Press.

Kuchitsu, K., and S. J. Cyvin. 1982. Representation and experimental determination of the geometry of free molecules. In *Molecular Structures and Vibrations*, ed. S. J. Cyvin. Amsterdam: Elsevier.

Kuchitsu, K., M. Nakata, and S. Yamamoto. 1988. Joint use of electron diffraction and high-resolution spectroscopic data for accurate determination of molecular structure. In *Stereochemical Applications of Gas-Phase Electron Diffraction, Part A*, ed. Hargittai, I., M. Hargittai. New York: VCH Publishers.

Appendix A: Bibliographies of Equilibrium Structures

Earlier compilations of molecular structures can be found in the following references:

Sutton, L. E. *Tables of Interatomic Distances and Configuration in Molecules and Ions.* Special Publication No. 11. London: The Chemical Society, 1558; and *Supplement 1956–1959.* Special Publication No. 18, 1965.

Harmony, M. D., V. W. Laurie, R. L. Kuczkowski, et al. Molecular structures of gas phase polyatomic molecules determined by spectroscopic methods. *J Phys Chem Ref Data* 8 (1979): 619–733.

Huber, K. P., and G. Herzberg. *Molecular Spectra and Molecular Structure: Constants of Diatomic Molecules.* New York: Van Nostrand-Rheinhold, 1979.

Structure data of molecules have been collected in a series of volumes of Group II of the new series of Landolt–Börnstein (http://www.springermaterials.com). Volume II/28 A–D is a supplement to Volume II/25 A–D. Volume II/25 also incorporates all the data of the previous Volumes II/7, II/15, II/21, and II/23 after appropriate revision. These critically evaluated compilations contain experimentally determined structures. The tabulations are frequently supplemented to bring them up to date:

Graner, G., E. Hirota, T. Iijima, et al. Structure data of free polyatomic molecules. In *Landolt-Börnstein New Series I*, vol. 25 A-D. edited by K. Kuchitsu. Berlin: Springer-Verlag, 1998–2003.

Hirota, E., T. Iijima, K. Kuchitsu, D. A. Ramsay, J. Vogt, and N. Vogt. Structure data of free polyatomic molecules. In *Landolt-Börnstein, Numerical Data and Functional Relationships in Science and Technology (New Series), Group II*, vol. 28 A-D. edited by K. Kuchitsu, N. Vogt, and M. Tanimoto. Berlin: Springer, 2006/2007.

Hirota, E., K. Kuchitsu, T. Steimle, M. Tanimoto, J. Vogt, and N. Vogt. Structure data of free polyatomic molecules. In *Landolt-Börnstein, Numerical Data and Functional Relationships in Science and Technology (New Series), Group II*, vol. 30. edited by K. Kuchitsu, N. Vogt, and M. Tanimoto. Berlin: Springer, 2011.

The equilibrium structures of diatomic molecules determined by spectroscopy are also found in a series of volumes of Group II of the new series of Landolt–Börnstein. Volume II/24 contains the most recent ones (up to 1998):

Demaison, J., H. Hübner, and G. Wlodarczak. Molecular constants mostly from microwave, molcular beam, and sub-Doppler laser spectroscopy. In *Landolt-Börnstein, Numerical Data and Functional Relationships in Science and Technology (New Series), Group II*, vol. 24A. edited by W. Hüttner. Berlin: Springer, 1998.

A review of equilibrium structures determined by gas-phase electron diffraction is given by

Spiridonov, V. P., N. Vogt, J. Vogt. Determination of molecular structures in terms of potential energy functions from gas-phase electron diffraction supplemented by other experimental and computational data. *Struct Chem* 12 (2001): 349–76.

A database called Molecular Gasphase Documentation (MOGADOC) contains references for about 10,000 molecules, which have been studied by microwave spectroscopy or gas electron diffraction. The database also comprises about 7,800 numerical datasets with internuclear distances, bond angles, and dihedral angles. Among them, there are about 900 entries with equilibrium structures. The database can be searched by textual, structural, and numerical retrievals. It is produced and distributed by the Chemieinformationssysteme at the University of Ulm (http://www.uni-ulm.de/strudo/mogadoc/), see also the following references:

Vogt, J., and N. Vogt. *J Mol Struct* 695 (2004): 237–41.
Vogt, J., N. Vogt, and R. Kramer. *J Chem Inform Comput Sci* 43 (2003): 357–61.
Vogt, N., E. Popov, R. Rudert, R. Kramer, and J. Vogt. *J Mol Struct* 978 (2010): 201–4.

Appendix B: Sources for Fundamental Constants, Conversion Factors, and Atomic and Nuclear Masses

CONTENTS

B.1 FUNDAMENTAL CONSTANTS

The Committee on Data for Science and Technology (CODATA) internationally recommended values of the fundamental physical constants may be found at http://physics.nist.gov/cuu/Constants/index.html.

The constants used in this book are reproduced in the table given at the bottom of this page.

With some of these values, the conversion factor $I \times B$ relating rotational constant B to moment of inertia I is obtained as follows:

$$I \times B = \frac{\hbar}{4\pi} = 505\ 379.005(50) \text{ u } \text{Å}^2 \text{ MHz}$$

Note that authors may have used variant values to some extent in their original work.

Quantity	Symbol	Value	Unit
Speed of light in vacuum	c, c_0	299 792 458	m·s^{-1}
Planck constant	h	$6.626\ 068\ 96(33) \times 10^{-34}$	J·s
$h/2\pi$	$\hbar$	$1.054\ 571\ 628(53) \times 10^{-34}$	J·s
Elementary charge	e	$1.602\ 176\ 487(40) \times 10^{-19}$	C
Electron mass	m_e	$9.109\ 382\ 15(45) \times 10^{-31}$	kg
Proton mass	m_p	$1.672\ 621\ 637(83) \times 10^{-27}$	kg
(Unified) atomic mass unit	u	$1.660\ 538\ 782(83) \times 10^{-27}$	kg
$1\,\text{u} = m_u = \dfrac{1}{12} m(^{12}c) = 10^3 \dfrac{\text{kg mol}^{-1}}{N_A}$			

B.2   ATOMIC MASSES

A table of the most recent atomic masses can be found at the LBNL Isotopes Project Nuclear Data Dissemination Home Page at http://ie.lbl.gov/toi.html. The 2003 data have been compiled by Audi, G., A. H. Wapstra, and C. Thibault. The AME2003 Atomic Mass Evaluation, *Nuclear Physics A* 729 (2003):337–676.

A large excerpt of this table is reproduced in the appendix "Atomic Masses" on the CD-ROM.

This information may also be found in the following book:

Cohen, E. R., T. Cvitas, J. G. Frey, et al., eds. *Quantities, Units and Symbols in Physical Chemistry.* Berlin: Springer, 2007.

B.3   NUCLEAR MASSES

Nuclear masses can also be found in the so-called Green Book of IUPAC. The second edition, I. Mills, T. Cvitas, K. Homann, N. Kallay, K. Kuchitsu, *Quantities, Units and Symbols in Physical Chemistry*, Oxford: Blackwell Science, 1993, can be downloaded freely from http://old.iupac.org/publications/books/author/mills.html.

Author Index

Subject Index

This index also includes molecules. Each molecule is listed under its formula ordered according to the Hill system. The molecules are arranged in alphabetical order of the symbols of the elements, with the exception of carbon and hydrogen atoms in organic compounds where they are written in that order.

A

Accidental resonance, *see* resonance
Accuracy, 30, 33, 34, 37, 52, 107
    of r_0 structure, 137
    of r_s structure, 146
    of structure, 35, 154, 198
Adiabatic correction potential, 196, 198
Airy function, 165
Analysis of residuals, 39, *see also* residuals
Angular momentum
    total, 92, 95, 104, 108, 255
    rotational, 92, 93, 104
    vibrational, 93, 94, 108
Anharmonic force field, *see* force field
Anharmonic resonance, *see* resonance
Anharmonic constant, 111
ArClH, Ar·HCl, 211, 224
ArClI, Ar·ICl, 225
ArMg, MgAr$^+$, 181
Ar$_2$, 178, 179
Asymmetric rotor, 56, 100, 132
Asymmetry parameter, 101
Atomic mass, 104, 105, 127, 149, 152, 153, 266
Atomic orbital (AO), 17
Autocorrelation, 48, 52
Average structure, r_v, *see* structure
Axis of symmetry, *see* symmetry axis
Axial rotational constant, 114, 124
Axis rotation, 147, 153, 158

B

Badger rule, 227
Band constants, 163, 164, 185

Base physical quantities, 58
Basis set
    atomic natural orbital (ANO), 24, 77
    correlation consistent, 24, 71–83
    Gaussian, 23
    one-particle, 17
Bastiansen–Morino shrinkage effect, 237
BeF, 173
Bending mode, 102, 225
Bias, 31, 48, 50
Birge–Sponer plot, 174, 177–179, 181, 228
Biweight function, 43, 44
Bohr, 90, 177
Bohr–Sommerfeld quantization condition, 166, 167, 169
Boltzmann distribution, 234–236, 257
Boltzmann law, 104
Bond dissociation energy, 159
Bond lengthening from quadrupole coupling, 226
Bond length, 167, 168, 198, *see also* equilibrium bond length and structure
Bond moment, 211
Born–Oppenheimer approximation, 2, 53, 56, 63
Born–Oppenheimer breakdown (BOB), 159, 160, 183, 184, 195, 196, 198
Bose–Einstein statistics, 207, 208
Boson, 207
Boundary condition, 162, 166
Boyle's law, 48

C

Ca$_2$, 170, 173, 192
CArClN, Ar·ClCN, 216
Cardinal number, 24, 25
CCl$_2$O, 139, 151
Center of mass, 35, 91, 92, 127–129, 132, 141
Centrifugal distortion, 40, 46, 49, 61, 62, 98, 100, 163, 183, 213–216
    correction to the rotational constants, 106
    planarity relations, 107
    quartic constants, 73, 98, 100, 101, 214
CFOUR, 62, 68